Plumbing

Level 4

FIFTH EDITION

Pearson

NCCER National Center for Construction Education and Research

NCCER
President and Chief Executive Officer: Boyd Worsham
Vice President of Innovation and Advancement: Jennifer Wilkerson
Chief Learning Officer: Lisa Strite
Senior Curriculum Leader: Chris Wilson
Plumbing Project Manager: Lauren Henehan
Technical Writing Manager: Gary Ferguson
Technical Writer: Phil Elmore
Art Manager: Bree Rodriguez
Technical Illustrators: Judd Ivines, Liza Wailes
Production Artist: Chris Kersten
Permissions Specialist: Sherry Davis
Lead Editor: Alexandria Willbond
Editors: Zi Meng, Lauren Tygrest, Terry ONeil, Melina Redondo
Project Coordinator: Colleen Duffy

Pearson
Commercial Product Manager – Associations: Andrew Dunaway
Senior Digital Content Producer: Shannon Stanton
Associate Project Manager: Monica Perez-Kim
Content Producer: Alexandrina Wolf, Mary Roth
Executive Marketing Manager: Mark Marsden
Designer: Mary Siener
Rights and Permissions: Kaitlyn Coddington
Composition: Integra Software Services
Printer/Binder: Lakeside Book Company
Cover Printer: Lakeside Book Company
Text Typefaces: Palatino LT Pro and Helvetica Neue

Cover Image
Cover photo provided by Wendy Becovitz-McKinzie of Ferrer Mechanical & Electrical.

10 9 8 7 6 5 4 3 2 1

1 2024

Pearson

NCCER National Center for Construction Education and Research

ISBN-10: 0-13-821472-7
ISBN-13: 978-0-13-821472-2

PREFACE

To the Trainee

Most people are familiar with plumbers who come to their homes to unclog a drain or install an appliance. Plumbers also install, maintain, and repair many different types of pipe systems. For example, some systems move water to a municipal water treatment plant and then to residential, commercial, and public buildings. Other systems dispose of waste, provide gas to stoves and furnaces, or supply air conditioning. Pipe systems in generation plants carry the steam that powers turbines. Pipes also are used in manufacturing plants, such as wineries, to move material through production processes.

Plumbers and their associated trades constitute one of the largest construction occupations, holding about 500,000 jobs. In fact, the occupation is expected to see up to 4% growth by 2032. Plumbers are also among the highest paid construction occupations.

New with *Plumbing Level Four*

This revised fifth edition of *Plumbing Level 4* has been reformatted to provide a better experience to both the Plumbing trainee and the instructor. *Plumbing Level 4* contains code updates, design improvements, and content updates to several technologies, including information related to pools, spas, mobile homes, and medical gas applications. Additional information from the *NASCLA Contractors Guide to Business, Law, and Project Management* has been incorporated.

Several new joining technologies and their associated equipment have been added. Coverage of plenums, water expansion tanks, and sealing elements—approved and not approved for use with compressed air systems—has also been enhanced throughout the curriculum, creating more concise and trade-appropriate learning outcomes.

We wish you success as you progress through this training program. NCCER appreciates and welcomes its customers' feedback. If you have any comments on how NCCER might improve this textbook, please complete the Online Curriculum Feedback Form using the QR code on this page. You may submit yours by emailing **support@nccer.org**. When doing so, please identify feedback on this title by listing *#PlumbingL4* in the subject line.

Our website, **www.nccer.org**, has information on the latest product releases and training.

NCCER Standardized Curricula

NCCER is a not-for-profit 501(c)(3) education foundation established in 1996 by the world's largest and most progressive construction companies and national construction associations. It was founded to address the severe workforce shortage facing the industry and to develop a standardized training process and curricula. Today, NCCER is supported by hundreds of leading construction and maintenance companies, manufacturers, and national associations. The NCCER Standardized Curricula was developed by NCCER in partnership with Pearson, the world's largest educational publisher.

Some features of the NCCER Standardized Curricula are as follows:

- An industry-proven record of success
- Curricula developed by the industry, for the industry
- National standardization providing portability of learned job skills and educational credits
- Compliance with the Office of Apprenticeship requirements for related classroom training (*OSHA CFR 29:29*)
- Well-illustrated, up-to-date, and practical information

NCCER also maintains the NCCER digital credentials, which provides transcripts, certificates, and wallet cards to individuals who have successfully completed a level of training within a craft in NCCER's Curricula. *Training programs must be delivered by an NCCER Accredited Training Sponsor to receive these credentials.*

For information on NCCER's credentials and the NCCER digital credentials, contact NCCER Customer Service at 1-888-622-3720 or visit **www.nccer.org**.

Digital Credentials

Show off your industry-recognized credentials online with NCCER's digital credentials!

Through Credly's platform, NCCER is providing online credentials. Transform your knowledge, skills, and achievements into digital credentials that you can share across social media platforms, send to your network, and add to your resume. For more information, visit **www.nccer.org**.

Cover Image

Modular air handling units like the one in the cover photo show the diversity of the types of equipment plumbers can install. Although this unit provides the HVAC for a large educational building, all the heated, chilled, and potable water pictured inside this unit is installed by plumbers. Ferrer Mechanical & Electrical is dedicated to ensuring the highest quality of education and training for its employee owners in the residential, commercial, industrial, and institutional industries.

Ferrer Mechanical & Electrical is a 100% employee-owned mechanical contractor located in Indianapolis. Founded in 2012, Ferrer provides service and construction for the HVAC, plumbing, and electrical industries. The high level of craftsmanship and skill shown by the employee owners in this installation is just one example of why Ferrer has become a leading MEP contractor in Central Indiana.

DESIGN FEATURES

Content is organized and presented in a functional structure that allows trainees to access the information where they need it.

Source: Courtesy of Apricus Solar Co., Ltd.

Objectives

Successful completion of this module prepares you to do the following:

1. Describe the principles of hydronic and solar heating systems.
   a. Describe water heaters, boilers, and collectors.
   b. Describe hot water circulation.
   c. Describe radiators and radiant loops.
2. Describe the basic types of hydronic and solar heating systems and their components.
   a. Describe the basic types of hydronic systems.
   b. Identify the basic types of solar heating systems.
3. Describe the procedures for roughing-in, installing, and testing hydronic and solar heating system piping.
   a. Describe how to rough-in the piping.
   b. Describe how to install boilers and heating units.
   c. Explain how to install controls.
   d. Explain how to test and balance the system.
   e. Describe corrosion prevention techniques.

Performance Task

Under supervision, you should be able to do the following:

1. Lay out a hydronic or solar heating system.

Overview

Hydronic systems use heated water or steam to heat rooms or an entire building through either radiant heat or forced heat. Most hydronic systems are powered by electricity or gas; some systems, called solar heating systems, are powered by the sun's heat. Hydronic and solar heating systems typically consist of a water heater or boiler (or, in solar heating systems, a collector), a circulation system, radiators, and controls. When installing hydronic or solar heating systems, plumbers first rough-in the piping; then install the heater, boiler or collector, and controls; and, finally, test, balance, and start the system.

CODE NOTE

Codes vary among jurisdictions. Because of the variations in code, consult the applicable code whenever regulations are in question. Referring to an incorrect set of codes can cause as much trouble as failing to reference codes altogether. Obtain, review, and familiarize yourself with your local adopted code.

Digital Resources for Plumbing

Scan this code using the camera on your phone or mobile device to view the digital resources related to this craft.

SCAN ME

Trainees can navigate *Plumbing Level 3* using color coded tabs on the upper right hand corner.

The Objectives list the skills and knowledge trainees need in order to complete the module successfully.

The Performance Tasks give you an opportunity to apply your knowledge to real-world tasks.

Section Openers provide a visual organizational structure for the information. Objectives and Performance tasks are broken out for each section.

3.0.0 Roughing-In, Installing, and Testing Hydronic and Solar Heating System Piping

Objective

Describe the procedures for roughing-in, installing, and testing hydronic and solar heating system piping.

a. Describe how to rough-in the piping.
b. Describe how to install boilers and heating units.
c. Explain how to install controls.
d. Explain how to test and balance the system.
e. Describe corrosion prevention techniques.

Performance Task

1. Lay out a hydronic or solar heating system.

Zone control valve: In a hydronic system, a valve that allows the temperature in various parts of the building to be preset.

For a hydronic system, begin by establishing the location of the boiler. Plan the pipe runs between the boiler and the heating units. Inspect the planned path of the pipe runs for obstacles, structural members, and other concerns. Determine the location of different heating zones within the building. Locate **zone control valves** at each of these zones. These are valves that allow the owner to set the temperature in various parts of the building. Next, determine

Trade Terms appear on the page adjacent to the text where they are first presented.

Step-by-step presentations and math equations help make the concepts clear and easy to grasp.

Step 1 Determine the capacity of each sink in cubic inches:

$$\text{length} \times \text{width} \times \text{height} = \text{capacity in cubic inches}$$

Step 2 Convert the capacity from cubic inches into gallons per minute (gpm) using the following formula:

$$\text{capacity in cubic inches}/231 = \text{capacity in gpm}$$

Step 3 Account for displacement, which is the sink's actual usable volume, using the following formula:

$$\text{capacity in gpm} \times 0.75 = \text{actual capacity in gpm}$$

Step 4 Add the actual capacities of the sinks together. This is the total capacity that the grease interceptor should be able to handle in one minute.

SCAN ME

QR codes link trainees directly to digital resources that highlight current content.

Important information is highlighted, illustrated, and presented to facilitate learning.

Placement of images near the text description and details such as callouts and labels help trainees absorb information.

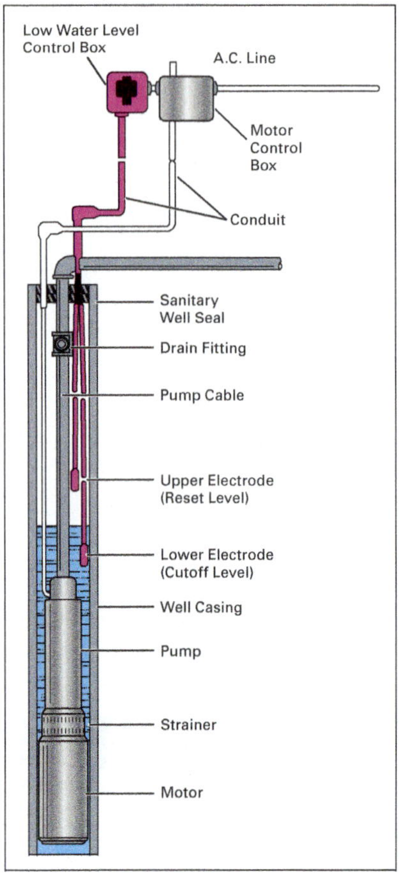

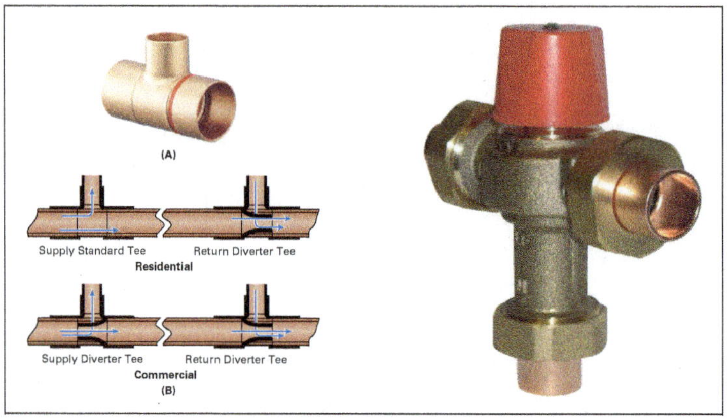

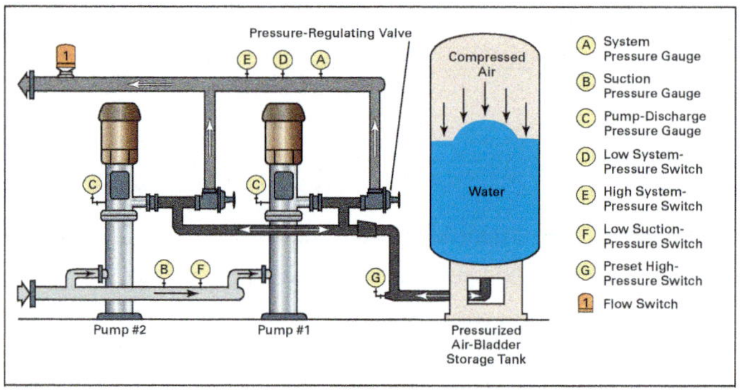

New boxes highlight safety and other important information for trainees. Warning boxes stress potentially dangerous situations, while Caution boxes alert trainees to dangers that may cause damage to equipment. Notes boxes provide additional information on a topic.

WARNING!

Some indirect systems use single-wall heat exchangers, which are more efficient than double-wall versions. Many codes do not permit single-wall heat exchangers because they are prone to leaking toxic liquids into the system over time. Codes require that each line carrying nonpotable liquids be marked. Identify each outlet on a nonpotable water line.

CAUTION

Because of contamination concerns, some jurisdictions prohibit hydropneumatic tank systems. Others may require special testing and certification to ensure the tank's structural soundness. Check your local code for restrictions on these systems in your area.

NOTE

Plumbers use the term corrosive-resistant waste piping to refer to pipe that is designed specifically to handle corrosive wastes. The term does not imply that such pipe is designed to resist other types of corrosion—for example, electrolysis, galvanic corrosion, scale, or rust.

Going Green looks at ways to preserve the environment, save energy, and make good choices regarding the health of the planet.

Going Green

Solar Hot Water Systems

City, state, and federal government programs are encouraging homeowners and businesses to use solar hot water heating. Solar energy is a renewable resource and creates few greenhouse gas by-products. Since 1992, the federal government has issued tax credits to businesses that use solar power. Many states offer similar incentives for homeowners and businesses.

Some systems convert sunlight into electricity. Others collect solar energy and use it to heat fresh water supplies and swimming pools directly. These direct systems, called solar thermal systems, are the least expensive form of solar energy for applications connected to power grids. Solar power is gaining popularity as people become more environmentally conscious. Plumbers can expect to install an increasing number of solar hot water systems in the near future.

These boxed features provide additional information that enhances the text.

Think About It

Making Solar Heating Cost-Effective

Solar heating systems often have high initial costs but lower operating costs than other types of heating systems. Operating costs can be reduced even further if other energy conservation steps are taken. Insulate piping, walls, and ceilings to reduce heat loss. Size the system to account for seasonal changes in demand and available sunlight. The client should not expect a solar heating system to provide 100 percent of the building's heating needs. Install a backup system for cloudy days.

Federal tax incentives for solar heating were eliminated in the mid-1980s. Many cities and states have followed suit. Some jurisdictions still provide rebates and tax incentives for homes that use solar heating systems. Ask an accountant about applicable local laws. Explain all the variables and options to a client before you install a solar heating system.

Review questions at the end of each section and module allow trainees to measure their progress

3.0.0 Section Review

1. Pressurized expansion tanks often come precharged with _____.
   a. helium
   b. water
   c. nitrogen
   d. air
2. When installing a boiler in a hydronic system, the return line is fitted with _____.
   a. a purging valve
   b. a T/P relief valve
   c. high-temperature adhesive
   d. $^3/_4$" noncorroding straps
3. A device that controls temperature and pressure more efficiently than a differential thermostat is the _____.
   a. thermistor
   b. thermosiphon
   c. aquastat
   d. limit switch

Module 02405 Review Questions

1. Hydronic and solar thermal heating systems can by classified as radiant systems or _____.
   a. gravity systems
   b. forced heat systems
   c. heat-sink systems
   d. blower-assisted systems
2. When 1 pound of steam condenses back to water, it gives off _____.
   a. 40 Btus
   b. 520 Btus
   c. 970 Btus
   d. 1,240 Btus
3. Steam heating systems are usually found in houses built _____.
   a. before World War I
   b. after 1950
   c. before World War II
   d. after 1985

NCCERconnect

This interactive online course is a unique web-based supplement in the form of an electronic book that provides a range of visual, auditory, and interactive elements to enhance training. Visit **www.nccerconnect.com** for more information!

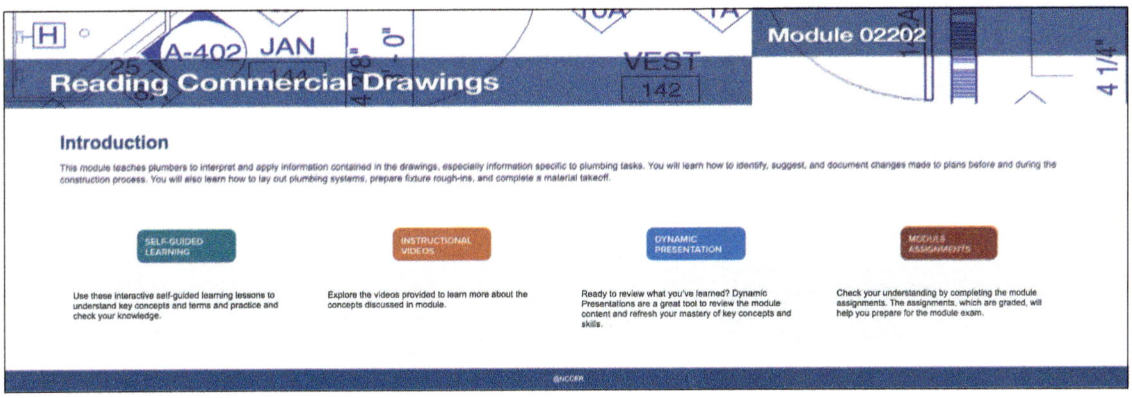

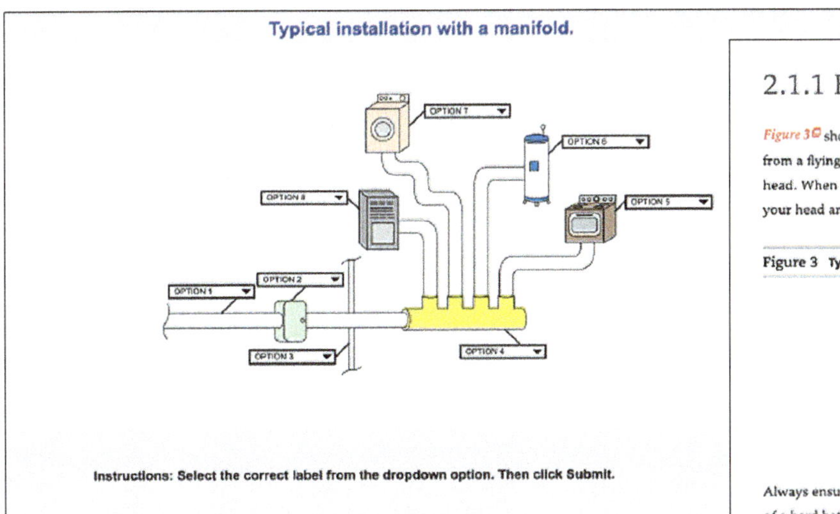

2.1.1 Hard Hats

Figure 3 shows a typical hard hat. The outer shell of the hat protects your head during a fall or from a flying object. The webbing inside the hat maintains a space between the shell and your head. When wearing a hard hat, adjust the headband, not the hard hat, so that the webbing fits your head and there is at least 1" of space between your head and the shell.

Figure 3 Typical hard hat.

Always ensure that the hard hat you use has not exceeded its life expectancy. The life expectancy of a hard hat is dependent upon factors such as when the hat was manufactured, and the make and model of the hard hat. For example, the manufacture date shown in *Figure 3* indicates the hard hat was made in September 2019. If the hard hat manufacturer indicates the lifespan of the hat is five years, then the hard hat needs to be replaced in September 2024.

ACKNOWLEDGMENTS

This curriculum was revised due to the vision and leadership of the following sponsors:

Campbell Plumbing
Ferrer Mechanical and Electrical
International Code Council, Inc.
Southern Crescent Technical College

This curriculum would not exist were it not for the dedication and unselfish energy of those volunteers who served on the Authoring Team. A sincere thanks is extended to the following:

Billy Elliott, Jr.
Mark Fasel
Richard Anderson
Wendy Becovitz-McKinzie
Mark Andrews

NCCER PARTNERS

To see a full list of NCCER Partners, please visit **www.nccer.org/about-us/partners**.

SCAN ME

CONTENTS

Module 02401 Business Principles for Plumbers

Module 46100 Introduction to Leadership

Module 02403 Water Pressure Booster and Recirculation Systems

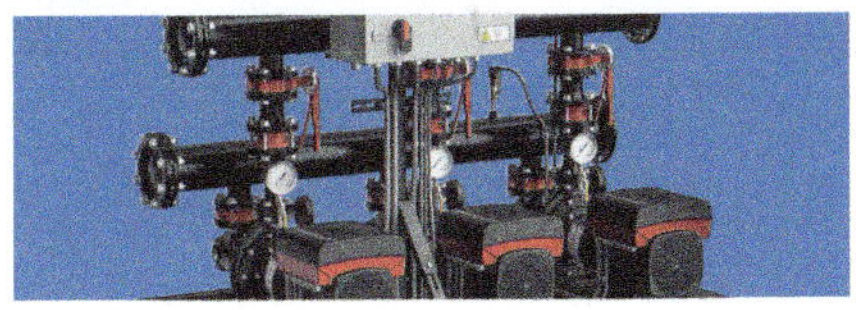

Module 02404 Indirect and Special Waste

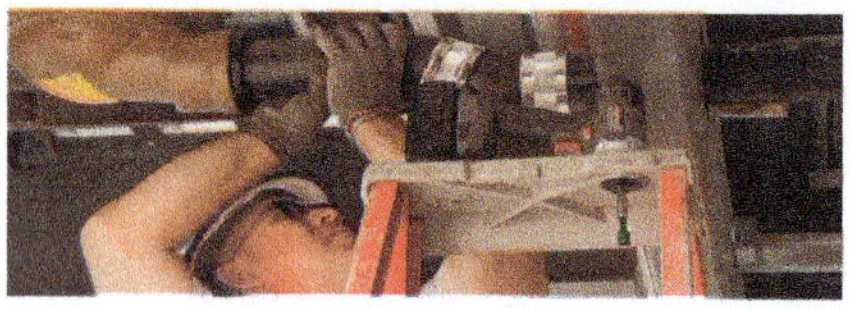

Module 02405 Hydronic and Solar Heating Systems

Module 02406 Codes

Module 02408 Private Water Supply Well Systems

Module 02409 Private Waste-Disposal Systems

Module 02410 Swimming Pools and Hot Tubs

Module 02411 Plumbing for Mobile Homes and Travel Trailers

Module 02412 Introduction to Medical Gas and Vacuum Systems

MODULE 02401

Business Principles for Plumbers

Source: Photo Smoothies/Shutterstock

Objectives

Successful completion of this module prepares you to do the following:

1. Identify cost control measures.
   a. Interpret a balance sheet and a profit and loss statement.
   b. Interpret how cost control measures affect the profit and loss statement.
   c. Interpret how on-the-job task organization affects profit and loss.
2. Identify the information required to prepare a material takeoff.
   a. Identify the information required for insurance and liability.
   b. Identify the information required for the estimating process.
   c. Identify the information required for updates and revisions.

Performance Task

Under supervision, you should be able to do the following:

1. Prepare a material takeoff as part of an estimate.

Overview

A successful business must make a profit, compete successfully for work, keep costs under control, and work efficiently. Plumbers should understand basic business accounting and practice effective techniques for bidding jobs, controlling costs, and organizing projects. Project management involves keeping control of orders, purchases, and deliveries; safe use and storage of materials, tools, and equipment; and issuing change orders. Task-planning skills enable plumbers to complete work on time and within budget.

CODE NOTE

Codes vary among jurisdictions. Because of the variations in code, consult the applicable code whenever regulations are in question. Referring to an incorrect set of codes can cause as much trouble as failing to reference codes altogether. Obtain, review, and familiarize yourself with your local adopted code.

NCCER Industry-Recognized Credentials

If you are training through an NCCER-accredited sponsor, you may be eligible for credentials from NCCER. The ID number for this module is 02401. Note that this module may have been used in other NCCER curricula and may apply to other level completions. Contact NCCER at 1.888.622.3720 or go to **www.nccer.org** for more information.

You can also show off your industry-recognized credentials online with NCCER's digital credentials. Transform your knowledge, skills, and achievements into credentials that you can share across social media platforms, send to your network, and add to your resume. For more information, visit **www.nccer.org**.

Digital Resources for Plumbing

SCAN ME

Scan this code using the camera on your phone or mobile device to view the digital resources related to this craft.

1.0.0 Cost Control Measures

Performance Tasks

There are no Performance Tasks in this section.

Objective

Identify cost control measures.

a. Interpret a balance sheet and a profit and loss statement.
b. Interpret how cost control measures affect the profit and loss statement.
c. Interpret how on-the-job task organization affects profit and loss.

All successful businesses start out with a set of clear goals and objectives. For plumbing businesses, the most important goal is to provide the highest-quality work possible at a reasonable cost to the client. Keeping clients happy is an important part of being in business, because happy clients will hire you again and recommend you to others. However, a business cannot run simply on goodwill. A successful business must also make a profit, compete successfully for work, keep costs under control, and work efficiently. That's a tall order for any business, but good businesses succeed because their employees understand how essential these goals are.

In this section, you will learn techniques for ensuring the success of your plumbing business through the use of cost control measures. You will review the basic principles of accounting to learn how companies stay in business. You will learn how to calculate and submit offers for new work, and you will examine effective ways to control costs and organize projects. All these techniques are essential for ensuring that a plumbing business can meet all its goals and be successful.

1.1.0 Interpreting the Balance Sheet and the Profit and Loss Statement

Like many people, you probably use an app or paper and pen to keep a running balance of your checking account. You understand the importance of having enough money in the bank to cover your expenses. You also usually try to end up with some money left over. The extra money goes to things that improve the quality of your life, like movies, vacations, or a new car. Businesses do the same thing but on a larger scale. Money, or income, comes in from projects and is used to pay for expenses, including salaries, equipment, and materials. The money that remains is called **profit**.

Profit: Money left over after expenses are subtracted from income.

Without profit, a business will not survive. In the construction trades, like any other business, owners track profits by compiling and analyzing financial records. This process is called **accounting**. Many people feel intimidated by accounting, but you don't need a degree in accounting to understand how it works. Basic accounting is a good skill for a plumber to have. Like a toolbox, you can take the knowledge with you wherever you decide to work. If you learn the basics of accounting, you will be able to take on positions of greater responsibility in the plumbing trade.

Accounting: The process of compiling and analyzing financial records.

The two basic tools of accounting are the **balance sheet** and the **profit and loss statement**. A balance sheet tracks what a business has and what it owes others, while a profit and loss statement lists all income and expenses over a period of time. Each of these basic tools is described in more detail in the following sections. There are also many books available on accounting for small businesses and the construction trades. When on the job, ask your employer how their accounting system works. Every system is a little different, but all accounting systems follow the same basic rules.

Balance sheet: A form used to track assets and liabilities.

Profit and loss statement: A list of all income and expenses over a given period.

1.1.1 The Balance Sheet

Anything a business owns is called an **asset**, while anything the business owes is called a **liability**. Assets include the following:

- Cash
- Inventory
- Property
- Real estate holdings

Asset: Any type of property owned by a business.

Liability: In accounting, money owed by a business. In insurance, a legal responsibility.

Liabilities consist of **debt** and **equity**. Debt is money the business owes to banks, other businesses, and individuals. Mortgages, taxes, loan repayments, and salaries are examples of debt. Equity is the value of the business if all the debts are paid. A business needs to have enough assets to pay its liabilities. This concept can be expressed as a simple equation:

Debt: Money owed by a business for payment of direct costs and indirect costs.

Equity: The value of the business if all debts are paid.

$$\text{Assets} \geq \text{Liabilities}$$

When the records show the assets are at least equal to the liabilities, the business can pay all its debt. This is the goal of accounting. When assets and liabilities are equal, the account is in **balance**. That is why one of the basic tools of accounting is called the balance sheet. Think of a balance sheet as a snapshot of the business's finances at one moment in time.

Balance: The condition of an account when assets and liabilities are equal.

Here is a simple example of how a balance sheet works. Imagine you are the owner of Bay City Plumbing and Heating, and you want to determine your company's assets, debt, and equity. Begin by looking at your records. Your bank statement shows $19,000 in the bank. You are awaiting $1,200 in payment from one client and $700 from another. Money owed to you but that has not yet been paid is called **accounts receivable**. Bay City has $3,900 worth of plumbing supplies in stock. Adding up the value of all the equipment, tools, and vehicles, you find they are worth $50,000. Now add up the various amounts to determine Bay City's assets. *Figure 1* shows the total assets equal $74,800.

Accounts receivable: Money that is owed to a business but has not been paid.

	A	B	C
1	Cash	$19,000.00	
2	Accounts Receivable	$1,900.00	
3	Plumbing Supplies	$3,900.00	
4	Equipment, Tools, Vehicles	$50,000.00	
5	**Total Assets**	$74,800.00	
6			
7			

Figure 1 Sample asset calculation for balance sheet.

Now determine Bay City's debts. Looking over the bills, you see Bay City owes $1,300 for a new set of tools. The company has $4,400 left to pay on a business development loan. The company will have to pay $1,100 in taxes for the year. Insurance payments are $2,800 for the year. Salary totals $44,000 for the year. Now add up the debts. *Figure 2* shows the total debts equal $53,600.

	A	B	C
1	Accounts Owned	$1,300.00	
2	Loan	$4,400.00	
3	Taxes	$1,100.00	
4	Insurance	$2,800.00	
5	Salary	$44,000.00	
6	**Total Debt**	$53,600.00	
7			

Figure 2 Sample debt calculation for balance sheet.

To determine the equity, start with the basic formula you just learned:

$$\text{Assets} \geq \text{Liabilities}$$

Remember liabilities consist of debt and equity, which translates as:

$$\text{Assets} \geq \text{Debt} + \text{Equity}$$

This means you can calculate both debt and equity by using simple substitution. Then, you can apply this formula to find the equity of Bay City Plumbing and Heating:

$$\begin{aligned} \text{Equity} &= \text{Assets} - \text{Debt} \\ \text{Equity} &= \$74{,}800 - \text{Debt} \\ \text{Equity} &= \$74{,}800 - \$53{,}600 \\ \text{Equity} &= \$21{,}200 \end{aligned}$$

Of course, most real-life cases are more complicated, but you can use the same basic formula to find the assets, debt, and equity of any company.

1.1.2 The Profit and Loss Statement

The other basic tool of accounting is the profit and loss statement (*Figure 3*). This is a summary of income and expenses over a given period. If a balance sheet is like a snapshot, then a profit and loss statement is like a short film.

	A	B	C
1	Gross Income	$30,300.00	
2	Materials	$10,120.00	
3	Salary	$3,300.00	
4	Taxes	$1,100.00	
5	Loan Payment	$175.00	
6	**Net Income**	$15,605.00	
7			

Figure 3 Sample profit and loss statement.

The profit and loss statement indicates how well the business has done over time. It does this by tracking the business's income and expenses, each of which are on separate accounts. The profit and loss statement tells how much profit the business made during a month, a quarter, or a year.

Profit is the money left after all expenses have been paid. The relationship can be expressed as another simple equation:

$$\text{Profit} = \text{Income} - \text{Expenses}$$

Net income: Another term for profit.

Profit, also called **net income**, is what allows a business to keep operating. Contractors use profit to buy new equipment, attract investors, and stay in business between projects. If a project goes over its budget, a contractor will have to make up the difference from its profit. If that happens too often, the contractor could go out of business. It is a delicate balance.

Gross income: All money earned before expenses are subtracted.

Consider this example: Imagine again that you are the owner of Bay City Plumbing and Heating. Looking over the paperwork for April, you find the company was paid $30,300 for five separate projects. This is called the **gross income**. Gross income is all the money made before related expenses are subtracted.

Next, add up the expenses. The company spent $10,120 on materials and equipment for the projects. Records show the company paid $3,300 in salary. The company made its yearly tax payment of $1,100. The monthly loan payment was $175. To determine the profit, subtract the expenses from the gross income of $30,300. The net income therefore equals $15,605.

Balance sheets and profit and loss statements show how well a business is doing. This information does more than help you pay taxes and suppliers; it helps you plan for the future of the business. If you see expenses in a certain area are higher than you expected, you can check whether there is a problem.

If the profit margin is lower than you would like, consider advertising for more business or raising your rates. Unless you look at the money coming in and going out, your business will eventually fail. Keeping track of the flow of money will help your company stay healthy and grow over time.

The Chart of Accounts

Businesses usually set up a separate account for each type of transaction. The accounts are grouped together in a logical order. Each account is given a number. The following example is a simplified chart of accounts:

100	**Assets**
110	Current Assets
120	Long-Term Assets
130	Other Assets
200	**Liabilities**
210	Current Liabilities
220	Long-Term Liabilities
230	Accounts Payable
300	**Net Worth**
310	Capital Stock
320	Retained Earnings
330	Equity Account
400	**Income**
410	Completed Contracts
420	Sales Returned
430	Discounts on Sales
500	**Direct Costs**
510	Materials Purchased
520	Direct Labor
530	Freight and Delivery

Take time to learn how the chart of accounts is structured where you work. It will help you understand how the business is organized.

1.2.0 Understanding the Impact of Cost Control on the Profit and Loss Statement

Imagine you are part of a crew of four plumbers working at a construction site. Your crew's task is to install soil pipe. Everything is going well until suddenly the team runs out of wyes. The nearest plumbing supply store is about 20 minutes away. The team leader instructs a plumber to go and buy enough wyes to finish the job. The plumber returns in an hour with the wyes. The team finishes installing the soil pipe and moves on to the next task.

The story seems innocent, but look at it through the eyes of a project manager. Because no one made sure there were enough wyes to complete the job, the following unnecessary costs were added to the project budget:

- Four man-hours of work
- Four hours' pay
- Mileage costs on a company vehicle
- Potential delays for other trades

In addition, the project schedule was pushed back an hour. That could mean your team starts its next task late, or it could mean there will be less time to deal with other issues in the future.

Project budgets are designed to allow workers to do the best job in the shortest practical time, for the lowest cost to the client, and with a profit that allows the contractor to stay in business. Plumbers on the team are responsible for ensuring that the project is kept on schedule and within budget. Costs can quickly spiral out of control if any of the following happen:

- Materials are ordered late or incorrectly.
- Materials are stored incorrectly, insecurely, or too far away.
- Tools and equipment are used improperly or damaged.
- Company vehicles are damaged or lost because of unsafe operation or negligence.
- Work that is not specified in the contract is done.

Plumbers are responsible for preventing these problems. If you see a problem, take steps to correct it. You will learn how to start and finish tasks in the right order and on time later in this module. The following sections discuss how to keep a project within budget.

1.2.1 Preplanning

Once a company has been awarded the contract for a project, the company must complete several steps before construction can begin. These steps are called **preplanning**. Preplanning steps are required to ensure that cost control measures are in place from the very beginning of a project, to prevent cost overruns that can eat into the company's profits. During this phase, subcontractors are selected, **purchase orders** are issued, construction schedules are developed, and the estimate is reviewed in depth. The project team holds discussions on staffing, equipment, and cost control, and then **mobilization** begins. Each of these steps is discussed in detail in the following sections.

Preplanning: The phase immediately following the awarding of a contract but before the start of construction, during which subcontractors are selected, purchase orders are issued, construction schedules are developed, and estimates are reviewed in depth.

Purchase order: An order form submitted to a supplier.

Mobilization: The final stage of preplanning, during which all items required to start construction are moved to the jobsite, arrangements are made with the other trades to coordinate jobsite activities, and all jobsite details are organized in preparation for construction.

Bid: A formal offer to do work according to an estimate.

Selecting Subcontractors

A **bid** is a formal offer to do the work according to the estimate. Just as the general contractor obtains a contract by submitting a bid to the project owner, a subcontractor obtains work by submitting a bid to the general contractor. Using a description of the work to be done, the plumbing subcontractor develops an estimate based on takeoffs, prices obtained from suppliers, vendors, the company's cost reports, and labor rates for the crew as well as workers who will be needed to support the crew's work. The bid should also include warranty information, if required by the state.

When the general contractor receives bids from subcontractors, the bids are reviewed carefully, making certain they are complete. Items usually reviewed include the following:

- Scope of work
- Cost
- Staffing resources
- Schedule

The general contractor then awards the contract to the subcontractor that proposes the most cost-effective mix of those items.

Unexpected Expenses

The cost of construction materials can increase substantially after the contracts are signed and before the material is purchased. This was especially evident during the Covid-19 pandemic, when costs of many materials and fuel soared. Price increases can cause significant problems for contractors unless the possibility is addressed in the bid. Contractors may opt to include an economic price adjustment clause in the contract to protect themselves in case costs increase. These types of clauses allow for an adjustment to the prices of concrete, pipe, or other materials when the costs of those materials go up or down.

Another type of unexpected expense is the cost of reinspection should an installation fail to meet the minimum requirements of the local plumbing code. Typically, a permit comes with a certain number of pre-paid inspections. Beyond that, the contractor is responsible for the cost of additional inspections. It is wise to budget for these potential expenses to avoid surprises later.

Issuing Purchase Orders

The contractor orders most materials for a project using purchase orders. A purchase order, often called a PO, is an order form submitted to a supplier. Purchase orders list the materials being ordered, the number needed, and the unit and total prices. Follow the contractor's procedures when issuing a purchase order (*Figure 4*). Distribute copies to the project file and to the superintendent. Send the purchase order to accounting only after comparing it to the packing slip and the invoice, and verifying they match. Purchase orders ensure that bills are paid only after the materials have been received. Often, the process is automated.

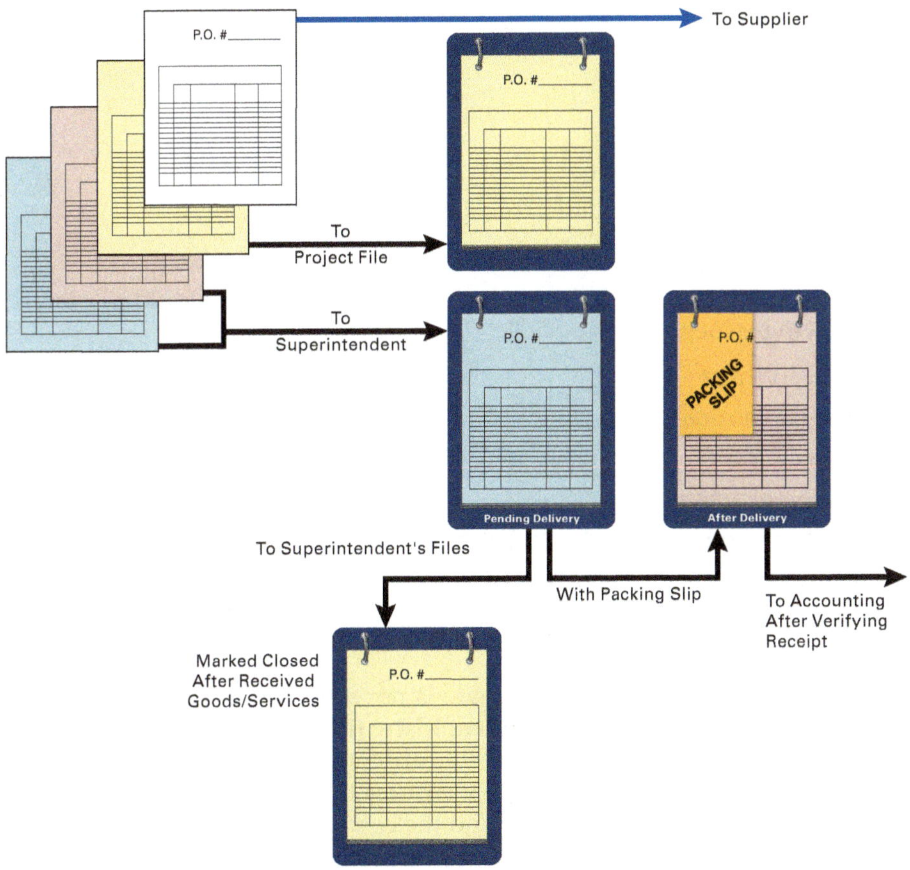

Figure 4 Typical work flow for a purchase order.

Developing a Schedule

The project schedule should be developed with cooperation and input from the following contributors:

- Members of the company's project team
- Major subcontractors
- Key suppliers
- Others who can contribute to a realistic schedule

Taxes

Plumbing tests nationwide often feature several questions about taxes. Before taking the test, be sure to learn how your local tax system works. Talk to an accountant to learn more about how businesses pay taxes. A little preparation on this topic can help improve your test scores.

Warranties

Many states require contractors to provide warranty information for construction services in the contract. The standard limited warranty in the United States for plumbing work is one year with an exception for drain cleaning, which is often not more than 6 months.

Smartphone Apps for Construction

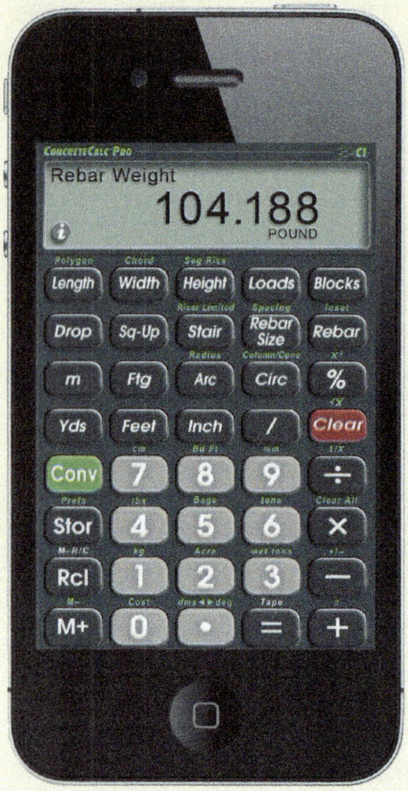

Although the craft of plumbing has been around for centuries, not every tool that a plumber uses is old. Smartphones are an increasingly popular form of communication, and they also offer a great deal of versatility for craftworkers. Smartphone cameras can be used to document on-the-job activities or potential safety violations. Best practices can be communicated to crew members using video clips. Construction calculators can be downloaded, providing craftworkers with the same (or even greater) capabilities than a handheld calculator.

The smartphone version of Calculated Industries, Inc.'s popular Construction Master® Pro, for example, lets craftworkers calculate and convert dimensions, plot right-angle conversions, find area and volume, and determine measures such as board feet, cost per unit, and even the angles of an equal-sided polygon. Calculated Industries, Inc., also makes an app version of its ConcreteCalc™ Pro as shown here, which masons can use to calculate stair dimensions, the length and weight of rebar, and the number of brick loads and mortar bags needed for a job.

Note that company and jobsite policies may not allow workers to carry smartphones while on the job. Cellphone use can distract you and prevent you from noticing safety hazards. Camera-equipped phones may raise privacy concerns on jobsites where photography is not allowed. Be sure to familiarize yourself with the applicable policies and procedures before bringing a smartphone or other cell phone to the jobsite.

Source: Calculated Industries, Inc. App shown at Courtesy of Calculated Industries, Inc

Throughout the scheduling process, the highest priority should be placed on the following issues:

- Long lead times
- Critical operations
- Sequencing
- Duration of activities
- Equipment deliveries and installation
- Staffing resources

Reviewing the Estimate

The preplanning review of the estimate is not a search for errors but a review to be sure all costs have been identified. Many times, errors both in the contractor's favor and against it are uncovered during the review. Even if no errors are found, a thorough project-team review ensures that everyone on the project has a clear understanding of the project and the costs involved.

Many contractors add a step that requires the construction team to prepare a construction estimate to compare against the original bid. This helps identify any lapses or doubling up of items that were in the original estimate.

Identifying Staffing Resources and Equipment Requirements

Labor and construction equipment requirements for all construction phases should be identified during review of the estimate and preparation of the schedule.

Determining the number of craftworkers from the various trades required at each stage of work gives the project team a basis for the following:

- Determining the availability of personnel
- Identifying possible worker shortages and surpluses
- Controlling labor cash flow

Reviewing equipment requirements ensures that cranes, lifts, generators, and other construction equipment are available when needed, used wisely and efficiently on site, and removed from the jobsite when no longer required.

Establishing a Cost Control System

Every step of the project, from beginning to end, is based on cost management. This makes a cost control system essential to all aspects of the proper management of any project.

- The owner considers cost when developing the project concept.
- The architect or engineer considers cost throughout the design phase.
- Cost is the basis for accepting and rejecting bids.
- Cost plays a central role in whether the project is profitable for those involved.

During the preplanning meeting, a review of the cost system, the extent of control, and the need for control must be established and understood.

Mobilizing

Mobilization includes the following activities:

- Moving all items required to start construction to the jobsite
- Coordinating jobsite arrangements with the other trades on the job
- Organizing all jobsite details in preparation for construction
- Setting up office and storage trailers
- Locating temporary facilities
- Establishing storage areas
- Assigning specific responsibilities to members of the project team

At the beginning of the project, find out when and from where you will receive the materials for each task. These factors will affect how you schedule your time. If some materials are scheduled to be delivered at the start of the project, begin with the tasks that use those materials. Schedule the other tasks to take place when the necessary materials arrive. Find out if plumbing suppliers in your area have orders delivered to them. Ask them when an order must be placed for it to arrive on schedule and when you can expect to pick it up.

Contractors often order materials from more than one supplier. Choose suppliers based on the location of the project and the price of the materials. Suppliers that have their own warehouses are more likely to have the materials you need in stock. Otherwise, they will need to order supplies for the project, potentially causing delays if the supplier has problems getting the materials on time. Large orders can be delivered directly to the jobsite. Before arranging for a delivery, ensure that the project budget allows for the cost of delivery.

Special requests and emergencies can eat into a budget quickly. Before a project starts, establish procedures for dealing with unexpected delays. Establish an account with a nearby supplier to meet emergency supply needs. Note the different suppliers for all the materials in the project schedule so that you can contact suppliers quickly in case of problems.

Establishing Storage Areas

In addition to the steps listed above, another important preplanning step is to establish areas where materials will be stored. Project costs are affected by the location of the stored plumbing materials on the jobsite. Whenever practical, have materials placed near where they will be needed. This step will help reduce labor costs during the installation.

Lean and LEED Certifications

Don't confuse the terms *Lean* and *LEED*. They are complementary, but they are not the same. LEED (Leadership in Energy and Environmental Design) is a set of standards that building designers, builders, owners, and operators follow to ensure that a building has been constructed using environmentally sound practices. The LEED standard was developed in 1998 by the US Green Building Council (USGBC). The Lean approach to business management focuses on identifying and eliminating wasteful practices and processes that cost time and money. The principles of Lean manufacturing are based on the Toyota Production System (TPS) developed by the car manufacturer in the late 1940s and now widely used in industries around the world. Construction professionals can obtain certifications as LEED and Lean professionals from approved accreditation organizations.

Figure 5 A field office.
Source: Photo courtesy of constructionphotographs.com

Materials stored on site must be secured, as they can be stolen. Store smaller items in a company vehicle or in the field office (*Figure 5*). The cost of replacing stolen materials and the time lost waiting for the replacements can eat away at a project budget. Store materials in the building only after doors with locks are installed. Account for materials on a regular basis. Depending on the project, this can be monthly, weekly, or daily. Track materials using an inventory form (*Figure 6*). Add materials to the inventory form when they are delivered, and subtract them when they are used.

Inventory					
Date	Orders Received	Issued from Inventory	Returned to Inventory	Balance	Physical Count

Figure 6 Sample inventory form.

1.2.2 Ordering and Taking Delivery of Materials

Time spent standing in line at a supplier to place an order and then waiting to collect the materials from storage could be better spent on the job. Consider ordering supplies online or by telephone. Learn the style of your company; as a company representative, you must present a good image when talking with suppliers and clients in person and on the phone.

Ensure that telephone orders are clear and complete. Before you place the order, make a list of the materials to be ordered and the questions you have. Speak slowly and clearly with a friendly voice. Include the quantities, sizes, and grades of the materials. Ask the supplier about the availability of each of the items on the list. Finally, ask the supplier to read back the order. This will cut down on errors.

If the person you are trying to reach is not available, leave your phone number and a brief message. If you cannot leave a message, find out when the person will be available. Keep a phone log of important calls, and note all business calls in a job diary. Whenever possible, try to get a written confirmation of an order, as this will help eliminate mistakes and misunderstandings.

Many suppliers accept emailed and faxed orders as well. When faxing an order, be sure to write clearly and use white paper. Some colored papers, especially yellow, do not fax well.

Invoice: An itemized list of materials for purchase or rent.

When a supplier sends a bill for the delivery of materials, the contractor checks the bill against a copy of the **invoice** for that shipment. An invoice is an itemized list of materials specifying prices and the terms of the agreement (*Figure 7*). An invoice is the only record a contractor has that confirms the bill can be paid. Also, the contractor will not know if the materials have been delivered on time. Without this information, the project could fall behind schedule, causing costs to escalate quickly. If you receive a shipment of materials, follow the correct procedures for handling the invoice.

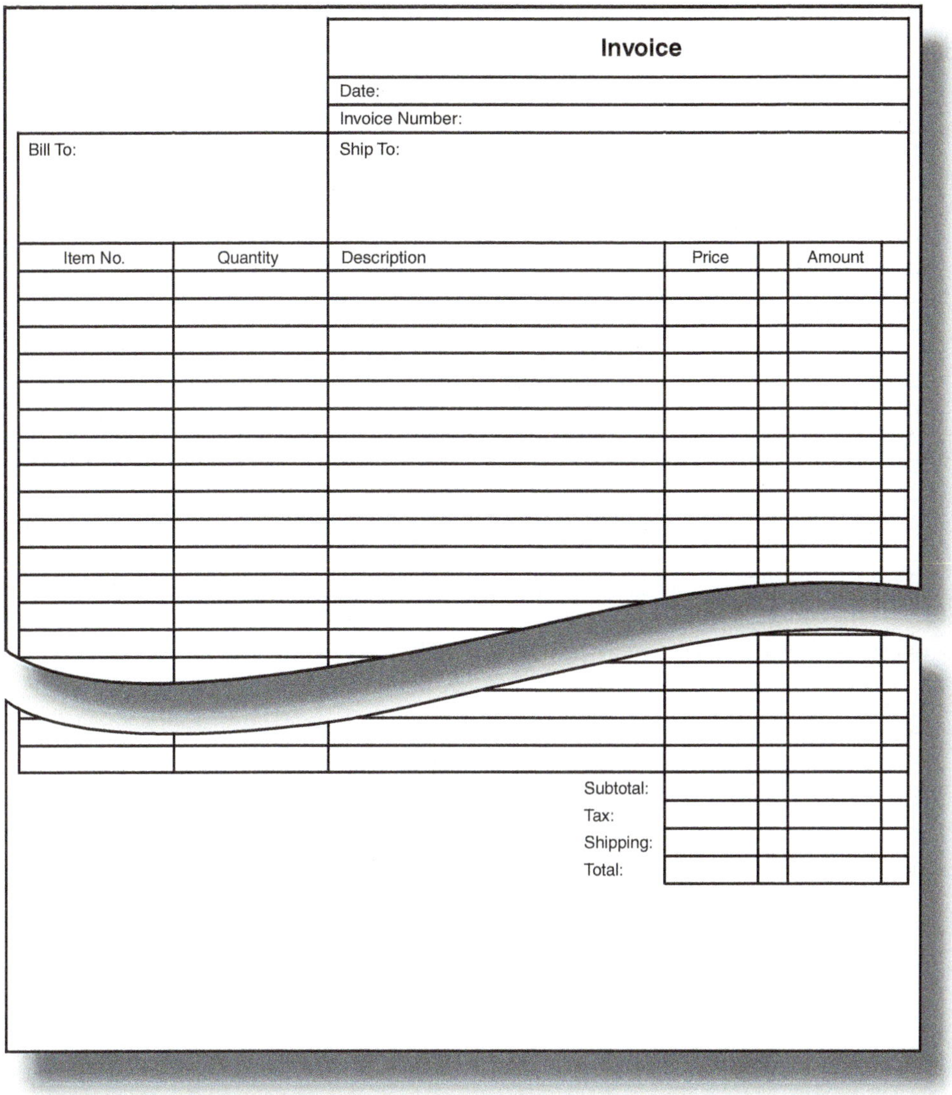

Figure 7 Sample invoice.

Ensure that the shipment contains the correct amount and type of materials, and that the materials are all in good condition. With fixtures, open each box to ensure that the fixtures are not damaged. Do not sign for the delivery until you are sure that everything that was ordered has been delivered in good condition. After signing the invoice, give a copy to your supervisor. If you purchased the materials with your own money, obtain a receipt. Your employer will use the receipt to reimburse you, and receipts are also used by contractors to control the project budget.

1.2.3 Proper Use of Tools and Equipment

Always use the proper tools and equipment on the job. This is not only a good habit but also sound business practice. Using the correct pipe and fitting the first time means you will not have to waste time and money in rework. Using the correct tools will save you time and energy. Finally, use tools properly. A damaged tool can cause injury and slow down the work. Keep records of tools and equipment to ensure that they are accounted for (see *Figure 8*).

Clean, dry, and oil tools before storing them. Proper maintenance prevents rust from developing and ensures that the tool stays in proper working order. Repair or replace damaged tools promptly.

Establish a storage area for tools and equipment. Return each tool to its proper place when you are finished using it. Store your own tools in a toolbox, and organize your toolbox so you can find tools quickly.

Equipment Record			
Description:			
Identification Number:		Classification:	
Size:	Model:	Style:	Engine Number:
New/Used:		Serial Number:	
Purchased From:			
Date of Purchase:		Terms of Purchase:	
Price:			
Tax:			
Delivery Charges:			
Installation Charges:			
Other Charges:			
Total:			
Date Sold:		Sold To:	
Gross Sale Price:			
Tax:			
Total Sale Price:			

Figure 8 Sample equipment record for a project.

1.2.4 Company Vehicles

Operate the company's vehicles safely and courteously. Remember that as the driver, you are advertising the company. This is especially true if the vehicle has the company's name on it. Reckless or discourteous driving will attract attention, but it will not be the kind of attention the company likes to get. Lock the vehicle when it is parked, and ensure that all toolboxes are secure. If you are responsible for a vehicle overnight, park it in a safe place. This can substantially reduce the risk of theft.

Use a checklist to account for all items in the vehicle; this checklist should include all tools, materials, and equipment. A checklist also serves as a record of the materials that have been used. Consult the checklist when resupplying the vehicle.

Keep the vehicle clean both inside and out. If you are responsible for maintaining the vehicle, follow the company schedule, and keep records of all oil changes and tune-ups. Take the vehicle in for regular service when scheduled, and keep track of the mileage. If you are using a company credit card, save all the sales slips and submit them to your supervisor regularly. Find out what the contractor expects you to do if the vehicle breaks down, and whom to contact and what to do in case of an accident.

1.2.5 Change Orders

Change order: An agreement to perform work in addition to what was originally agreed to in a contract.

The success of the project depends on plumbers performing their assigned tasks. Do not take on additional tasks outside the scope of what you are required to do. If you do, the company may lose money on the project. If additional work needs to be done, the client issues a **change order**. Change orders (*Figure 9*) are agreements to perform additional work outside the scope of the original contract. Change orders are supplements to the contract and are issued for the following:

- Additional tasks not foreseen in the original contract
- Tasks to be deleted from the contract
- Changes to tasks outlined in the contract
- Work to correct problems discovered during the job

Rework: To redo a previously completed task.

Sometimes, a change order requires the team to redo a task that it has already completed. For example, a change order could require moving a bathroom fixture group because the client changed the floor plan. This type of activity is called rework. **Rework** can be frustrating because it requires the team to tear out something that took time to install. Do not take this type of work personally; it is not a reflection on the quality of the team's workmanship. Do not let the need for change cause the team to lower its standards of quality. Perform the rework as if it were a brand-new task. Maintaining a positive attitude shows people that you take pride in your work and care about your job.

Change Order

This CHANGE ORDER is an integral part of the CONTRACT dated __________ between ____________________ (contractor) and ____________________ (client) for the following work: ____ to be performed at the following location: ______________________________. The following changes are the only changes allowed to be made under this CHANGE ORDER. This CHANGE ORDER signifies that the changes outlined below shall henceforth become a part of the original CONTRACT and may not be altered again without written authorization and consent of the undersigned parties.

Changes are as Follows:

____

____

____

____

These changes will: ____ increase ____ decrease the original contract amount. The terms of payment for these changes are as follows: ______________________________. The amount of change in the contract price will be as follows: _______________($).
The new total contract price shall be as follows: ________________ ($).

The undersigned parties hereby agree that the changes outlined in this CHANGE ORDER are the only changes allowed to be made. Verbal agreements shall not be considered valid. Further alterations to the CONTRACT shall not be allowed unless additional written authorization is first obtained. Written authorizations must be agreed to and signed by all parties.

This CHANGE ORDER constitutes the complete and entire agreement between all parties to alter the original CONTRACT.

______________________ Client	______________________ Contractor
______________________ Date	______________________ Date
______________________ Client	
______________________ Date	

Figure 9 Sample change order.

1.3.0 Implementing On-the-Job Task Organization

The division of project responsibilities is an essential part of on-the-job task organization. Project responsibilities are typically divided among a contractor's staff (*Figure 10*). Common divisions of responsibility are as follows:

- Either the owner of the contracting business or the project manager is responsible for ensuring that the entire project stays on schedule and within budget, and that it meets the client's needs.
- The job superintendent is responsible for coordinating the work at the jobsite.
- The plumbing foreman is responsible for coordinating the installation of the plumbing if the project involves more than one trade (such as a new building); the foreman is also responsible for assigning work to each member of the plumbing crew.
- The plumber is responsible for specific tasks, such as locating the main waste stack or installing bathroom fixtures.
- The apprentice is expected to serve on field crews for several years (typically three to four years) to obtain sufficient on-the-job experience to meet an apprentice's educational requirements, and also to learn the trade by working alongside more experienced co-workers.

This division of responsibilities is part of the process of **task planning**. Task planning allows plumbers to undertake and complete tasks in the right order and on time.

Task planning: The process of organizing activities and materials so that tasks are completed in the right order and on time.

Check Your Math

Electronic aids such as calculators, smartphone apps, and online estimators increase the accuracy and speed of the estimating process. However, it is still important to double-check all calculations. Calculators and computer programs can still produce incorrect estimates if the information entered is incorrect. As computer programmers say, "Garbage in, garbage out."

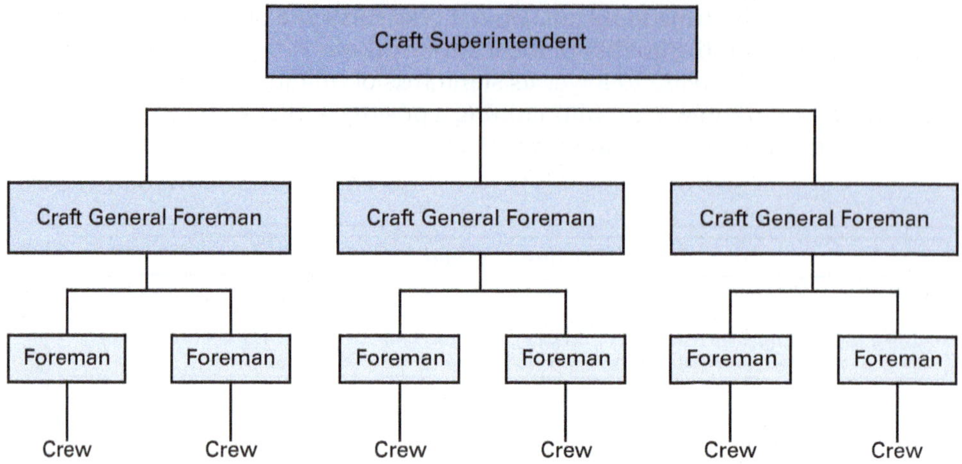

Figure 10 Sample organizational chart for a project.

Estimating Software

Back in the office, plumbers can also use special computer software, such as AccuBid by Trimble®, to develop estimates for a project. The AccuBid system includes a built-in database of materials with unit prices, labor estimates, and cost tracking codes for each item. Plumbers can use this software to prepare takeoffs on the fly. Like a spreadsheet, the estimating software reflects changes in real time so the plumber can see the effects of changes immediately, without having to recalculate by hand.

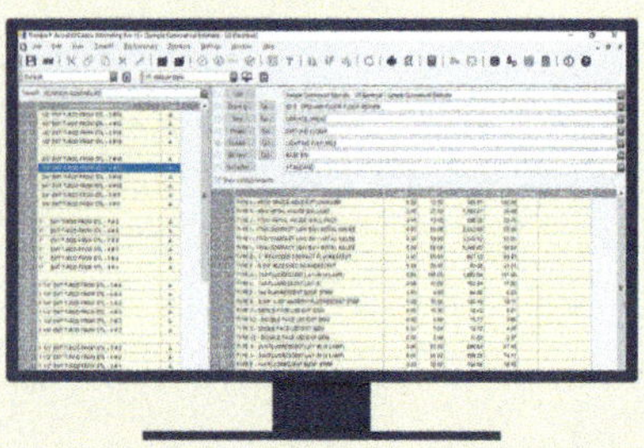

Source: Courtesy of Trimble Navigation Limited

Electronic Aids for Estimating

In the field, plumbers use specialized calculators to estimate materials. In addition to the standard calculator functions, these calculators can perform mathematical operations on dimensional measurements. Dimensional measurements can be easily converted into other English or metric units. Estimating calculators, such as the Pipe Trades Pro™ by Calculated Industries, include specialized functions designed to perform many typical calculations used in estimating the following:

- An area, given the length and width
- A volume, given the length, width, and height
- Cost estimates and per-unit costs
- The weight of materials given the volume, or the volume for a given weight
- The square-up or diagonal length of a rectangle, given the length and width

Source: Courtesy of Calculated Industries, Inc

Every task has a logical position in the overall project schedule. Each task must occur before, after, or at the same time as other tasks. **Critical path management (CPM)** is a popular way to visualize how tasks are arranged. CPM uses diagrams that show the relationships between tasks and the estimated time each will take (see *Figure 11*).

Critical path management (CPM): A task-planning process in which diagrams show the relationships between tasks and estimated completion times.

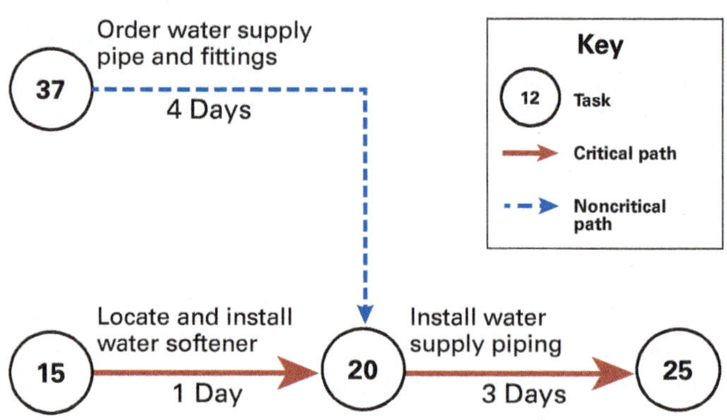

Figure 11 Sample critical path management (CPM) diagram.

Critical activities: Tasks in the critical path management process that could cause significant delay if not completed on time.

Using CPM, activities that could significantly delay the project if not completed on time are labeled **critical activities**. They are placed along the main schedule line, or critical path, in the diagram. Other, less critical, activities are shown in parallel to the critical activities. They intersect the critical path at the point where they must be performed for the project to stay on schedule. By adding up the durations of all the activities along the critical path, you can determine the duration of the entire project.

Before beginning each task, ensure that you understand what you are expected to do and what is required to accomplish the task. This includes knowing where your work is to begin and end, and what types of materials are to be installed. Divide the task into individual steps, and decide the sequence in which you will perform the work. Teams need to know the following before starting a task:

- The start and end dates of the project
- The team size for each task
- The dates materials will be delivered
- The schedules of the other trades
- Potential bottlenecks

Contractors rely on **daily work reports** to keep track of the project's schedule (*Figure 12*). The job superintendent writes up a daily work report at the end of each day's work. The daily report shows the contractor how much work was completed that day. If there were problems or delays, the contractor can decide what steps to take to keep the project on schedule. Consult the daily work reports to spot delays that could affect your team.

Daily work report: A report written by a job superintendent describing activities at a jobsite.

Daily Work Report

Project: ______ Weather: ______
Job number: ______ Temperature: ______
Superintendent: ______ Wind: ______

Work force/subcontractors:
______ Foreman ______ Plumbers
______ Pipefitters ______ (Other)

Equipment on the job: ______

Remarks: ______

Visitors:

Time:	Name:	Representing:	Remarks:

Equipment needs, rentals, problems: ______

Work completed (describe): ______

Work in progress (describe): ______

Remarks: ______

Superintendent's signature: ______ Date: ______

Figure 12 Sample daily work report for a project.

Obtain all the materials, tools, equipment, and supplies that you will need to perform the task. Perform the task according to plans, specifications, and code requirements, referring to the drawings and specifications, if needed. If the installation includes fixture rough-ins, ensure that you have the correct manuals. Confirm the finished elevations of drains installed in the ground.

One of the most important parts of task planning is to think ahead. At least once a day, stop and ask yourself the following questions:

- What materials, equipment, and information will I need during the next couple of days?
- Are all the supplies, special tools, and equipment available?
- Will I need help to perform these tasks?
- Are there any potential conflicts between my work and that of other trades?
- Do I need the cooperation of other trades to perform my work efficiently?
- What other work could I do if the task is interrupted?

If supplies, tools, and equipment are not available, inform your supervisor so the supervisor can get them to you quickly. Look ahead to ensure that you collect all the information you will need. Plan your task so you can work around other trades. If their work interferes with what you want to do, turn to other tasks waiting to be done. Good task planning ensures that all time is used wisely.

Effective Communication

A successful project depends on clear communication. Always speak with precision and provide the necessary information. When someone asks you a question, listen closely and answer promptly. These may seem like basic skills, but they are worth practicing all the time.

Getting along with your co-workers is an important part of communication. You will be working with other plumbers, workers in other trades, supervisors, plumbing suppliers, and inspectors. All these people deserve your respect. Cooperation leads to productivity, and the more productive the crew, the better the work. Address personal problems or conflicts outside of regular work hours. Otherwise, they will decrease your productivity and slow down your colleagues. Do not let a problem or misunderstanding on the job go unresolved, because it will harden into resentment and mistrust. Ask a supervisor for advice in solving the problem, if necessary.

1.0.0 Section Review

1. Profit is _____.
   a. income plus expenses
   b. income minus expenses
   c. assets plus debt
   d. assets minus debt

2. To account for all items in a company vehicle, _____.
   a. leave all items where you found them
   b. store them in a secure locker
   c. take photos of the items
   d. use a checklist

3. Typically, apprentices are expected to serve on field crews for _____.
   a. one to two years
   b. three to four years
   c. five to six years
   d. seven to eight years

2.0.0 Preparing a Material Takeoff

Objective

Identify the information required to prepare a material takeoff.

a. Identify the information required for insurance and liability.
b. Identify the information required for the estimating process.
c. Identify the information required for updates and revisions.

Performance Task

1. Prepare a material takeoff as part of an estimate.

Many commercial and municipal projects choose plumbing companies through a competitive process called bidding. In the bidding process, each contractor estimates how much time and money it will take to complete the project. The contractors then submit these estimates to the client as bids. When the client settles on the winning bid, the contractor and the client sign a **contract**. This is a legal agreement for the work to be done on the project.

Contract: A legal agreement between a contractor and a client for work to be done.

Public and commercial projects do not solicit bids in the same way. Public projects are bid under formal rules. They must be advertised and open to all qualified bidders. Commercial projects can be offered to a few selected bidders. They can also be offered to a single contractor of the owner's choosing. This is called a **sole source** bid.

Sole source: A type of bid in which a project is offered to a contractor of the client's choice.

Contractors calculate costs for a bid through a process called **estimating**. People often use the word estimating to mean guessing. An estimate is a forecast of something that has not happened yet, but it is more than a guess; plumbing estimates are careful calculations. They are based on information provided by the client and on the contractor's experience. They consider a wide variety of factors:

Estimating: Calculating the costs to complete a specific project, which are submitted as part of a bid.

Direct costs such as the following:

Direct cost: An expense for an activity or an item related directly to a project.

- Materials
- Labor
- Tools
- Equipment
- Worker insurance

Indirect costs such as the following:

Indirect cost: An expense for an activity or an item that supports the administrative and overhead activities of a business.

- Salaries and benefits
- Office rent
- Permits
- Licensing and registration fees with state and/or local municipalities
- Utilities
- Liability insurance
- Office supplies
- Advertising
- Property and equipment depreciation
- Profit

Direct costs apply to items and activities that are directly related to the project. Indirect costs are the items and activities that support the contractor's business, and they are sometimes called **overhead**. Payroll taxes are an important overhead expense. Payroll taxes are money withheld from employee paychecks to cover health insurance, Social Security, and other individual obligations.

Overhead: The costs of maintaining a business that are not applicable to a specific job. Office and administrative expenses are common examples of overhead.

Profit, as discussed in the section *The Profit and Loss Statement*, is the amount of money left after all the expenses, including overhead, have been paid. Profit allows the contractor to stay in business. If estimates are not accurate and realistic, the contractor will not make a profit.

In this section, you will learn how to estimate and bid on a plumbing job. To prepare an accurate estimate, you will use skills that are already familiar to you. Estimating involves reading blueprints and creating material takeoffs, which you have learned how to do. You will use your basic math skills to calculate the time, materials, and labor involved. You will also use effective writing skills to clearly communicate the cost estimate. The process can be intimidating at first, but with practice you will become familiar with the steps involved in estimating and bidding on a wide variety of jobs.

2.1.0 Preparing Insurance and Liability Information

You probably know that insurance protects your car, home, and health in case of an accident. As a contractor, you will need insurance too. Contractors are responsible for the safety of their employees. They are also responsible for ensuring that the client's property is not damaged while the plumbing work is being performed. These are only two of the many things a contractor is responsible for on the job. The legal term for such responsibility is liability. With insurance, the term *liability* has a different meaning than it does with accounting.

Insurance is one of the most important indirect costs on a project. Find out early on what the insurance requirements are for the project. Ensure that you have the proper coverage. For the estimate, you will need to determine the **premium**. This is what you pay the insurance company to cover the project. You can get this information from your insurance specialist. If you do not have all the necessary coverage, find out if it will be possible to get the coverage before the project's start date. Some clients prefer that the contractor have the proper insurance before submitting a bid for the work. Be sure to read the client's bid information carefully.

Premium: What the contractor pays the insurance company to ensure that the project is covered.

Clients often require contractors to guarantee that all work will be completed correctly, on time, and within budget. The guarantee usually requires that the contractor pay a penalty to the client if the work is poor, late, or more expensive than agreed upon. Such financial guarantees are called **bonds**. Performance bonds and payment bonds are common. A performance bond guarantees the work will be done according to the terms of the contract. A payment bond guarantees the contractor will pay subcontractors brought in to do specific tasks. Banks issue bonds to contractors, so a contractor needs to have a good reputation to secure a bond.

Bond: A financial guarantee that work will be completed according to the terms of a contract.

Some projects result in the creation of hazardous wastes. Pollution liability coverage covers the management of hazardous wastes. Hazardous wastes include the following:

- Acids and alkyds
- Adhesives, epoxies, and solvents
- Cleaners and degreasers
- Fuels, motor oil, and hydraulic fluids
- Insulation and asbestos
- Primers, paints, and thinners
- Scrap metal and soiled rags

Federal rules say that contractors have **cradle-to-grave responsibility** for hazardous waste. This means the contractor is responsible for the wastes from their creation to their disposal. Ensure that the estimate includes the cost of an approved waste hauler, which must dispose of the waste using proper methods. Ensure that the work crew is trained to work with hazardous materials.

Cradle-to-grave responsibility: A contractor's legal responsibility for the management of hazardous wastes from creation to disposal.

2.2.0 Preparing Estimates

Begin an estimate by reviewing records of previous jobs. Most contractors keep detailed records of past work for reference and accounting. Records of past projects should contain the following information:

- Material, equipment, and labor costs
- Labor productivity rates by task
- Effective crew sizes, and combinations of supervisors, apprentices, journeymen, and laborers

Use this information as a guide when developing a new estimate. Remember that the material, equipment, and labor costs in old project records may be out of date. Be sure to use the most recent figures.

Legal and contract requirements may affect how estimates are developed. Use the following steps as a guideline. Note that the steps will vary depending on the contractor.

Step 1 Using the project's drawings and specifications, perform a takeoff of all the materials needed for the job. Write this information on a takeoff worksheet (*Figure 13*).

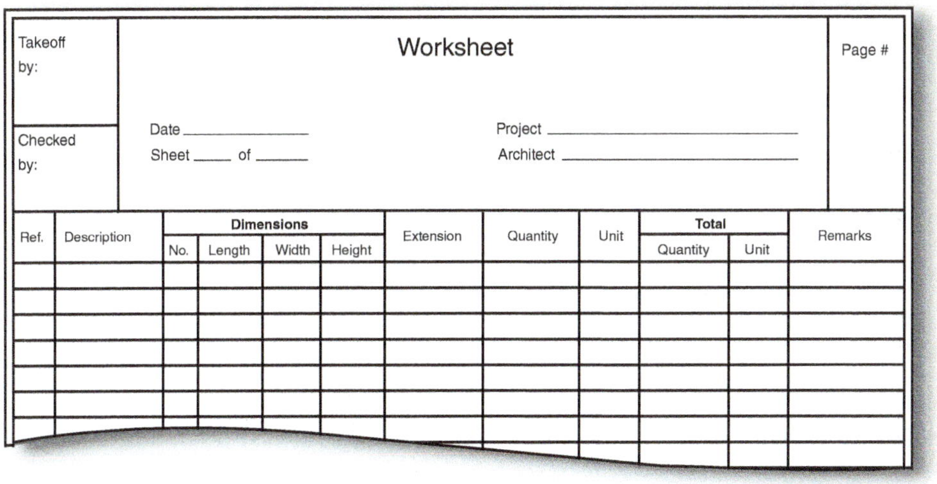

Takeoff by: | Worksheet | Page #

Checked by: | Date ______ Sheet ____ of ____ | Project ______ Architect ______

Ref.	Description	Dimensions				Extension	Quantity	Unit	Total		Remarks
		No.	Length	Width	Height				Quantity	Unit	

Figure 13 Sample takeoff worksheet.

Step 2 Determine the size of the crew needed for the job. Calculate how long it will take the crew to perform each task. These calculations are known as **production figures**. Refer to figures from past jobs. Published estimating manuals are also available, but they should be used only if figures from past jobs are not available.

Production figure: An estimate of the time required for a crew to perform a task.

Step 3 Multiply the labor hours for each task by the hourly labor rates. For subcontractor labor, have the subcontractor calculate the costs per task.

Step 4 Calculate the equipment costs. If any equipment needs to be rented, be sure to use up-to-date rental costs.

Step 5 Transfer the total material quantities, material costs, labor costs, and equipment costs to a summary sheet (*Figure 14*). Include the **unit cost** for the materials used. This is the total cost divided by the number of units of material.

Unit cost: In estimating materials used in a job, the total material cost divided by the number of units used.

Step 6 Transfer the material, labor, and equipment costs for each activity to a **recapitulation sheet**, also called a recap sheet (*Figure 15*). A recapitulation sheet serves as a task-by-task summary of an entire project.

Recapitulation sheet: A summary of material, labor, and equipment costs for each task in a project. Also called a recap sheet.

Step 7 Calculate the indirect costs, including insurance, taxes, overhead, and profit. Consult an accounting specialist for this information. Add the indirect costs to the direct costs on the recapitulation sheet. The resulting figure is the estimated total job cost.

Step 8 When the estimate is complete, submit it to the client as a bid.

The two most common mistakes plumbers make when performing an estimate are omitting all or part of an item and making an arithmetical error.

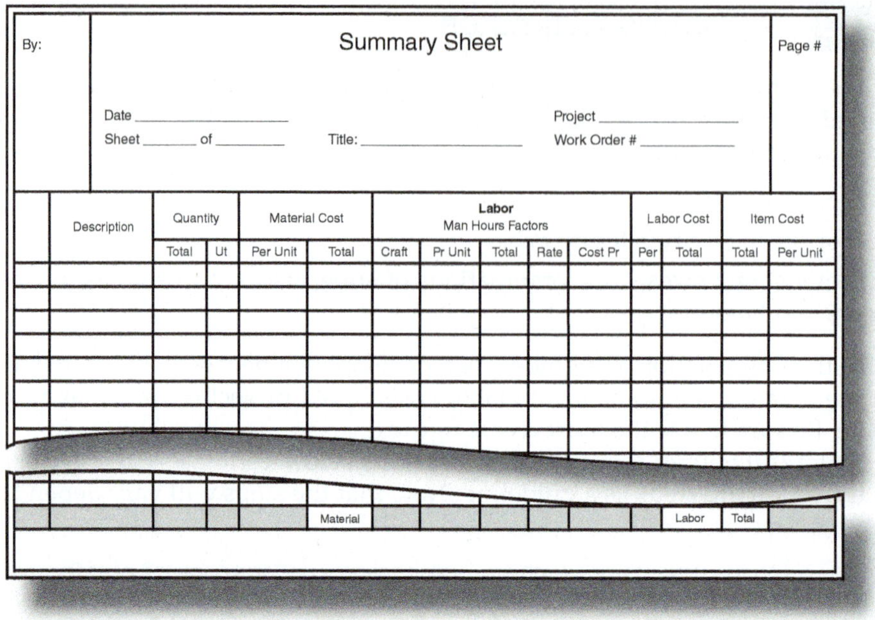
By:
Summary Sheet
Page #
Date ______
Sheet ______ of ______
Title: ______
Project ______
Work Order # ______

Description	Quantity		Material Cost		Labor Man Hours Factors					Labor Cost		Item Cost	
	Total	Ut	Per Unit	Total	Craft	Pr Unit	Total	Rate	Cost Pr	Per	Total	Total	Per Unit
				Material							Labor	Total	

Figure 14 Sample summary sheet.

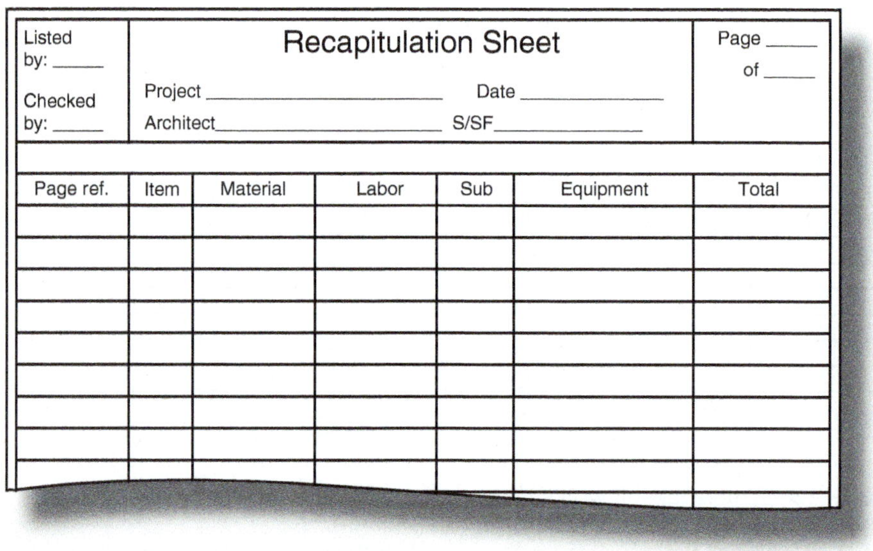
Listed by: ______
Checked by: ______
Recapitulation Sheet
Project ______ Date ______
Architect ______ S/SF ______
Page ______ of ______

Page ref.	Item	Material	Labor	Sub	Equipment	Total

Figure 15 Sample recapitulation sheet.

Always recheck the material takeoffs against the construction drawings. Check all the arithmetic at each stage. Review the summary worksheet against the individual item worksheets, then review the recapitulation sheet against the summary sheet. This way, you have a better chance of catching errors.

2.3.0 Preparing Updates and Revisions

Project schedules and budgets are continually updated throughout the life of a project. Updates allow the contractor to ensure that the costs, materials, and schedule are being followed as the project moves along. It is important to monitor the progress of a project. If expenses are not in line with the progress of the job, the contractor may need to revise the cost structure and budget. If a task is taking too long, the contractor may need to find ways to speed it up. This requires revisions to the schedule. For these reasons, project managers in the field continually track the actual costs against the estimated costs and provide updates. Doing so allows the contractor to determine if the project is staying within budget. Otherwise, the contractor may find out about problems too late to correct them.

Effective Writing

Cost estimates are the heart of a bid, but numbers alone do not sell a contractor. A bid, like anything a contractor provides to a client or potential client, must be clearly written. A poorly written bid, no matter how right the price, is likely to make the client choose another contractor.

At the end of the project, save the takeoffs, summaries, and recapitulation sheets. They can be used as reference materials for future projects. Contractors can also use them to find out how profitable a company has been up to that time. This allows contractors to plan on future expenses, because they know how much they have made.

2.0.0 Section Review

1. The guarantee that a contractor will pay subcontractors brought in to do specific tasks is called a _____.
   a. premium
   b. cradle-to-grave responsibility
   c. contract
   d. payment bond

2. When preparing an estimate, the first step is to prepare a _____.
   a. work flow
   b. takeoff
   c. summary sheet
   d. recapitulation sheet

3. During the course of a project, if the contractor finds that expenses are not staying in line with the job's progress, the contractor may need to _____.
   a. revise the cost structure and budget
   b. add more workers to the crew
   c. lay off staff
   d. confer with the architect or designer to find ways to scale back the project

Module 02401 Review Questions

1. The most important goal of a plumbing business is to _____.
   a. keep costs under control
   b. provide the highest-quality work possible at a reasonable cost to the client
   c. make clients happy so they will recommend you to friends
   d. keep prices competitive to obtain work

2. The two basic tools of accounting are the balance sheet and the _____.
   a. profit and loss statement
   b. budget document
   c. cash-flow report
   d. accounts-receivable ledger

3. The value of a business if all its debts are paid is its _____.
   a. profit margin
   b. capitalization
   c. book value
   d. equity

4. Amounts of money that are owed to the company but have not yet been received are considered _____.
   a. accounts receivable
   b. pending credits
   c. due bills
   d. accounts payable

5. Gross income is the money a company has made _____.
   a. after taxes are paid
   b. in a specified period
   c. before deducting expenses
   d. on a particular project

6. When ordering materials, choose suppliers based on the price of the materials and _____.
   a. delivery schedules
   b. warehouse location
   c. billing cycle
   d. location of the project

7. An itemized list of the prices of materials furnished and the terms of agreement is called a(n) _____.
   a. invoice
   b. bill of lading
   c. account payable
   d. packing list

8. When additional work outside the scope of the original contract is needed, the client issues a(n) _____.
   a. amendment
   b. change order
   c. contract extension
   d. rework ticket

9. The person responsible for coordinating the work at the jobsite is the _____.
   a. job superintendent
   b. general contractor
   c. plumbing foreman
   d. project manager

10. Undertaking and completing jobs in the proper order and on time is the purpose of _____.
    a. budgeting
    b. priority assignment
    c. task planning
    d. time study

11. To help keep track of project progress, the job superintendent provides the contractor with a _____.
    a. weekly progress summary
    b. task-specific checklist
    c. labor-and-materials utilization form
    d. daily work report

12. On a CPM diagram, the main schedule line is referred to as the _____.
    a. backbone
    b. primary track
    c. critical path
    d. control vector

13. A client request for a bid from only one contractor is called _____.
    a. directed bidding
    b. sole-source bidding
    c. assigned bidding
    d. preferred-contractor bidding

14. Office rents, insurance, and permit costs are examples of _____.
    a. indirect costs
    b. out-of-pocket costs
    c. controllable costs
    d. plumbing fees

15. The recapitulation sheet used in an estimate is a _____.
    a. compilation of anticipated labor hours
    b. detailed materials list
    c. task-by-task summary of the entire project
    d. listing of direct and indirect costs

Answers to odd-numbered review questions are found in the Module Review Answer Key at the back of this book.

Answers to Section Review Questions

Answer	Section	Objective
Section One		
1. b	1.1.0	1a
2. d	1.2.4	1b
3. b	1.3.0	1c
Section Two		
1. d	2.1.0	2a
2. b	2.2.0	2b
3. a	2.3.0	2c

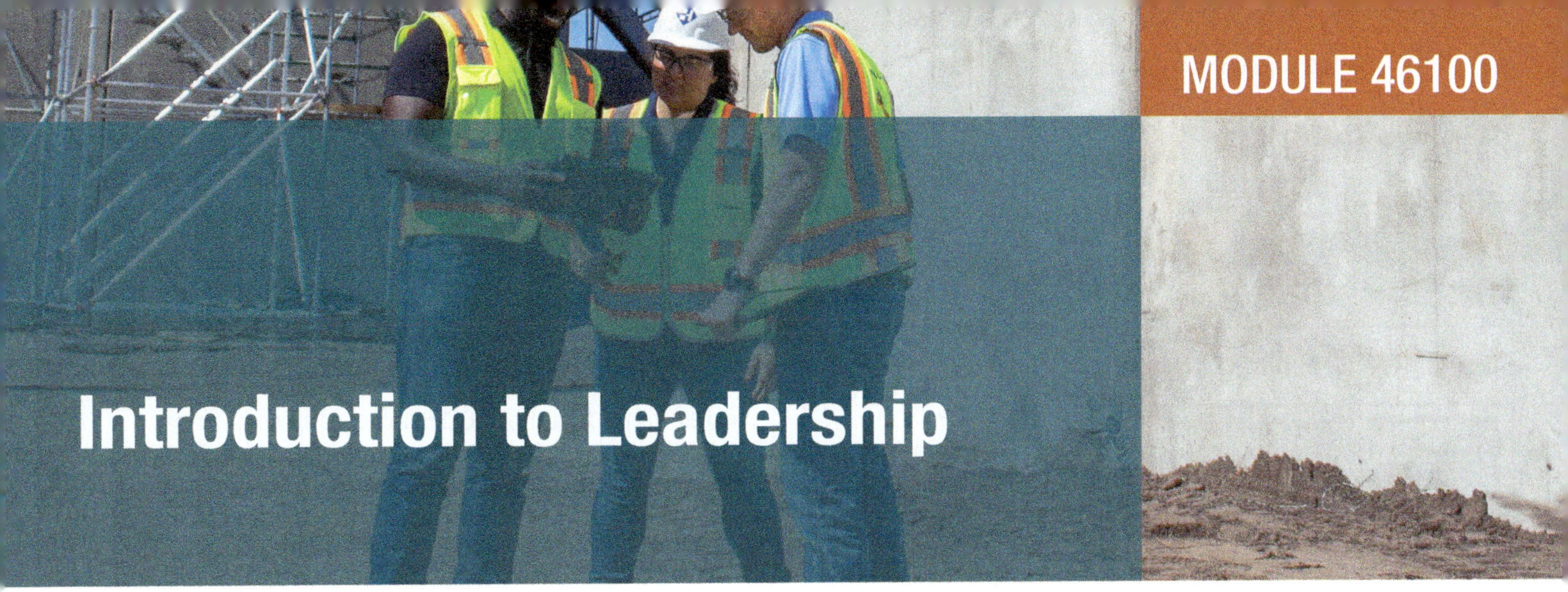

MODULE 46100

Introduction to Leadership

Objectives

Successful completion of this module prepares you to do the following:

1. Explain business and organizational structures in the construction industry.
   a. Understand organizational charts.
   b. Describe organizational roles and relationships, including job descriptions and responsibility, authority, and accountability.
2. Describe frontline leadership and communication skills.
   a. Describe the qualities and role of a leader, including leadership styles and ethical leadership.
   b. Explain the communication process and distinguish between verbal, nonverbal, written, and visual communication types.
   c. Compare problem-solving and decision-making.
3. Identify key skills for team leadership and motivation.
   a. Choose strategies to motivate individuals and a team.
   b. Describe key approaches to building and leading teams.
   c. Recognize how to approach leadership challenges, such as transitioning from peer to leader, addressing performance issues, and resolving conflict.
4. Identify a frontline leader's safety leadership responsibilities.
   a. Assess the impact of incidents.
   b. Explain the purpose of OSHA and describe the role of OSHA in administering worker safety.
   c. Describe the importance of safety and health programs.
   d. Explain the frontline leader's role in incident preparedness and response.
   e. Recognize the importance of supporting your crew's total health.
5. Demonstrate an understanding of construction project phases, project planning and scheduling, field reporting, and cost and resource management.
   a. Identify construction project phases.
   b. Summarize crew planning and scheduling responsibilities, including creating lookahead schedules.
   c. Describe the frontline leader's role in field reporting.
   d. Explain approaches to managing costs and controlling project resources, such as materials, tools, and equipment.

Performance Tasks

Under supervision, you should be able to do the following:

1. Organize and deliver a safety-related toolbox talk, applying communication skills.
2. Complete a job safety analysis.
3. Develop and present a lookahead schedule.

Overview

As a frontline leader, you must be able to communicate effectively, provide direction to your crew, plan and schedule work, manage costs, and ensure the jobsite is safe for everyone. Whether you are already a frontline leader or want to become one, this module will help you learn more about the requirements and skills needed to succeed.

Digital Resources for Plumbing

Scan this code using the camera on your phone or mobile device to view the digital resources related to this craft.

SCAN ME

NCCER Industry–Recognized Credentials

If you are training through an NCCER-accredited sponsor, you may be eligible for credentials from NCCER. The ID number for this module is 46100. Note that this module may have been used in other NCCER curricula and may apply to other level completions. Contact NCCER at 1.888.622.3720 or go to **www.nccer.org** for more information.

You can also show off your industry-recognized credentials online with NCCER's digital credentials. Transform your knowledge, skills, and achievements into badges that you can share across social media platforms, send to your network, and add to your resume. For more information, visit **www.nccer.org**.

1.0.0 Business and Organizational Structures

Performance Tasks

There are no Performance Tasks in this section.

Objective

Explain business and organizational structures in the construction industry.

a. Understand organizational charts.

b. Describe organizational roles and relationships, including job descriptions and responsibility, authority, and accountability.

Crew leader: A person appointed leader of a team or crew, usually an experienced craft professional who has demonstrated leadership qualities. Also known as *frontline leader*, *frontline supervisor*, *lead person*, or *foreman*.

Superintendent: An on-site supervisor who is responsible for one or more crew leaders or frontline supervisors.

Project manager: Person responsible for managing one or more projects.

When a crew is assembled to complete a job, one person is appointed the leader. Different companies give this role different titles, such as **crew leader**, *frontline supervisor*, *lead person*, or *foreman*. In this module, this role is called the *crew leader* or *frontline leader*. A crew leader or frontline leader is often an experienced craft professional who has demonstrated leadership qualities.

A **superintendent** is essentially an on-site field supervisor who is responsible for one or more crew leaders or frontline supervisors. A **project manager** may be responsible for managing one or more projects. This training will concentrate primarily on the supervisory role of the frontline leader. These individuals work with others on projects as part of an organization.

1.1.0 Organizations and Organizational Structures

Organization: A group of people working together.

Organizational chart: A diagram that shows how the various management and operational responsibilities relate to each other within an organization. Named positions are ranked top to bottom, from those functional units with the most and broadest authority to those with the least. Lines connect positions and indicate chains of authority and other relationships.

An **organization** is a group of people working together. Businesses, clubs, and associations are examples of types of organizations. On a jobsite, the relationships among the people within the company or project create a type of organization.

1.1.1 Organizational Charts

A formal organization will have an **organizational chart** that shows all of the positions in the organization and how those positions are related. A detailed organizational chart, or org chart, may also show the name of the person in each position. *Figure 1* shows an example of an organizational chart for a construction company. Each position on that organizational chart represents an opportunity for career advancement for a crew leader.

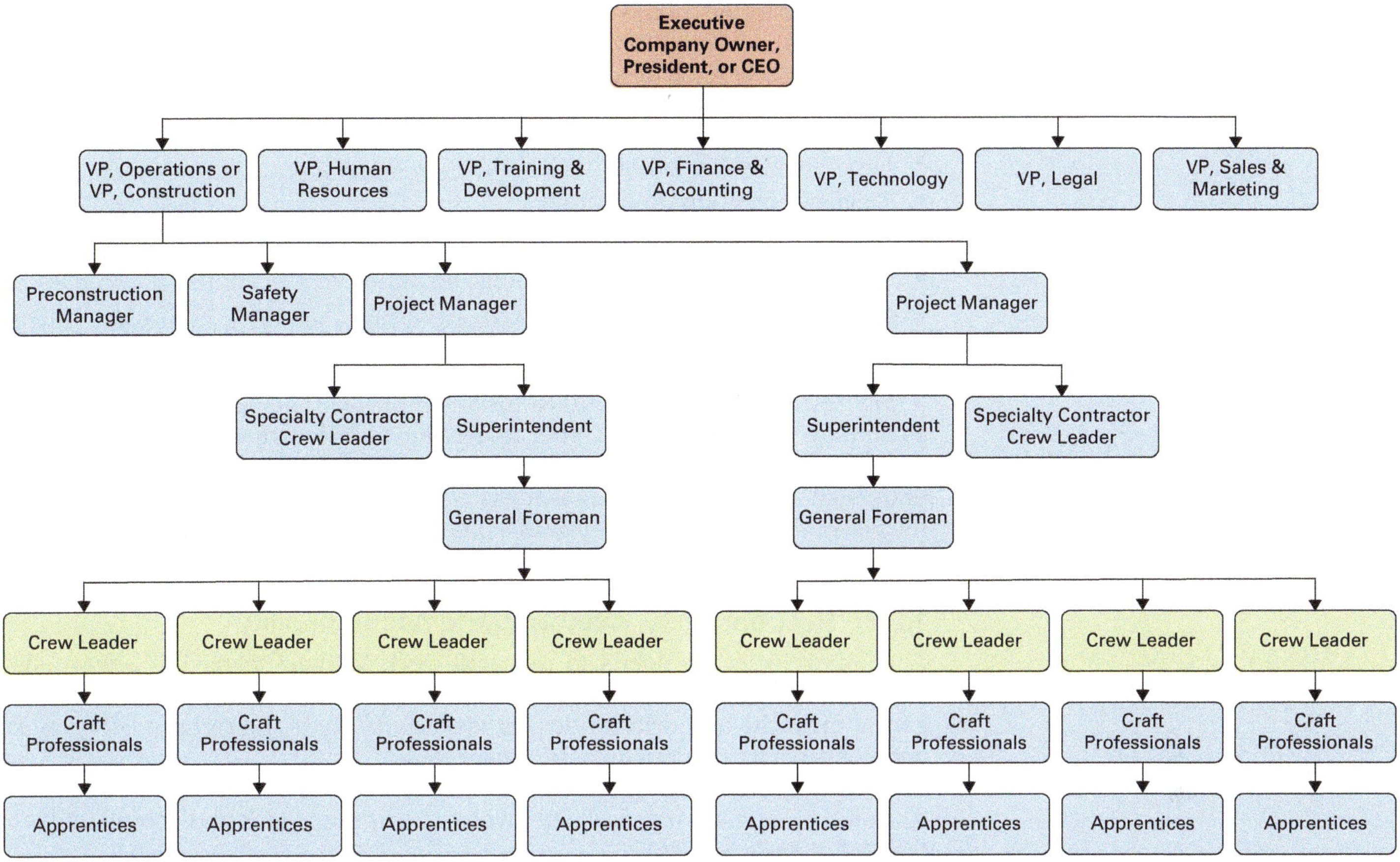

Figure 1 Sample organization chart for a construction company.

As shown in *Figure 1*, most companies have separate groups or departments that focus on running specific functions. Smaller companies may have one or two individuals who perform multiple functions. Larger companies will often create specific roles or departments to handle more complex projects and larger teams.

Companies may have additional roles or teams, such as architectural and engineering design functions. Depending on the size of these groups, they may become separate departments.

NOTE

Organizational charts vary widely based on the size of the company and the specific construction sector in which they operate. For example, an industrial company might have crew leaders who report to a plant manager. Some larger companies will have a superintendent per shift, who will report to a project superintendent, who reports to the project manager. If you are interested in learning more about your company's organizational chart, ask your supervisor or manager.

1.2.0 Organizational Roles and Relationships

As a crew leader, you will assign others the responsibility to perform specific jobs or tasks. Your crew and team members will perform best if the following is true:

- They have a clear job description that explains their job duties.
- They have the required skills and abilities to perform the task.
- They know who has **authority**, **responsibility**, and **accountability**.
- They have a clear understanding of expectations.
- They are trained on policies and procedures.
- They have all needed tools, equipment, and material.

Authority: The right to act or make decisions in carrying out an assignment.

Responsibility: The act of accepting a task and completing it to the best of one's ability.

Accountability: In relation to job completion, accepting responsibility and being personally answerable for the results.

1.2.1 Job Descriptions

As part of the hiring process, most companies give each employee a **job description**. A job description is a statement that describes the duties and responsibilities of a position with a company. A **duty** is a work task an employee is expected to perform. Responsibility involves accepting a task and completing it to the best of one's ability.

Job description: A description of the scope and responsibilities of a worker's job so that the individual and others understand what the job entails.

Duty: A work task an employee is expected to perform.

A job description should include enough detail so that an employee understands their job. It also should have any information needed to evaluate the employee's job performance. A job description should contain at least the following:

- Job title
- The supervisor to whom the position reports
- A general position description
- Specific duties and responsibilities
- Required and desired qualifications, including certifications and licenses
- Other information, such as working conditions, job type, hours, pay ranges, and more

Job descriptions set a standard for an employee. They outline the tasks that an employee should handle. After reviewing a job description, a new employee should understand all of the duties and responsibilities of the job. Having a clear understanding of the job makes it easier for new employees to work well with the rest of the crew while they gain experience. Clear job descriptions help crew leaders to judge performance and identify training needs for new employees.

1.2.2 Responsibility, Authority, and Accountability

A good crew leader needs to delegate work to their team, but always stay accountable for the outcome. When you assign a crew member work, you are giving them the authority to do that task. *Authority* is the right to act or make decisions in carrying out an assignment.

When the employee takes on a job or assignment, they take on responsibility. As noted previously, *responsibility* involves accepting a task and completing it to the best of one's ability.

In relation to job completion, *accountability* is accepting responsibility and being personally answerable for the results.

Each project or company has differences in the type and amount of authority, responsibility, and accountability a supervisor or worker has. As a crew leader, you must give sufficient responsibility and authority to allow employees to be successful at their job activities. You then must hold employees responsible for completing those assigned activities.

As shown in *Figure 2*, even though you may delegate authority and responsibility to crew members, the overall accountability for tasks assigned to a crew always rests with the frontline leader or crew leader.

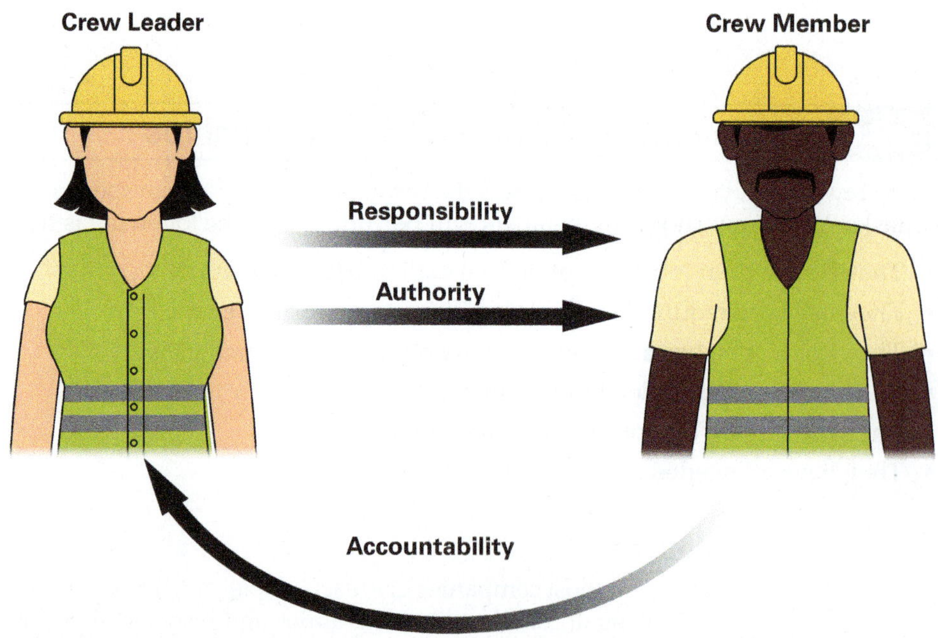

Figure 2 Accountability always rests with the frontline leader.

1.2.3 Policies and Procedures

Most companies have formal policies and procedures established to help crew leaders carry out their duties.

- A **policy** is a general statement establishing guidelines for a specific activity. Examples include policies on vacations, breaks, workplace safety, and checking out tools.
- **Procedures** are formal instructions to carry out and meet policies. For example, a procedure written to implement a policy on workplace safety would include guidelines for reporting incidents and following general safety practices.

Policy: A general statement establishing guidelines for a specific activity.

Procedures: Formal instructions to carry out and meet policies.

A crew leader must understand any company policies and procedures that relate to their work, especially safety practices. The **Occupational Safety and Health Administration (OSHA)** is a national public health agency that is part of the US Department of Labor. To promote a safe and healthy work environment, OSHA issues standards and rules for working conditions, facilities, equipment, tools, and work processes. When OSHA inspectors visit a jobsite, they often question crew leaders and employees about the company safety policies. During incident investigations, the inspector will verify if the responsible crew leader knew and followed the applicable company policy.

Occupational Safety and Health Administration (OSHA): A national public health agency that is part of the US Department of Labor. To promote a safe and healthy work environment, OSHA issues standards and rules for working conditions, facilities, equipment, tools, and work processes.

1.0.0 Section Review

1. Which of the following is shown on an organizational chart?
   a. How to organize materials on a jobsite
   b. How positions in an organization are related
   c. How prevailing wages are determined
   d. How to coordinate work with other crafts

2. Although you may delegate authority and responsibility for tasks to crew members, as a frontline leader, you have the overall _____.
   a. policy
   b. accountability
   c. job description
   d. authority

2.0.0 Leadership and Communication

Objective

Describe frontline leadership and communication skills.

a. Describe the qualities and role of a leader, including leadership styles and ethical leadership.
b. Explain the communication process and distinguish between verbal, nonverbal, written, and visual communication types.
c. Compare problem-solving and decision-making.

Performance Tasks

There are no Performance Tasks in this section.

In addition to being an experienced craft professional, a frontline leader also must demonstrate leadership qualities and strong morals. They must be able to communicate effectively, solve problems, and make good decisions. Frontline leaders are generally promoted up from a work crew. A worker's ability to accomplish tasks, get along with others, meet schedules, and stay within the budget have a significant influence on the selection process. The frontline leader must lead the team to work safely and provide a quality product.

This section explains the importance of developing and improving leadership skills as a frontline leader. It will cover the qualities and roles of a leader, as well as effective ways to communicate with coworkers and employees at all levels, solve problems, and make decisions.

2.1.0 The Qualities and Role of a Leader

There are many ways to define leadership. One simple definition of leadership is the ability of a person to influence others to achieve a common goal. A leader is someone who sees how to accomplish or improve things. A leader then rallies people to move toward that vision. Effective leadership requires communicating ideas effectively to engage others to act accordingly.

Leaders often have power based on a role or title. However, leadership is much more than power. An effective leader inspires others to act, while also directing the way that they act. A leader must have respect, so others follow their directions. A leader also must have the critical thinking skills to know the best way to use available resources.

Motivate: To make someone feel determined or enthusiastic or to make someone behave in a particular way.

Some people develop leadership qualities as part of their upbringing. Others may work hard to develop the qualities that **motivate** others to follow and perform. Leadership is a skill you can learn through experience and training, as well as hard work, focus, and practice.

2.1.1 Functions of a Leader

What is the purpose or function of a leader? The answer will vary based on the specific project, the work environment, the team they lead, and the tasks they must perform.

As shown in *Table 1*, a frontline leader has certain functions that apply to almost every work situation.

TABLE 1 Frontline Leadership Functions

Leadership Duties and Responsibilities	Leadership Functions
Planning the Work	Organize, plan, staff, direct, and control work
Safety and Compliance	Ensure that the team understands and follows company policies and procedures
Empathy	Be sensitive to the needs and qualities of a diverse workforce
Motivating and Coaching	Give team members the confidence to make decisions and take responsibility for their work
Conflict Resolution	Maintain a cohesive team by resolving tensions and differences
Professionalism	Represent the team or group and represent the company or organization
Accountability	Accept responsibility for the successes and failures of the team's performance

2.1.2 Leadership Traits and Behavior

Trait: A distinguishing quality of a person.

Behavior: The way a person responds to a specific situation.

As previously noted, leadership is a skill you can learn and develop through experience and training. As you continue to grow your leadership skills, remember that while each leader has some unique approaches, the most effective leaders share some common traits and behaviors. A **trait** is a distinguishing quality of a person. A **behavior** is the way a person responds to a specific situation.

Effective leaders share the following traits and behaviors:

- Follows company policies and procedures
- Has necessary technical knowledge

- Can plan, prioritize, and organize work
- Leads by example
- Is fair and ethical
- Solves problems, makes decisions, and assumes responsibility
- Communicates effectively
- Can promote an idea
- Is enthusiastic and shows strong initiative
- Is able to motivate individuals and teams
- Is loyal to their company and crew
- Extends respect and trust to all team members
- Can teach others and learn from others

First and above all, excellent leaders lead by example. They work and live by the standards that they set for their crew members. A leader will not ask someone to complete a task that they are not willing to do themselves.

Leaders tend to have a high level of drive and determination, as well as a persistent attitude. When faced with obstacles, effective leaders don't get discouraged. Leaders identify the potential problems and make plans to overcome them. They then work toward achieving the intended goal. In the event of failure, effective leaders learn from their mistakes and apply that knowledge to future situations. They also learn from their successes.

Respect is an important aspect of leadership. Leaders earn respect by being fair to employees and by listening to their complaints and suggestions. Using appropriate incentives and **rewards** to motivate a crew helps leaders earn respect. In addition, frontline leaders who have a positive attitude tend to gain the respect of their team and their peers. A **peer** is someone who is in a position or level in an organization equal or similar to yours.

Rewards: Tangible items given to an employee as recognition for good work.

Peer: Someone who is in a position or level in an organization equal or similar to yours.

Trust is an equally important part of being a great frontline leader. Trust is a group's belief that their leader is committed, capable, and will do what's best for all involved. To build trust, effective leaders create interactions and conversations with the team. They treat everyone on the team as an individual.

Leaders are good communicators who lay out the goals of a project to their crew members clearly. To do this, you may need to overcome language barriers, bias, or differences in personalities to ensure that each member of the crew understands the established goals of the project.

Effective leaders can motivate their team to work to their full potential. They will work to develop crew members' skills. They encourage them to improve and learn so they can contribute more to the team effort. Effective leaders strive for excellence from themselves and their team, so they work hard to provide the necessary skills and leadership.

It is important for leaders to have an expert knowledge of the activities they supervise. This is important because the crew members need to know that they have someone to turn to when they have a question or a problem, need some guidance, or must make modifications or changes on the job.

In addition, leaders must possess organizational skills. They know what needs to be accomplished and they use their resources to make it happen. Because they can't do it alone, leaders require the help of their team members to share in the workload. Effective leaders delegate work to their crew members. They also implement company policies and procedures to complete the work safely, effectively, and efficiently.

Finally, outstanding leaders have the experience, authority, and self-confidence that allow them to make decisions and solve problems. To accomplish their goals, leaders must be able to calculate risks, take in and understand information, set a course of action, make decisions, and assume the responsibility for those decisions.

2.1.3 Leadership Styles

Many terms are used to describe different leadership styles. As a frontline leader, you will never lead the exact same project twice. Having a leadership style that is also adaptable is important for you, the project, and your company's success.

Situational leadership: A leadership style that involves adjusting to the specific situation or task and the leadership needs of the group.

Situational leadership is a leadership approach that involves adjusting the way you lead to reflect the specific situation or task and the needs of the group. The main idea behind situational leadership is that there is no "one size fits all" or best style of leadership. Instead, you must adapt or adjust your leadership style to the situation.

As shown in *Figure 3*, situational leadership provides a model with four primary leadership styles:

- *Directing:* The leader tells a team what to do and how to do it.
- *Coaching:* The leader coaches individuals to improve skills. The leader encourages more back-and-forth between leaders and the team.
- *Supporting:* A leader supports the group, participating to help increase motivation. The leader offers less direction. Instead, the leader has team members take a more active role in coming up with ideas and making decisions.
- *Delegating:* A leader allows other people to own the task, and they facilitate it happening. Team members make most of the decisions and take most of the responsibility for what happens.

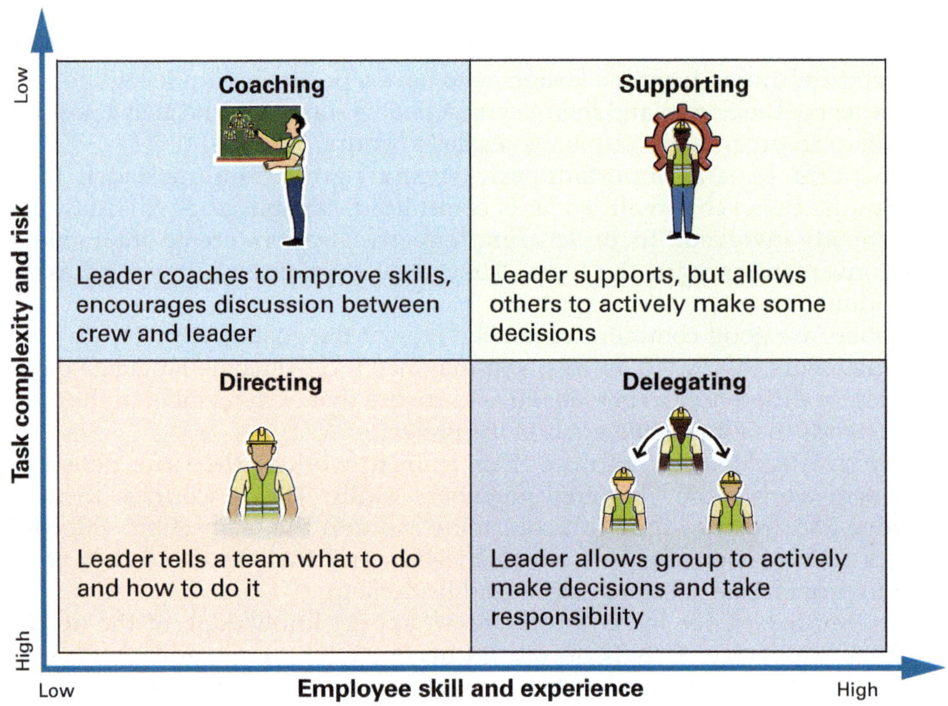

Figure 3 Situational leadership styles.

Knowing which leadership style to apply depends on several key factors:

- *Team relationships:* Consider your team, how well they work together, and how productive they are. A group that is not efficient or productive might benefit from a Directing style. That style emphasizes direction, organization, and clearly defined roles. On the other hand, a highly cohesive group might benefit from a Supporting style, which gives the team more inputs on decisions.
- *Team skills:* Consider the skills and abilities of the individuals completing the task. Giving a task to a team or team member who lacks the skills to complete the task is risky, is dangerous, and may end in failure.

- *Work tasks:* Consider the actual work or tasks at hand. You must have an expert knowledge of the tasks and activities to ensure they are done well, completely, and in the safest way possible. You may need to coach or train workers who do not have experience to help them complete the task.
- *Work situation:* Consider the situation you're facing. In some cases, the situation itself will override all other factors.

For example, if the crew does not have enough experience for the job ahead, you may want to use a *Directing* style. A Directing style works with people and teams who have limited experience or skill performing a task. It also works when team members are insecure or unmotivated to try a task. The Directing style requires your close supervision to spot signs of progress, which you can use to motivate ongoing development. If a job task or situation is very risky, a Directing style can help reduce the risk by providing clear and very specific direction.

If your team is working well and showing strong progress on work tasks, you might use a *Coaching* style. You still direct how and when specific tasks are completed, but you also discuss why the task matters and where it fits into the overall project or job. With this style, you recognize the enthusiasm, interest, and commitment of the team. You also reinforce the importance of learning and gaining task-related experience.

You may want to use a *Supporting* style if the team is reasonably skilled and work tasks are not complex, but commitment and attitude are an issue. The goal of the Supporting style is to create alignment. As a leader, you can discuss a person's commitment by asking open-ended questions. These questions should help you understand what is behind the performance challenges and create possible solutions. The Supporting style of leadership also works well when work is creative, as it is helpful to brainstorm and exchange ideas with crew members.

The *Delegating* style works best with an experienced, strong, committed crew on a well-defined project. The Delegating style of leadership is also effective when jobs have repetitive tasks that require few decisions. To use the Delegating style, however, your company must give you enough authority as the frontline leader to do the job.

Effective leadership takes many forms. The correct style for a particular situation depends on the team and the work you need to accomplish. Knowing the factors to consider can help you decide which type of leadership to apply in which situation. This flexibility allows you to meet each situation with the leadership style that brings out the best in your team and gives them the highest chance of success.

2.1.4 Ethics in Leadership

Ethics are the moral principles that guide a person's actions when dealing with others. *Figure 4* outlines three basic types of ethics. Frontline leaders must maintain strong ethics and the highest standards of honesty and legality. Every day, a leader makes decisions that have fairness and moral implications. When you make an unethical decision, it reflects poorly on you. It also impacts your team, other project team members, and the company as a whole.

As a leader, you will find yourself in situations where you will need to weigh the ethical results of a choice or decision. For example, imagine a crew member shows symptoms of heat exhaustion. Do you keep that employee working just because the superintendent says the project is behind schedule? If you are aware that your crew did not properly erect the reinforcing steel, do you stop the pour and correct the situation? In both cases, you must act ethically as a leader and make the right choice. If supervision ever asks a frontline leader to carry out an unethical decision, that individual must inform the next level of authority of the unethical nature of the issue.

Ethics: The moral principles that guide an individual's or organization's actions when dealing with others.

NOTE

Anyone who is aware of illegal activity can face legal concerns simply by not acting. This is true even if the activities were requested by one's supervisor.

Business or legal ethics	**Professional or balanced ethics**	**Situational ethics**
Adheres to all laws and published regulations related to business relationships or activities.	Carries out all activities in such a manner as to be honest and fair to everyone under one's authority.	Relates to specific activities or events that may initially appear to be a gray area. For example, you may ask yourself, "How will I feel about myself if my actions were going to be published in the news or if I need to justify my actions to my family, friends, and colleagues? Would I still do the same thing?"

Figure 4 Types of ethics.

2.2.0 Communication

Communication: The process of exchanging facts, information, and ideas. Communication comes in many forms, including verbal, written, visual, and nonverbal.

Successful crew leaders learn to communicate well with people at all levels of an organization. **Communication** is the act of accurately and effectively sharing facts, feelings, and opinions with another. Simply stated, communication is the process of exchanging information and ideas. Communication comes in many forms, including verbal, written, visual, and nonverbal.

2.2.1 Why Communication Matters

Effective communication on a jobsite or project is critical for many reasons (*Figure 5*). First and foremost, clear communication is a key part of safety. Everyone on a jobsite must receive and understand communications about safety policies and procedures. If this does not happen, it creates safety issues for everyone in the group.

Miscommunication is also expensive (*Figure 5*). One study indicated that over half of all rework is due to poor communication and bad information. If communication on a construction site fails, it can cause delays and quality issues for the

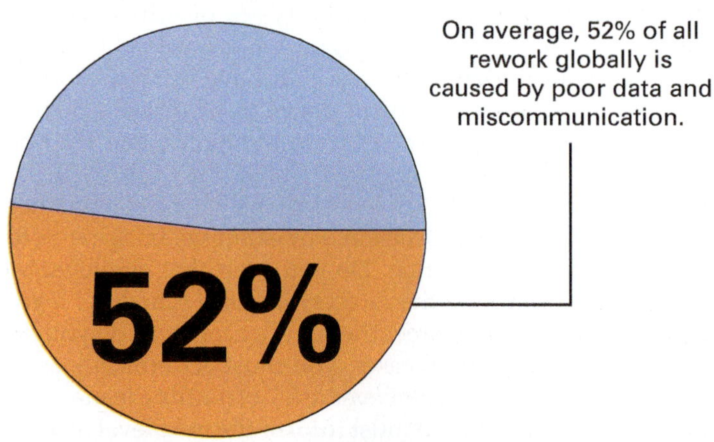

Figure 5 Industry impacts of miscommunication in the construction industry.

project. It can also result in damage to equipment or property. The good news is that communication is a skill that can be taught, practiced, and improved.

2.2.2 The Communication Process

Clear communication has three basic steps, as shown in *Figure 6*.

1. *Sender sends message to receiver.* A sender sends a message to a receiver, as a verbal, written, visual, or nonverbal communication. A message can be a set of directions, an opinion, or a personal message such as praise or a warning. Whatever it is, a message is an idea or fact that the sender wants the receiver to know.
2. *Receiver interprets message and repeats the message to provide feedback.* The receiver is the person who takes in the message. When the receiver gets the message, they figure out what it means by listening, viewing, or reading, and then repeat the message to provide feedback. *Feedback* is the receiver's communication back to the sender in response to the message. Feedback is a very important part of the communication process. It shows the sender how the receiver interpreted the message and if the receiver understood it as intended. In other words, feedback is a checkpoint to make sure the receiver and sender are on the same page.
3. *Sender validates that receiver correctly understood the message.* The sender confirms the receiver heard and understood the message.

This process is called **three-way communication**. Three-way communication is also sometimes called *repeat-back communication*.

Three-way communication is a simple way to make sure that communication is clear. If this process is easy, why is good communication so hard to achieve? When we try to communicate, many things can get in the way. These obstacles or **distractions** are anything that disrupts the communication flow between sender and receiver. Also called *noise*, distractions can include poorly worded messages, emotions, interpretations of a word, or the actual noise from a construction site. Distractions can happen at any point in the communication process.

Three-way communication: A form of communication with an initial message from a sender, which is then repeated by the receiver to confirm accuracy, and then validated by the sender. Also called *repeat-back communication*.

Distractions: Anything that disrupts the communication flow between sender and receiver. Also called *noise*.

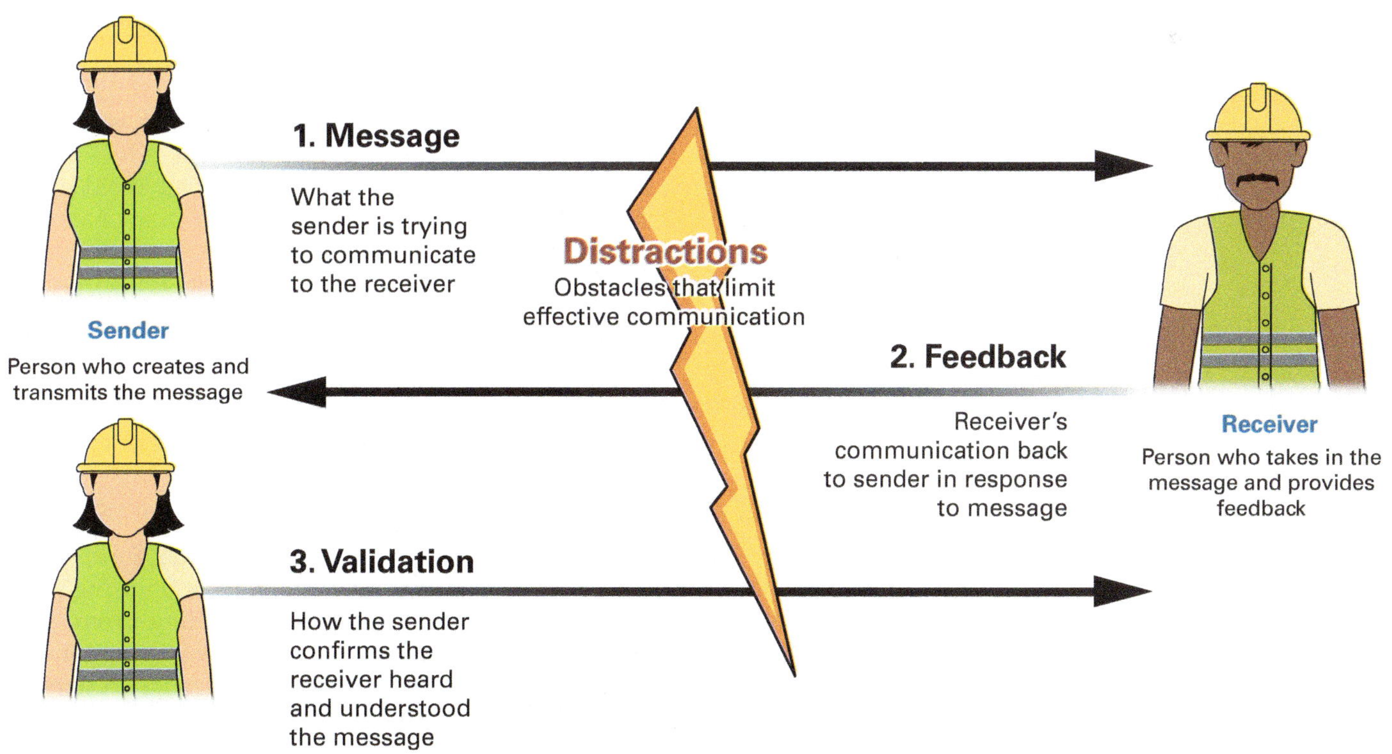

Figure 6 The three-way communication process.

Verbal	**Written**
• Spoken or oral communication • Examples: face-to-face conversations, telephone call, videoconference, voice chat	• Messages or information sent using words • Can be digital or physical • Examples: text message, email, training guide, weekly report, change request, purchase order
Communication Types	
Visual	**Nonverbal**
• Using visual elements, such as photos, graphics, charts, colors, and fonts to convey messages • Examples: sketches, diagrams, graphs, photos, construction drawings, signs	• Things the speaker and the receiver see or hear while communicating • Examples: appearance, personal environment, use of time, facial expressions, body movements, hand gestures, tone of voice, and eye contact

Figure 7 Types of communication.

As noted in *Figure 7*, there are four different types of three-way communication. The following sections review these different types of communication and discuss ways to increase the effectiveness of your communication, regardless of type.

2.2.3 Verbal Communication

Verbal communication: Messages that are sent using voice or other audible sounds. Also called *spoken communication* or *oral communication*.

Verbal communication includes messages that are sent face-to-face or by telephone, videoconferencing, voice chat, or other methods. Verbal communication is also called also *spoken* or *oral* communication.

With verbal communication, the way you speak is as important as the message itself. As a leader, you should assess your audience before you speak. Even in a meeting as standard as a daily huddle, using clear verbal communication sets your crew up for a successful day. Consider each person's knowledge of the topic and any language, gender, or cultural differences. Do not talk down to anyone in a group. Everyone, whether in a senior or junior position, deserves respect and courtesy.

You should also be sure to use words and phrases that the receiver can understand. For example, avoid technical language, acronyms, or slang if the receiver does not know these terms. In addition, speak at a reasonable pace, which gives the audience time to understand one point at a time.

If you are receiving a message, be sure to listen actively and avoid distractions that interfere with the delivery of the message. There are many barriers to effective listening, particularly on a busy construction jobsite.

As a receiver, you can show that you are actively listening. You can accomplish this without saying a word. Examples include maintaining eye contact, nodding your head, and taking notes. If you are not clear on the message, ask questions to get additional explanation or information. You can also provide feedback by using your own words, or *paraphrasing*, to repeat a message back to the sender. This lets the sender know that you understood the message.

2.2.4 Written or Visual Communication

Written communication: The process of sending messages or information using words, whether digitally or physically.

Visual communication: The process of using visual elements, such as photos, graphics, colors, and fonts, to convey messages.

Much of the information on a job or project is in written or visual form. **Written communication** is the process of sending messages or information using words, whether digitally or physically. Examples include everything from a text message to an email to a printed training guide. A lot of project documentation is written, including weekly reports, requests for changes, purchase orders, and contracts. Many of these items are written so they can form a permanent record for business and historical purposes.

Visual communication is the process of using visual elements to convey messages. Photos, diagrams, graphics, charts, and colors are all types of visual communication. Visual communication is widely used because it is an effective way to convey information quickly. Traffic signs, for example, use standard visual symbols because they are clear and well-understood. Many messages on a job also use visual communication. Examples include the project plans, drawings, or a video of a worksite incident. Visual communication can help workers who speak different languages or have limited language skills.

Written and visual communication used together are even more clear. For example, tags and signs on a project (*Figure 8*) use both written words and visual elements to help make them clear to all workers on a jobsite.

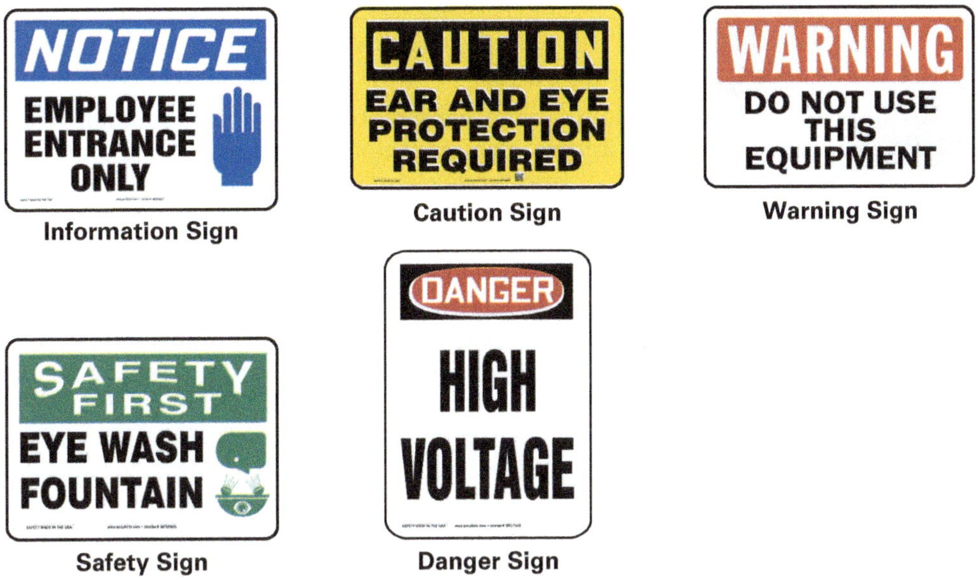

Figure 8 Communication tags and signs combine written and visual communication.
Source: Courtesy of AccuformNMC.com LLC

When writing or creating a written or visual message, you first should assess the receiver (the reader). The receiver must be able to read the message and understand the content. If they cannot, the communication process will be unsuccessful. Always consider the actual meaning of words or diagrams and how others might interpret them. *Table 2* and *Table 3* provide checklists to help make your written and visual communications effective.

TABLE 2 Checklist for Effective Written Communication

Did You . . .?	Yes/No
Assess your audience. Consider existing knowledge. Consider any language, gender, or cultural differences.	
State the purpose of the message clearly.	
Present the information in a logical manner.	
Provide an adequate level of detail.	
Avoid emotion-packed words or phrases.	
Avoid making judgments, unless asked to do so.	
Avoid using unfamiliar acronyms or abbreviations.	
Avoid using technical language unless you know your audience knows those terms.	
Make sure that the communication is clear and your handwriting is readable.	
Proofread your work. Check for typos, spelling errors, and grammatical errors. Do not assume spell-check or autocorrect got it right.	
Be prepared to provide a verbal or visual explanation if needed.	

TABLE 3 Checklist for Effective Visual Communication

Did You . . .?	Yes/No
Avoid complex graphics. (Simple is better.)	
Use colors and graphical elements to highlight key information.	
Ensure that the visual is large enough to be useful.	
Present the information in a logical order.	
Provide an adequate level of detail and accuracy.	
Provide a legend or written description if you include symbols.	
Be prepared to provide a written or verbal explanation of the diagram if needed.	

2.2.5 Nonverbal Communication

Nonverbal communication: Things the speaker and the receiver see or hear while communicating. Examples include facial expressions, body movements, hand gestures, tone of voice, and eye contact.

Unlike verbal or written communication, **nonverbal communication** does not involve spoken or written words or images. Nonverbal communication refers to things the speaker and the receiver see or hear while communicating. This includes appearance, personal environment, use of time, and body language. Examples include facial expressions, body movements, hand gestures, tone of voice, and eye contact.

Nonverbal communication can provide an external sign of someone's inner emotions. It often occurs at the same time as verbal communication. You can express feelings in many ways without realizing it. You may show you are happy by smiling and giving someone a warm handshake. You may show you are angry by raising your voice.

People observe your physical, nonverbal cues when you are communicating. In fact, nonverbal signals can have more impact than your verbal communication. Be aware of nonverbal cues to avoid miscommunication based on your posture, facial expression, or other signals. For example, if you have your arms crossed in a defensive posture while speaking, the person might assume you are angry, even if you are not.

On a jobsite, nonverbal communication is also a matter of safety. For example, a major aspect of rigging safety is having clear communication between a crane operator and the designated signal person on the ground. This communication relies on common signals, including nonverbal hand signals. The Emergency Stop signal used in rigging operations is a nonverbal communication. Anyone on the ground within sight of the crane operator can use this signal in the event of an emergency (*Figure 9*).

Figure 9 The Emergency Stop signal is an example of nonverbal communication.

Did You Know?

The Major Impact of Nonverbal Communication

Over the years, many studies have tried to measure how much of all communication is nonverbal. The specific numbers vary by study but indicate that between 55 to 93 percent of all communication is nonverbal. This means that your nonverbal signals send an even more powerful message than your verbal communication. Always remember that people might not remember what you say, but they do remember how you make them feel. Nonverbal cues have a major impact on how you make people feel when you communicate.

Did You Know?

Communication Is Number One

Research shows that the most common thing a supervisor does daily is communicate. They may do this as either a sender or receiver of messages. Eighty percent of a typical supervisor's day is spent communicating through writing, speaking, listening, or using body language. Of that time, studies suggest that around 20 percent is written communication, while 80 percent involves verbal communication.

2.2.6 Additional Communication Considerations

As a leader in the field, the way that you communicate has significant impact on the safety, quality, and productivity of your group. Regardless of your communication method, remember that everyone communicates differently. In some cases, you might need to communicate in multiple ways. You might also need to adjust your level of detail to ensure that everyone understands the intended meaning. For instance, a visual learner may need a map to comprehend directions. Another team member might just need a verbal explanation of which turns to take to a destination. *Figure 10* shows an example of ways to tailor the message to your audience.

Verbal Instructions (Experienced Crew)	**Verbal Instructions (Inexperienced Crew)**
"Please drive to the supply shop to pick up our order."	"Please drive to the supply shop. Turn right here and left at Route 1. It's at 75th Street and Route 1. Tell them the company name and that you're there to pick up our order."

Written Instructions

1. Turn right at exit.
2. Drive 2 miles to Route 1. Turn LEFT.
3. Drive 1 mile (past the tire shop) to 75th Street.
4. Look for supply store on right...

Visual Instructions (Diagram/Map)

7501 N. Highway 1
(800) 555-7567

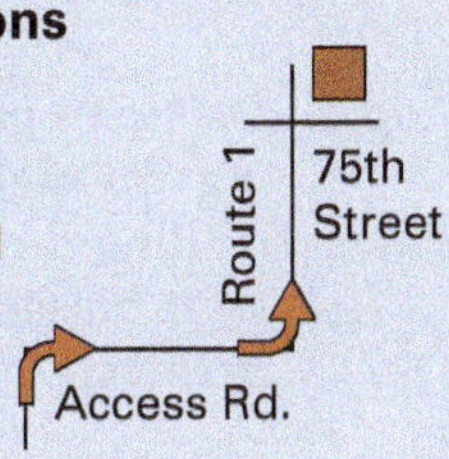

Figure 10 Assess your audience and then tailor your message.

The construction industry is a diverse and multicultural one. It is quite common to have a crew with people from different cultural backgrounds working together. Language and cultural differences can make communication even more challenging. Even if employees speak the same language, different understandings of words or symbols can create confusion or cause offense.

An important part of effective communication is to remember that people are different and to adjust your communication style to meet the needs of the person or people receiving your message. This means field leaders need to invest the time to learn how to communicate well with each individual. This requires you to know each crew member, recognize how each person learns, how each person delivers feedback, how each person works with others, and who they are as individuals, not just coworkers.

2.3.0 Problem-Solving and Decision-Making

Problem-solving and **decision-making** are a large part of every field leader's daily work. They are a part of life for all supervisors, especially in fast-paced, deadline-oriented industries like construction.

What is the difference between problem-solving and decision-making? Problem-solving is a process used to identify possible solutions to a situation. Decision-making refers to the action of making a choice. Decision-making involves starting or stopping a plan or choosing a different plan of action as

Problem-solving: A process used to identify possible solutions to a situation.

Decision-making: The action or process of making a choice. Decision-making involves starting an action, stopping one, or choosing a different plan or action.

appropriate for the situation. A decision may be complicated and involve many steps, but it is simply a choice.

The key difference between problem-solving and decision-making is that solving problems is a process, while making decisions is an action based on experience or information learned during the problem-solving process.

2.3.1 Problem-Solving

The ability to solve problems is an important skill in any workplace. It is especially important for craft professionals, whose workday is often not predictable or routine. Your group will look to you as a leader to solve problems regularly.

Figure 11 shows a five-step process for solving problems, which you can apply to any situation.

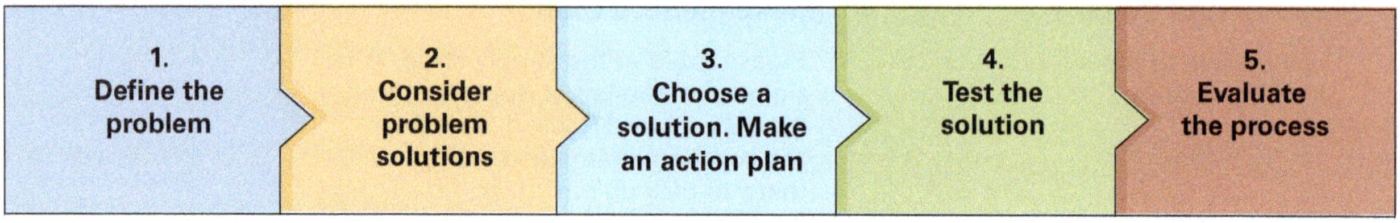

Figure 11 The problem-solving process.

Step 1 *Define the problem.* Defining a problem is not as easy as it sounds. Thinking through the problem often uncovers additional problems. Also, drilling down to the facts of the problem may mean setting aside your own biases or presumptions toward the situation or the individuals involved.

Step 2 *Consider possible solutions to the problem.* A problem may have more than one solution. Think through a number of solution options. As you do this, get inputs from people most affected by the problem, those who must correct the problem, and those who will be most affected by any potential solution.

Step 3 *Choose the solution that seems best and make an action plan.* After you consider several options to solve the problem, you must think through each possible solution and pick the best one. Again, you do not have to do this alone. Ask for feedback from others who have a stake in solving the problem well. Keep an open mind. The best solution might be taking pieces of two different solutions and combining them to create a new solution.

Step 4 *Test the solution to determine whether it actually works.* A solution might sound perfect before you try it but turn out not to be effective when you do. Alternatively, you might discover from trying to apply a solution that it is acceptable with a little change. If a solution doesn't work, think about how you could improve it, and then test your new plan.

Step 5 *Evaluate the process.* After your solution is in place, review the steps you took to discover and implement the solution. Could you have done anything better? The next time you are faced with a similar situation, you can use this experience to help solve the problem more effectively.

This problem-solving process can apply to almost any situation. *Figure 12* walks through a scenario to show how problem-solving works in practice. A

Problem Scenario

You are leading a team assigned to a new shopping mall project. The project will take about 18 months to complete. The only available parking is half a mile from the jobsite. The crew must carry heavy toolboxes and safety equipment from their cars and trucks to the work area at the start of the day, and then carry them back at the end of their shifts.

1. Define the problem	2. Consider problem solutions	3. Choose a solution Make an action plan	4. Test the solution	5. Evaluate the process
1. Problem: Workers are wasting time and energy hauling their tools and equipment to and from the worksite.	**2. Problem solution options:** Think about different ways to solve the problem. Several workers have proposed solutions: • Install lockers for tools and equipment closer to the worksite. • Have workers drive up to the worksite to drop off their tools and equipment before parking. • Bring in another construction trailer.	**3. Solution choice and action plan:** The worksite superintendent doesn't want an additional trailer in your crew's area. You and the team agree the shuttle service is the best option. It solves the time and energy issue *and* workers can keep their tools with them. The project supervisor plans to arrange a large van and driver to provide the shuttle service. **Note that this is a decision point in the problem-solving process.**	**4. Test the solution:** At first, the solution works, but a new problem arises. Everyone starts and leaves at the same time. The van can only hold half of the team and their equipment. To solve this problem, the supervisor schedules shuttle pickups every 15 minutes and adjusts work schedules so workers board the van at two different times. **Note that this is also a decision point in the problem-solving process.**	**5. Evaluate the process:** This process gave management and workers a chance to express an opinion and discuss the various solutions. Everyone feels pleased with the process and the solution.

Figure 12 Applying the problem-solving process, with decision points.

key point to note is that it's important to receive input from anyone in your group affected by the problem or the solution. Each solution will have pros and cons, so it's important to receive input from the workers affected by the problem. For example, workers will probably object to any plan (like the drop-off plan) that leaves their tools vulnerable to theft. Gaining the inputs of others will help you gain the trust and respect of your team.

2.3.2 Decision-Making

Figure 12 also shows where decisions are made as part of the problem-solving process. Decision-making is a choice made by using one's judgment. This judgment might be based on experience or on information learned during a problem-solving process. In general, you will have to make three types of decisions as a field leader, as shown in *Figure 13*.

No matter what the type, some decisions are simple or can be made based on past experiences. Other decisions are more difficult to make. These decisions require more careful thought about how to carry out an activity by using problem-solving techniques.

Note that decisions do have impacts, which you will need to consider. If you choose to use a shuttle to transport workers to a jobsite in shifts, as shown in *Figure 12*, using the shuttle may create a safety liability while workers are transported.

Decision Type	Description	Examples
Yes or no decisions 	Decisions in which the answer is yes or no. Also called *whether* decisions, because you have to decide whether to take an action or not.	• Whether to hire a van to use as a shuttle service • Whether or not to promote a team member • Whether to alert a supervisor to an unsafe condition • Whether to proceed with a concrete pour
Options decisions 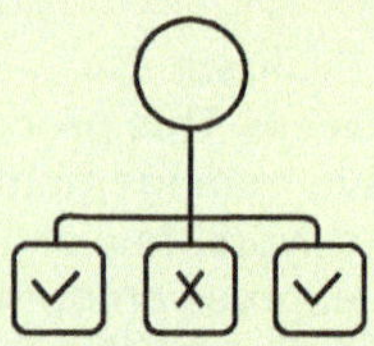	Decisions that require you to choose from a group of options. Also called *which* decisions, because you have to choose which option to take.	• Which of the three options will best address the problem of workers having to haul tools and equipment to and from the worksite? • Which toolbox talk would be most important to deliver to the team today? • Which team member should I have drive to the supply shop to pick up the order?
Contingent decisions 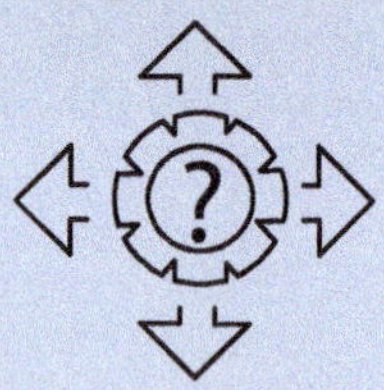	Decisions already made but on hold, until other specific events happen. These decisions can be very powerful because you can "pre-decide" what you will do in the event those events occur.	• If Carlo is late again two more times, I will set up a performance conversation. • If the materials arrive early, we will store them in the back of the jobsite. • If the team completes this task on time, I will find a clear way to recognize and reward the success.

Figure 13 Types of decisions.

2.0.0 Section Review

1. Which of the following is *not* a trait or behavior of effective frontline leaders?
   a. They lead by example.
   b. They are good communicators who lay out project goals to their crew.
   c. They motivate the team to work to their full potential.
   d. They have limited knowledge of the activities they supervise.

2. Which of the following is *not* a good practice to ensure effective communication?
   a. Speaking clearly at a reasonable pace
   b. Using highly technical language, acronyms, and slang
   c. Using visual cues, such as sketches or diagrams, to clarify instructions
   d. Using three-way communication to validate that a message was clear

3. While problem-solving is a process used to identify one or more possible solutions to a problem, the action of making a choice is called _____.
   a. decision-making
   b. a problem statement
   c. a decision statement
   d. action planning

3.0.0 Team Leadership and Motivation

Performance Tasks

There are no Performance Tasks in this section.

Objective

Identify key skills for team leadership and motivation.

a. Choose strategies to motivate individuals and a team.
b. Describe key approaches to building and leading teams.
c. Recognize how to approach leadership challenges, such as transitioning from peer to leader, addressing performance issues, and resolving conflict.

In your role as a frontline leader, you will have incredible impact on the individuals on your team. Motivating individuals and building successful teams are important and challenging tasks for any leader. You will also face leadership challenges as a new or developing leader, including managing performance issues, handling conflict, and leading team members who were once your peers. In this section, you will learn strategies and approaches to succeed as a team leader.

3.1.0 Motivating Individuals

The ability to motivate others is a key skill that leaders must develop. To motivate is to make someone feel determined or enthusiastic or to make someone behave in a particular way. On a jobsite, a leader uses authority and influence to encourage individuals to put in their best effort to accomplish the project goals. Lack of motivation affects job quality and productivity, reducing profits and growth for companies, businesses, contractors, or organizations that rely on people to complete the work.

As a supervisor or crew leader, it is important to understand what motivates each individual on your team. Different things motivate different people in different ways. This means there is no single best approach to motivating all of your crew members, because what motivates one crew member may not motivate another. In addition, what works to motivate a crew member once may not motivate that same person again in the future.

How can you know what motivates the members of your project team? Effective communication skills, specifically active listening, are a key part of getting to know each person on your team. You can also learn more about what motivates an employee by looking at factors in the employee's performance. Examples of factors that may indicate an individual's motivation include the level or rate of unexcused absenteeism, the number of complaints, and the quality and quantity of work produced.

Generally, the factors that motivate individuals are the same as those that create job satisfaction. These include the following:

- Clear expectations and goals from leaders
- Training to help individuals be successful
- Consistent sense of **accomplishment** and growth
- Recognition and rewards
- A good work environment and culture

Accomplishment: An individual's need to set challenging goals and achieve them.

A crew leader's ability to provide strong support in these areas increases the likelihood of high **morale** within a crew. Morale refers to an individual's emotional outlook toward work and the level of satisfaction gained while performing the jobs assigned. High morale means that employees will be motivated to work hard, and they will have a positive attitude about coming to work and doing their jobs. A motivated team is also generally a more productive and safer team, which can have a positive effect on project success.

Morale: An individual's emotional outlook toward work and the level of satisfaction gained while performing the jobs assigned.

3.1.1 Clear Expectations and Goals

As a leader, it is important for you to clearly communicate your expectations. A lack of clear expectations can leave your team feeling lost and demotivated. To get the best performance from your team and to motivate them, you need

> **Did You Know?**
>
> **Money Doesn't Always Buy Motivation**
>
> Many leaders consider money to be a good motivator. Money can drive motivation, but it often is only a temporary solution. If you give someone a raise, they tend to work harder for a period of time. However, research shows that this behavioral change does not last. Further, after someone has reached a level of financial security, factors other than money may come into play. Over time, too much reliance on money for motivation can negatively impact a worker's commitment. Remember, employees are individuals who are motivated by different things, including pride in their work, work environment, and more. For those people, a great workplace may mean more to them than better pay.

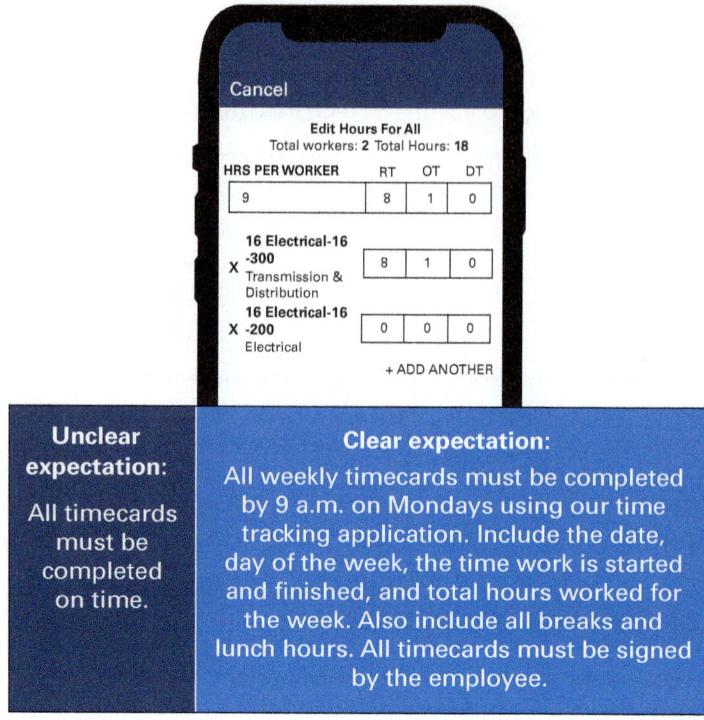

Figure 14 Examples of unclear and clear expectations.

to set clear goals for completing tasks. You also need to ensure that everyone on the team understands the deadlines for those tasks. *Figure 14* includes an example of clear expectations.

Set achievable short-term and long-term goals, including tasks to be done and expected timeframes. Having short-term goals gives you a chance to provide regular feedback, positive praise, and rewards. However, because short-term goals can change, also setting long-term goals gives the team additional opportunities for success. Do not set workers up for failure, as this leads to frustration, and frustration can lead to reduced productivity.

In addition to setting clear goals and expectations, you must also deliver timely feedback. People want and need feedback and positive reinforcement for their contributions and efforts. However, telling someone they did a good job last year or criticizing them for work from a month ago is meaningless. When sharing timely feedback, always include specifics to help the individual know what they did well or where they could improve.

You always should share negative feedback in private. When delivering negative feedback, approach the problem as a coaching moment, working with the individual to understand their challenges and working with them to correct the problem. Every failure or miss can be an opportunity for you to help that individual grow, if the conversation and feedback are constructive.

Positive feedback can be shared privately or publicly. When you praise a person in front of their team, it may also encourage other members of the team. Praise can be a valuable tool to reinforce behaviors that you want. (Not all people like to be publicly praised, so consider that before you offer public praise.)

3.1.2 Accomplishment and Growth

Accomplishment refers to an individual's need to set challenging goals and achieve them. A sense of accomplishment motivates most people. That is why setting and working toward recognizable goals tends to make employees more productive. A person with a sense of achievement is likely to be better motivated and help to motivate others.

A person who can align a task to their sense of meaning is much more likely to complete it. Frontline leaders can help their team feel a sense of accomplishment by keeping the work challenging and interesting. Workers who are completing the same task every day or performing work above or below their skill level can quickly become demotivated. For many individuals, overcoming the challenge of new or difficult assignments is much more satisfying than performing repetitive, boring tasks.

Remember, however, that too many or significant changes in work may have a negative impact on morale, because people often prefer some consistency and predictability in their work. Crew members react to change in many different ways and reasons. For some individuals, change creates anxiety and uncertainty, which can result in a negative attitude toward the work. To help manage change with your crew, remember to share why the change is occurring and explain what the employees and company will gain.

As a leader, you must ensure the individuals on your team feel that their skills and abilities are valued and make a difference. Employees who do not feel valued tend to have performance and attendance issues. Leaders should work to ensure the following is true for every crew member:

- They feel like an important part of the team with a skill set that contributes to the project success.
- They know that every position helps them build skills and experience that will last throughout a career.
- They have a chance to learn new things and develop their skill sets.
- They see opportunities to grow and progress within the company.

Professional and personal growth are other critical factors in motivation. **Professional growth** refers to gaining new skills and work experience that can help you reach a goal in your career. It is important that employees know that they aren't limited to their current jobs. Let them know that they have a chance to grow with the company and to be promoted as acknowledgment for excelling in their work when such opportunities occur.

Professional growth: Gaining new skills and work experience that can help one to reach a goal in one's career.

Whenever possible, show your team that you value their professional growth by giving them more advanced responsibilities when they are ready to perform them safely. People who are presented with challenging tasks often are motivated to work harder and improve. This approach is a win–win since the construction project benefits from a higher level of productivity, while the worker experiences professional growth.

Effective leaders encourage each of their crew members to work to their full potential. In addition, they share information and skills with their employees to help them advance within the organization.

Personal growth is about improving your talents and potential, both in and out of the workplace. For example, a team member may find it very rewarding to master a new competency on the job. Experiencing this type of personal growth at work helps to motivate, reduce boredom, and spark enthusiasm for additional growth.

Personal growth: Improving one's talents and potential, both in and out of the workplace.

Personal growth also takes place outside the workplace. Crew leaders should encourage their employees to identify goals for themselves both professionally and personally and then support their efforts to accomplish those goals. As an example, training can be a way to improve both professional and personal skills, such as time management, communication, financial management, and more.

3.1.3 Training and Upskilling

Another key factor in motivating individuals—and ensuring productive and safe projects—is providing sufficient training and opportunities for **upskilling**. Upskilling is a type of training that provides someone with more advanced skills through additional education. Upskilling provides employees with the tools and knowledge they need to perform their current job more effectively, more efficiently, or in a more modern way.

Upskilling: A type of training that provides someone with more advanced skills through additional education.

As people enter the construction workforce, many will have completed training programs but may not have the specific practical experience needed to complete a task on a project site. When a new employee joins your crew, you will need to take an active role in confirming their skills and helping with on-the-job training if there are gaps. A frontline supervisor on a construction jobsite only has a short time to evaluate their crew members on what tasks they can do or how much experience the crew member actually has.

In the first week, with any new employee, you should ensure that you train the new employee on any jobsite processes and allow time to do a task verification (*Figure 15*). After you have established the skills the individual has, you can work on a plan to leverage those skills while establishing new ones. If you identify gaps, you can create a training plan for that individual to address those areas.

3.1.4 Recognition and Rewards

Recognition: Acknowledgement of an employee's accomplishments through verbal and nonverbal communications.

Positive feedback is a form of **recognition**. Recognition acknowledges an employee's accomplishments through verbal and nonverbal communications. Individuals want to have good work appreciated, applauded, and acknowledged by others. You can accomplish this by thanking employees, offering an award for Employee of the Month, or giving tangible or monetary awards.

The following are tips to remember for giving recognition and praise:

- Be available on the jobsite so that you see good work as it is happening.
- Provide positive feedback publicly or in private.
- Give recognition and praise only when truly deserved. People quickly recognize insincere praise and will lose respect for the person delivering it.
- Encourage improvement by showing confidence in the ability of the crew members to do above-average work.
- Use goal setting as a tool to build confidence and create opportunities for praise.

Rewards are tangible recognition for good work. Rewards can be monetary, such as opportunities for raises, bonuses, or other incentive pay. Other types of rewards might include company gear, vouchers for new boots, gift cards, a team outing at the end of the week, or the best spot in the parking area for a month.

3.1.5 Good Work Environment and Culture

Having a good work environment and culture is critical for creating strong employee motivation and high morale. As a leader, you can create a good work environment and culture for your team in several ways.

Treat employees well. Be considerate, kind, and respectful. Treat people fairly. Mentor your employees. Coaching and supporting employees boosts their self-esteem, their self-confidence, and their motivation. Remember that small things matter. Having clean lunch areas, good housekeeping, and a comfortable workplace communicates that you respect the group.

Figure 15 When a new employee starts, be sure to evaluate the crew member's skills.
Source: kali9/E+/Getty Images

Communicate actively. As a leader, you should make yourself available for questions, opinions, and ideas. Actively engaging and communicating with workers plays a crucial role in motivating your team, as it shows that you appreciate and value their feedback. Involve employees in discussions so that they feel their opinions are valued. This, in turn, leads to a sense of pride and active participation. Making yourself available also prevents team members from feeling isolated from leadership.

When you communicate, be positive. Approaching your leadership responsibilities by threatening employees with negative consequences can result in higher employee turnover instead of motivation.

As much as possible, maintain a sense of humor, especially toward your own failings as a human being. No one is perfect. Your employees will appreciate that and be more motivated to encourage you.

Help individuals manage stress. In a job that can be demanding, fast-paced, and tiring at times, managers in the construction industry must take measures to ensure that their team is not exposed to undue stress. Work with your supervisors to ensure you have a safe and reasonable work environment. This means not only physical spaces, but also policies and procedures. For example, allowing workers to take time off to deal with personal matters or emergencies shows that the company prioritizes their well-being. Engage with your supervisors to help address resources and to organize shifts in such situations to ensure that project progress and construction costs are not negatively impacted.

Lead by example. Employees will be far more motivated to follow someone who is willing to do the same things they are required to do. Being a good leader and motivator means leading by example. Employees will not be motivated to work if upper management does not display the same level of commitment to the job that they are asked to have. For example, if you ask everyone to work late, but then leave early to go see a movie, this will be noticed and may create feelings of resentment and decrease team morale. As a leader, you can inspire your team by displaying the behaviors you would like to see from them, including working hard, showing personal and professional integrity, and having a can-do attitude.

3.2.0 Leading Teams

As a crew leader, you will need to lead a team of individuals from a range of backgrounds and experiences and create a high-performing team. To do so, you will need to delegate work to that team and implement policies and procedures that support the team. You will also need to address challenges like language barriers and **harassment**.

Harassment: Unwelcome conduct that is based on race, color, religion, sex, national origin, age, disability, or genetic information. Harassment includes verbal or physical conduct that impacts a person's work performance or creates an intimidating, hostile, or offensive work environment.

3.2.1 Delegating to Your Team

Once the various activities that make up the job have been determined, the crew leader must identify the person or persons who will be responsible for completing each activity. This requires that the crew leader be aware of the skills and abilities of the people on the crew. Then, you must put this knowledge to work by matching the crew's skills and abilities to specific tasks needed to complete the job.

After matching crew members to specific activities, you must then delegate the assignments to the responsible person(s). Generally, when delegating responsibilities, the crew leader verbally communicates directly with the person who will perform or complete the activity.

When delegating work, remember to do the following:

- *Delegate work to a crew member who can do the job properly.* If it becomes evident that the worker doesn't perform to the desired standard, either teach the crew member to do the work correctly or turn it over to someone else who can (without making a public spectacle of the transfer).
- *Make sure crew members understand what to do and the level of responsibility.* Be clear about the desired results, specify the boundaries and deadlines for accomplishing the results, and note the available resources.

- *Identify the standards and methods of measurement for progress and accomplishment.* Also identify the consequences of not achieving the desired results. Discuss the task with the crew member and check for understanding by asking questions. Allow the crew member to contribute feedback or make suggestions about how to perform the task in a safe and quality manner.
- *Give the crew member the time and freedom to get started.* Don't make them feel pressured by too much supervision. When making the work assignment, be sure to tell the employee how much time there is to complete it and confirm that this time is consistent with the job schedule.
- *Check in on progress.* Be sure to periodically check in on the crew member's progress. This will allow you to see how the task is going and allow the crew member to ask questions. Checking in along the way will also help you clear up any misunderstandings early and avoid rework.
- *Examine and evaluate the result.* After the task is complete, ask the crew member for their comments on how the work went. Provide your own feedback. Having this discussion will help you know what kind of work to assign that crew member in the future. It will also provide a means of measuring your effectiveness in delegating work.

There may be a time when someone else in the company issues written or verbal instructions to a crew member or your crew without going through you as the crew leader. If this happens, work with that individual to understand why they took that approach. Then, help that person understand why it is important for them to work through you for this type of instruction, whenever possible, so that you can manage the work of the entire crew safely and productively.

3.2.2 Implementing Policies and Procedures

Every company establishes policies and procedures that crew leaders are expected to implement and employees are expected to follow. Company policies and procedures are essentially guidelines for how the organization does business. They can also reflect organizational philosophies, such as putting safety first or making the customer the top priority. Examples of policies and procedures include safety guidelines, credit standards, and billing processes.

The following tips can help you effectively implement policies and procedures:

- Learn the purpose of each policy. This will help you follow the policy and apply it appropriately and fairly.
- If you're not sure how to apply a company policy or procedure, check the company manual or ask your supervisor.
- Always follow company policies and procedures. Remember that they combine what's best for the customer and the company. In addition, they provide direction on how to handle specific situations and answer questions.

NOTE

Try to obtain a supervisor's policy interpretation in writing or print out their email response so that you can append the decision to your copy of the company manual for future reference.

Crew leaders may need to issue directions to their crew members. A *direction* is a form of communication that initiates, changes, or stops an activity. Directions may be general or specific, written or verbal, formal or informal. It is up to you how you issue directions, but the policies and procedures of the company may govern your choice.

When issuing directions, do the following:

- Make them as specific as possible. Avoid being general or vague unless it is impossible to foresee all the circumstances that could occur in carrying out the directions.
- Recognize that you do not have to write directions for simple tasks unless your company requires this.
- Write directions for more complex tasks, tasks that will take considerable time to complete, or tasks that are permanent (standing) orders.

- Consider what is being said, the audience to whom it applies, and the situation under which it will be implemented to determine the appropriate level of formality for the direction.

3.2.3 Addressing Language Barriers

Leading a diverse team also means addressing language barriers. Millions of US workers speak languages other than English. Jobsites where two or more languages are used are common. In addition, even among English speakers, accents and dialects can present communication barriers.

As you have learned, any barrier in communication keeps people from being able to understand each other clearly. If a person does not fully understand the task a frontline supervisor gives them, the task may be completed incorrectly, which can cause delays, cause additional expenses, or create an unsafe situation. As a supervisor of a crew, you need to know that each crew member understands the work to be done. The following tips will help when communicating across language barriers:

- Be patient. Give workers time to process the information in a way they can comprehend.
- Avoid humor. Humor is easily misunderstood. The worker may misinterpret what you say as a joke at the worker's expense.
- Do not assume an individual is not intelligent simply because they don't understand what you are saying.
- Speak slowly and clearly. Avoid the tendency to raise your voice.
- Use face-to-face communication whenever possible. Other forms of communication can be more difficult when a language barrier is involved.
- Use the plans and drawings or make sketches to ensure people understand clearly.
- For critical written instructions, ensure the individual's literacy level matches the level of difficulty and language of the document. A person may speak clearly in a second language, but struggle with written understanding.
- Communicate in multiple ways or adjust your level of detail or terminology to ensure that everyone understands the meaning as intended.
- Confirm understanding. Ask for feedback with open-ended questions that require more than yes-or-no answers. Use three-way communication and have an individual repeat back what they heard to ensure mutual understanding and prevent miscommunication. If a worker is not fluent in English, ask the person to demonstrate their understanding through other means.

Many organizations offer language classes for supervisors and employees to help create more effective communication on jobsites. Whenever possible, take advantage of these training opportunities and encourage your crew to do so.

3.2.4 Preventing Harassment

As a frontline supervisor of a team, you must be aware of harassment as a potential challenge. As defined by the US Equal Employment Opportunity Commission (EEOC), harassment is unwelcome conduct that is based on race, color, religion, sex, national origin, age, disability, or genetic information. Harassment includes verbal or physical conduct that impacts a person's work performance or creates an intimidating, hostile, or offensive work environment.

The following are examples of activities that might qualify as harassment:

- Telling offensive jokes
- Calling someone offensive names
- Making verbal or physical advances or assaults

- Making threats or intimidating others
- Displaying offensive objects or pictures

Harassment becomes unlawful when a person must endure the offensive conduct to remain employed or the conduct is severe or common enough to create an intimidating, hostile, or abusive work environment. While not all acts of harassing behavior may be serious enough to be considered a violation of the law, all such behavior destroys teamwork and negatively affects jobsite safety and productivity.

Anyone can be a victim of harassment. Historically, many have thought of harassment as an act perpetrated by someone in power over another. However, a harasser can be a supervisor, a supervisor in another area, a coworker, or any other person in the work environment.

The EEOC enforces sexual harassment laws within industries. When investigating allegations of harassment, the EEOC looks at the whole record, including the circumstances and the context in which the alleged incidents occurred. Supervision may hold the crew leader responsible if they were aware of sexual harassment and did nothing to stop it. You should not only take action to stop harassment, but also should serve as a good example for the rest of the crew.

Prevention is the best way to eliminate harassment in the workplace. Your employer will have a harassment policy that clearly communicates zero tolerance for harassment and outlines a complaint or grievance process. This process should include taking immediate and appropriate action when an employee complains. This anti-harassment policy should apply to all employees (including managers, supervisors, and coworkers) and nonemployees (such as suppliers, owners, vendors, and others). Creating an open and inclusive work environment will discourage harassment.

As a leader, you must understand and uphold your employer's policies and procedures on harassment. If you believe that someone has violated the harassment policy, immediately notify your supervisor. If you are not comfortable reporting the incident to your immediate supervisor, you may report the complaint to any employee in a supervisory position or Human Resources. If you are not sure how to proceed in a situation, reach out to your Human Resources partners for guidance.

3.3.0 Leadership Challenges

As the leader of a team, you will need to manage leadership challenges and opportunities every day on the job. While the specific challenges will vary based on your team, the project, and other conditions, you must learn to address these when they arise. Such challenges can include navigating your initial transition from peer to leader, managing performance issues with your crew, or resolving conflicts that arise on the jobsite.

3.3.1 Transitioning from Peer to Leader

Making the transition from crew member to a crew leader can be difficult, especially when the new position involves overseeing a group of former peers. When you move from being a peer to a leader, your relationship with your former team members changes in important ways.

When you become a crew leader, you are no longer responsible for your work alone. You are now accountable for the work of a diverse team of individuals with varying skill levels, personalities, work styles, and cultural and educational backgrounds.

Crew leaders must learn to put personal relationships aside and work for the common goals of the entire crew. The crew leader can overcome these problems by working with the crew to set mutual performance goals and by freely communicating with them within permitted limits. Use their knowledge and strengths along with your own so that they feel like they are key players on the team. *Table 4* lists some additional suggestions on how to make the transition from peer to leader easier.

TABLE 4 Making the Transition from Peer to Leader

Strategy	Tactic
Reset boundaries with existing friends	Expectations and boundaries must be set with existing friends. Having a strong, professional working relationship is important, but you may need to reduce time spent with old peers outside of the project. Be clear and explain that, on the job, business comes before personal relationships.
Demonstrate that you have earned this leadership role	Quickly demonstrate good leadership for your team to earn employees' trust and respect. Be fair and consistent in everything you do. The perception of favoritism will tear a crew apart and negatively impact the outcome of any project.
Make the best decisions for the project and your company	Always make decisions that are best for the business. While it can be hard to make a decision that impacts someone who was a peer and is a friend, don't compromise to make a popular decision.

3.3.2 Managing Performance Issues

At the end of the day, if someone is simply not being accountable and completing their work, it might be time for a difficult conversation. In the workplace, a difficult conversation is one in which you have to manage emotions and information in a sensitive way to deal with a workplace issue. A difficult conversation may involve the following:

- Communicating business decisions
- Discussing poor performance or behavior
- Addressing conflict
- Handling complaints
- Ending employment or other changes in job status

As a leader and manager, you will need to have a difficult conversation at some point in your career. These conversations are never enjoyable, but you can handle them effectively by following some basic steps.

Identify the problem to resolve. Ensure that you have a clear understanding of the problem that you want to resolve. Before you start preparing for the conversation, ask: Why do you want to have the conversation? What do you want to achieve? Is the outcome you're looking for realistic?

Gather the information you need. Make sure you know the facts of the situation before approaching the employee. Bring any documents you will need to the meeting. For example, if you are discussing how an employee is late to work, you will want to have relevant policies and tardiness data at hand. If you are not sure about any internal workplace policies and procedures, you can always seek advice from a supervisor or Human Resources manager.

Think about the employee's point of view. Before you speak, you should think about the matter from your employee's point of view. Engage in the conversation with an open mind and truly seek to understand the situation from the employee's point of view.

Get the timing right. Try to schedule the conversation as soon as possible, but not if emotions are running high. Choose a time and place appropriate for the discussion, where you will not be rushed or interrupted. Generally, avoid having negative discussions at the end of the week and, if an employee needs time to process the conversation, offer to take a break and meet again later that day or the next.

Prepare what you are going to say. Write down the key points you need to cover, so that you can keep your conversation on track and stay in control. While it is important to prepare notes, avoid preparing a rehearsed script, as it may hamper your ability to listen effectively, react accordingly, or find alternative solutions.

Give the employee time to prepare. Make an appointment with the employee and provide some context to give them some time to prepare. Remember that knowing this conversation is coming may be stressful for the employee, so phrase the approach carefully and don't leave it too long. Examples of ways to phrase the appointment might be:

> "I'd like to talk about a series of quality issues I'm noting and get your point of view. Can you come see me tomorrow?"

> "I think we have different perceptions about what it means to have a clean work area. I'd like to hear your thoughts on this. Are you free this afternoon?"

Seek support if you need it. If you're not comfortable having the conversation one-on-one with the employee, enlist help from your manager or Human Resources person.

After the conversation, be sure to take the following steps:

- *Document the discussion.* Write down any agreements that were reached, and indicate the date, time, and who was present. This will be helpful if you need to revisit the discussion at a future date. Your *job diary*, which you will learn about later, is a useful tool here.
- *Follow up with the employee.* Confirm what you had agreed in writing and send these to the employee. Be sure to give the employee time to actually act upon the steps they agreed to.
- *Take some time to reflect on the matter.* Learn from the experience and think about what you could do differently next time. Debrief with your supervisor or Human Resources if you feel you need to discuss the matter.

Also, remember that you should never discuss this matter with other staff members. You must treat the employee the same as you had prior to the conversation and remain completely professional.

If you do need to have this type of difficult conversation, do it as soon as you start to have concerns or see early signals of conflict. The worse a situation becomes, the harder it can be to manage and resolve. In the meantime, your project and the rest of the crew are being affected.

As previously noted, the best communication strategy is to have an open-door policy or informal one-on-one conversations so that issues can be dealt with as naturally as possible. Be approachable and keep in touch with your team, so they are more likely to come to you with problems before they escalate.

3.3.3 Conflict Resolution

Conflicts will happen during a project. When you have many groups, including owners, general contractors, and specialty contractors all working together to complete a project, disagreements are likely to happen. Conflicts can arise within your crew as well. Each person or group might have a different view or opinion on how something should be done. Those differing opinions often lead to conflicts.

As a leader of the team, you will have many opportunities to look for positive and proactive means to resolve conflict. Resolving conflicts can lead to a better understanding or a new way of doing things that improve the overall project. If you handle conflict well, it will further earn the respect and trust of your crew. The following are some strategies to use when addressing conflict:

Recognize that conflict is normal. Not all conflict is bad. Conflicts on the construction site can be good. If the individuals engaged in the conflict are mature, conflict can improve processes and approaches by creating conversation about new ways to do work.

Do not ignore the conflict. Take action early to avoid further complications down the line. Even a small dispute that may appear inconsequential can balloon into a larger problem if it goes unresolved. If you allow the parties involved to work it out on their own, you still need to follow up and make sure that there aren't any lingering issues.

Understand the real issue behind the conflict. You must understand the real issues causing the conflict to determine the best solution. Conflicts are often caused by different interpretations of plans and specifications, change orders, scheduling, and poor communication. Work stress can also be an underlying issue. Conflicts, for example, may happen more often later in the day or in a work shift, due to fatigue. Use your active-listening skills to listen to everyone involved. This shows everyone that their opinions matter and that you are committed to working with them to resolve the issue fairly.

Remain neutral and focus on business, not people. Conflicts can escalate quickly from a simple disagreement to a shouting match. If appropriate, call for a short break before starting the discussion again. Remind everyone that resolving conflicts is about making the best business decision for the project, not personal feelings. Both parties should sit down and stay focused on developing a solution that is in the best interest of the project.

Compromise and work with the team. Conflicts rarely have a clear-cut right and wrong. As you work to resolve the conflict, you will need to encourage others to be willing to compromise and work together to solve the problem. The effort here must stay focused again on the desired business result: "What is the best decision for this situation?" Construction is all about cooperation and collaboration. However, if the conflict involves safety rules and regulations or adhering to building codes, any solution must follow the proper policy, procedure, or regulation.

Avoid conflict through clear communications. Poor communication on the jobsite often leads to conflicts. Make sure that you clearly communicate expectations with everyone on the project. Keeping everyone up to date with any changes to the project or schedule can help avoid conflict.

Communicate the resolution and follow up. Resolving conflicts may not result in a win for everyone. It is important that once resolution is reached, everyone gets an explanation and is clear on what is expected moving forward. They might not always agree with the decision. Having a clear expectation will help others accept the decision. Whenever an agreement is reached, follow up to determine if both sides are working as discussed or if issues still exist. You also will remind everyone that, although the conflict might have been difficult, the resolution produced a good outcome for the project.

If you cannot resolve a conflict on your own, always reach out to your supervisor for guidance. Your supervisor or another senior leader can make a decision and communicate it to the appropriate individuals.

Remember, the goal of any construction project is to see the successful completion of the job on time and within budget. When conflicts arise, it can threaten the goal of on-time and on-budget job completion. As the leader of your crew, you must learn to address and resolve conflicts, so that the project work can proceed. Resolving conflicts can be challenging. It all comes down to making the best business decision to keep the project moving forward.

3.0.0 Section Review

1. Which of the following is *not* a way to motivate a team?
   a. Providing training and helping individuals feel a sense of accomplishment and growth
   b. Setting expectations without clear short-term and long-term goals
   c. Offering recognition and rewards when appropriate
   d. Having a good work environment and culture

2. One sign of a high-performing team is _____.
   a. the leader cannot delegate work to the team
   b. harassment is common on the worksite
   c. everyone on the team follows policies and procedures
   d. language barriers prevent effective communication

3. As a frontline leader, steps to successfully resolving conflict may involve any of the following *except* _____.
   a. remaining neutral and focusing on the business, not on people
   b. compromising and working with the team
   c. ignoring the conflict
   d. understanding the real issues causing the conflict

4.0.0 Safety and Safety Leadership

Performance Tasks

1. Organize and deliver a safety-related toolbox talk, applying communication skills.
2. Complete a job safety analysis.

Objective

Identify a frontline leader's safety leadership responsibilities.

a. Assess the impact of incidents.
b. Explain the purpose of OSHA and describe the role of OSHA in administering worker safety.
c. Describe the importance of safety and health programs.
d. Explain the frontline leader's role in incident preparedness and response.
e. Recognize the importance of supporting your crew's total health.

Hazards: Conditions or activities that, if left uncontrolled, have the potential for harm.

Incidents: Unplanned events that result in property damage, fatalities, injuries, illnesses, or near misses.

Craft professionals routinely face **hazards**. Examples of these hazards include falling from heights, working on scaffolds, using cranes in the presence of power lines, operating heavy machinery, and working on electrically powered or pressurized equipment. Despite these hazards, applying preventive safety measures can significantly reduce the number of **incidents**.

Safety matters because people matter. As a crew leader, some of your most important responsibilities are to enforce the company's safety program, make sure that all workers are performing their tasks safely, and ensure that everyone on the jobsite remains safe and goes home healthy (*Figure 16*). You must recognize safety risks and work to eliminate or reduce them.

To be successful, the crew leader should do the following:

- Be aware of the human and business impacts of incidents.
- Understand all federal, state, and local governmental safety regulations applicable to your work.
- Be the most visible example of the best safe work practices.
- Be involved in training workers in safe work methods.
- Conduct training sessions.
- Get involved in safety inspections and incident investigations.

Crew leaders are in an important position to ensure that their crew members perform all jobs safely. Providing employees with a safe working environment by preventing incidents and enforcing safety standards will go a long way toward maintaining the job schedule, ensuring a timely job completion within budget, and keeping your team healthy.

Figure 16 Committing to safety is a key part of a crew leader's role.
Source: RainStar/E+/Getty Images

4.1.0 The Impact of Incidents

An incident is an unplanned event that results in property damage, a fatality, injury, illness, or a near miss. A near miss is an unplanned event that does not result in personal injury but may result in property damage or is worthy of recording.

Incidents including work-related injuries, sickness, and fatalities cause immense suffering for workers and their families. Resulting project delays and budget overruns can cause huge losses for employers, and worksite incidents damage the overall morale of everyone on a project or jobsite.

Businesses lose billions of dollars every year because of on-the-job incidents. As just one measure of the business impact of incidents, construction companies in the United States pay over $1 billion per week in **workers' compensation** for nonfatal injuries. Workers' compensation, often called *workers' comp,* is insurance that companies have to cover medical expenses, lost wages, and other costs for employees who are injured or become ill while completing work for their job. It also pays death benefits to families of employees who are killed on the job.

Workers' compensation: Insurance that companies have to cover medical expenses, lost wages, and other costs for employees who are injured or become ill while completing work for their job.

Preventing injuries means protecting the lives of workers, while reducing significant costs for their employers. It is estimated that the full impact of workplace injuries and fatalities cost companies approximately $151 billion per year. These costs affect the individual employee, the company, and the construction industry as a whole.

Organizations encounter both direct and indirect costs associated with workplace incidents.

- Direct costs are the money companies must pay out to workers' compensation claims and sick pay.
- Indirect costs are all of the costs a company must account for as the result of a worker's injury or death, such as time off from work.

The direct and indirect costs of incidents are comparable to the visible and hidden portions of an iceberg, as shown in *Figure 17.* The visible part of the iceberg represents direct costs, some of which are covered by insurance. The hidden part of the iceberg represents the indirect costs, which are a greater financial burden than the direct costs.

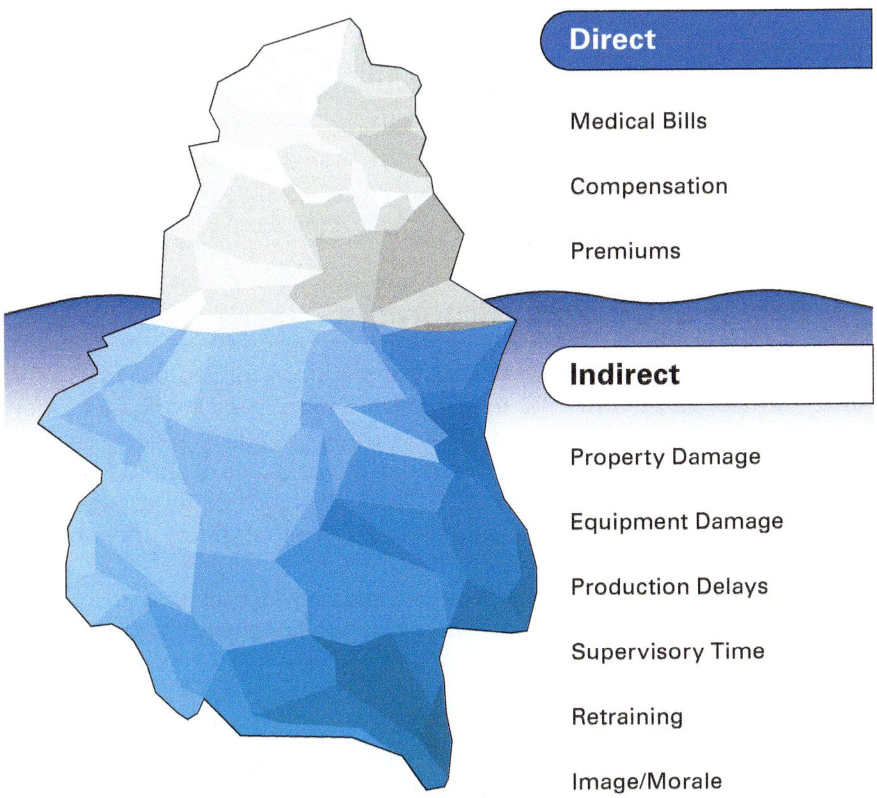

Figure 17 Incidents create both direct and indirect costs.

4.2.0 OSHA

> **Did You Know?**
> **OSHA 10 and OSHA 30 Authorized Certification**
> OSHA helps to make safety training available through their Outreach Training Program's authorized 10-hour and 30-hour safety courses. The goal of these courses is to promote workplace safety and health and to make workers more knowledgeable about workplace hazards and their rights. As a crew leader, you may already have completed the OSHA-authorized 10-hour training program, which is designed for entry-level workers. The OSHA-authorized 30-hour training program provides more in-depth training and is intended for supervisors or others with some safety responsibility. If you want to take an OSHA-authorized 10-hour or 30-hour training program, work with your supervisor to obtain this training.

To reduce safety and health risks and the number of injuries and fatalities on the job, the federal government has enacted laws and regulations, including the Occupational Safety and Health Act of 1970 (OSH Act of 1970). This law created the Occupational Safety and Health Administration (OSHA), which is part of the US Department of Labor. OSHA also provides education and training for employers and workers. Through the administration of OSHA, the US Congress seeks to ensure safe and healthful working conditions for workers by setting and enforcing standards and by providing training, outreach, education, and assistance.

OSHA has many functions. OSHA issues standards and rules for working conditions, facilities, equipment, tools, and work processes. It does extensive research into occupational incidents, illnesses, injuries, and deaths to reduce the number of occurrences and adverse effects. In addition, OSHA regulatory agencies conduct workplace inspections to ensure that companies follow the standards and rules. The agencies also have the right to take legal action if companies are not in compliance with OSHA standards and rules.

OSHA has established significant monetary fines for the violation of its regulations. *Table 5* lists the maximum OSHA penalty amounts as of 2022. In addition to the fines, there are possible criminal charges for willful violations resulting in death or serious injury. The attitude of the employer and their safety history can have a significant effect on the outcome of a case.

TABLE 5 OSHA Maximum Penalty Amounts

Type of Violation	Maximum Penalty
Serious	$14,502 per violation
Other-Than-Serious	
Posting Requirements	
Failure to Abate	$14,502 per day
Willful or Repeated Violation	$145,027 per violation

> **Did You Know?**
> **State OSHA Programs**
> Nearly half the states in the United States and some of its territories have state-run OSHA programs. These programs are set up under federal OSHA guidelines and must establish job health and safety standards that are at least as effective as the federal standards. The states/territories have the option of adopting more stringent standards or setting standards for hazards not addressed in the federal program. Of the 22 states and territories with nonfederal OSHA programs, six of them limit their coverage to state and local government employees.

4.3.0 Safety and Health Programs

Under the OSH Act of 1970, employers have a responsibility to provide a safe workplace. Your employer must set up a program to manage workplace safety and health and to reduce work-related injuries, illnesses, and fatalities. The program must be appropriate for the conditions of the workplace. It should consider the number of workers employed and the hazards they face while at work. On a project site, a hazard usually is a condition or activity that, if left uncontrolled, has the potential for harm.

One way employers help create a safe working environment is to create a **construction safety plan**. A construction safety plan is a document that outlines the procedures, rules, and regulations that are or will be put in place to protect workers during a construction project. The safety plan should include steps to prevent incidents and respond if they occur, including emergency medical services and processes to review and report incidents. For some projects, a **site-specific safety plan (SSSP)** will include a focused set of information that accounts for a jobsite's unique hazards. Safety plans should be created before a construction project is underway, but also should be updated on a regular basis as conditions on the site change.

Construction safety plan: A document that outlines the procedures, rules, and regulations that are or will be put in place to protect workers during a construction project.

Site-specific safety plan (SSSP): A safety plan that includes a focused set of information that accounts for a jobsite's unique hazards.

4.3.1 Education and Training

All employees should receive a safety policies and procedures manual and initial training when they are hired. A company's safety manual should cover the company's safety expectations and the procedures that must be followed to create a safe work environment and meet OSHA requirements. During orientation, appropriate company staff should guide new employees through the safety policies and procedures manual, focusing on the sections most important for their job. An employer must ensure that any barriers to worker understanding, such as language barriers, are removed or addressed.

In addition to formal training, safety is a topic you need to discuss with your crew every day on the jobsite. You need to help employees be aware of and avoid the hazards to which craft professionals are exposed. As a leader, you should plan to conduct frequent safety trainings and meetings with your crew, including holding toolbox talks (*Figure 18*). These meetings are important to building a strong safety culture and reinforcing your company's commitment to protecting your workers.

Toolbox talks are one way to effectively keep all workers aware and informed of safety issues and guidelines. Toolbox talks, sometimes referred to as *tailgate meetings* or *safety briefings*, are short, informal safety meetings held at the start of each day or shift. Toolbox talks are 5- to 10-minute meetings that review specific health and safety topics. *Table 6* provides a checklist for delivering effective safety briefings or toolbox talks.

Toolbox talks: Short, informal safety meetings held at the start of each day or shift. Sometimes referred to as *tailgate meetings* or *safety briefings*.

Always document your toolbox talks or other safety training. Even if certain OSHA standards do not require documentation of safety training, keep a record of each safety session signed by all attendees. In the record, include the date, a summary of the training session, the trainer, and attendees.

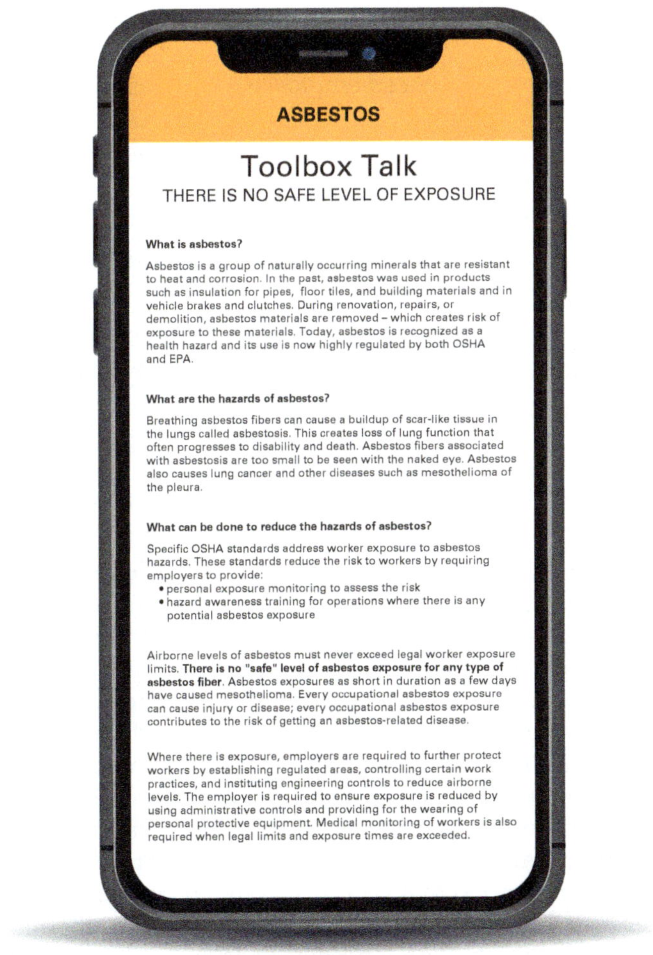

Figure 18 Safety management software can help companies deliver and track worker training. Some of the solutions help you schedule, track, and deliver toolbox talks.

TABLE 6 Delivering Effective Safety Briefings or Toolbox Talks

Did You . . .?	Yes/No
Schedule time in advance and create time to allow workers to participate.	
Speak clearly and directly at a natural pace.	
Try to find a spot free of noise and other distractions. If the workers cannot hear you talking, or are distracted by other activities in the area, they won't be focusing on your talk.	
Use props to help make your point and provide visual communication cues. For example, have a safety harness with you for your talk on fall protection.	
Reinforce safety basics, focus on high-risk scenarios, and inform workers about changes to the jobsite and working conditions that may have occurred since their last shift.	
Discuss any incidents or injuries that have occurred and how they could have been prevented.	
Focus the discussion on active work. For example, do not hold a toolbox talk on sitework when electrical work is the only thing being done that day.	
If appropriate, break out a toolbox talk into smaller groups to focus on individual crafts or work activities for certain individuals.	
Focus on the positive, too. Highlight safe working practices. Provide positive reinforcement and applaud workers following safety rules and regulations.	
Always give workers an opportunity to ask questions at the end of the toolbox talk.	

One-on-one or small group training is also a critical part of keeping everyone on the project safe. When you assign an inexperienced employee a new task, you must ensure that the employee can do the work in a safe manner. You can accomplish this by providing safety information or training for groups or individuals.

When assigning an inexperienced employee a new task, do the following:

- Define the task.
- Explain how to do the task safely.
- Explain what tools and equipment to use and how to use them safely.
- Identify the necessary personal protective equipment and train the employee in its use.
- Explain the nature of the hazards in the work and how to recognize them.
- Stress the importance of personal safety and the safety of others.
- Review Safety Data Sheets that may be applicable.

4.3.2 Hazard Identification, Assessment, Prevention, and Control

Crew leaders and workers play an important role in identifying and reporting hazards. Hazards are often identified in job safety or job hazard analyses, which also document how the hazards will be eliminated or mitigated. A **job safety analysis (JSA)**, or *job hazard analysis (JHA)*, is a technique that focuses on job tasks as a way to identify hazards before they occur. It focuses on the relationship between the worker, the task, the tools, and the work environment. After you identify uncontrolled hazards, you will take steps to control them by eliminating or reducing them to an acceptable risk level.

Job safety analysis (JSA): A technique that focuses on job tasks as a way to identify hazards before they occur. It focuses on the relationship between the worker, the task, the tools, and the work environment. Also known as *job hazard analysis (JHA)* or *activity hazard analysis (AHA)*.

Figure 19 shows an example of a form used to conduct a JSA. In a JSA, the task at hand is broken down into its individual parts or steps and then each step is analyzed for its potential hazards. Once a hazard is identified, certain actions or procedures are recommended that will correct that hazard. For example, during a JSA, it is determined that using a chain hoist to install a pump motor in a tight space would be safer than having a worker do it manually. By using the chain hoist, the chance that the worker's hand would get crushed during installation is reduced. Using the JSA process would help protect the worker from injury.

JOB SAFETY ANALYSIS

TITLE OF JOB OR TASK

TASK	START	END	HAZARDS	CONTROLS
1.				
2.				
3.				
4.				
5.				
6.				
7.				
8.				

Required Training: Required Personal Protective Equipment (PPE):

Job Name: ______________

Job Number: ______________

Supervisor: ______________

Date: ______________

Weekly Vehicle Checklist: ___ Tire Pressure ___ Transmission Fluid ___ Oil ___ Lights ___ Air Filter ___ Wkly Mileage

Name of Employees:

PRINT NAME	SIGN NAME	TOTAL HOURS

Figure 19 Job safety analysis form.

A job safety analysis can also be used during planning to ensure that safety is planned into the job. Taking this approach helps ensure that safety does not get compromised for the sake of schedule and productivity.

Crew leaders must protect workers from existing or potential hazards in their work areas. It is the crew leader's responsibility to determine what working conditions are unsafe and to inform employees of hazards and their locations. In addition, they should encourage their crew members to tell them about hazardous conditions. To accomplish this, crew leaders must be present and available on the jobsite.

4.3.3 Management Leadership and Employee Participation

A culture of commitment to safety and a strong safety and health program require management leadership and employee participation. As a frontline leader, the best way for you to encourage employee participation in safety and health is to lead by example. Your behavior sets standards for your crew members. If you cut corners on safety, you are telling your crew they may do so as well.

You should always be sure to do the following:

- Stay actively involved in your company's safety and health program.
- Exceed expectations when it comes to safety. Help others follow the rules and ask your crew if there are safer ways to do a task.
- Host safety training and meetings. Discuss safety at the start of each shift. Have regular toolbox talks. Include safety in all meetings, even if the primary purpose is schedule, cost, or production related.
- Be a visible participant on safety walks.
- Review and provide feedback on hazard analyses.
- Always follow up on safety concerns raised by the workers.

CAUTION

Team members often "follow the leader" when it comes to unsafe work practices. Sometimes, supervisory personnel engage in unsafe practices and take more risks because they are more experienced. However, inexperienced or careless workers who take the same risks won't be as successful in avoiding injury. As a leader, you must follow all safety practices to encourage your crew to do the same.

Your employer will use their policies and procedures for incident investigation and documentation. Be sure to follow any of these guidelines, when the situation arises.

4.4.0 Incident Preparedness and Response

While the goal is to have zero incidents on a worksite, you and your team must understand how to respond quickly to a safety incident. If an incident occurs, you should follow the safety policies and procedures in the construction safety plan. As a frontline leader, you are responsible for making sure everyone on your team understands the safety plan. You also may have reporting and investigation responsibilities after an incident.

4.4.1 Administering First Aid

First aid: Help given to a sick or injured worker on the worksite immediately after an injury occurs.

AED (automated external defibrillator): A device used to help those experiencing sudden cardiac arrest.

When someone is hurt or has a medical event, your first goal is ensuring the best outcome: minimizing the health impact to the person. **First aid** is help given to a sick or injured worker on the worksite immediately after an injury occurs. The primary purpose of first aid is to provide immediate and temporary medical care to employees involved in incidents, as well as employees experiencing non-work-related health emergencies, such as chest pain or breathing difficulty.

Many companies have very specific policies on administering first aid. If you are unsure of these policies, ask your supervisor and your safety team for guidance. You need to know how to handle these situations before an incident occurs. A quick response can be the difference between life and death. The victim of an injury or sudden illness at a jobsite may be at a remote location. The site may be far from a rescue squad, fire department, or hospital, presenting a problem in the rescue and transportation of the victim to a hospital.

Your employer must ensure prompt first aid treatment for any injured employees. To do this, they will provide access to readily available first aid kits. You and everyone on your team should be aware of the location and contents of first aid kits available on the jobsite. Many jobsites also have an **AED (automated external defibrillator)** that is used to help those experiencing sudden cardiac arrest (*Figure 20*). Many lives have been saved due to a prepared team who knew where the AED device was stored and how to use it.

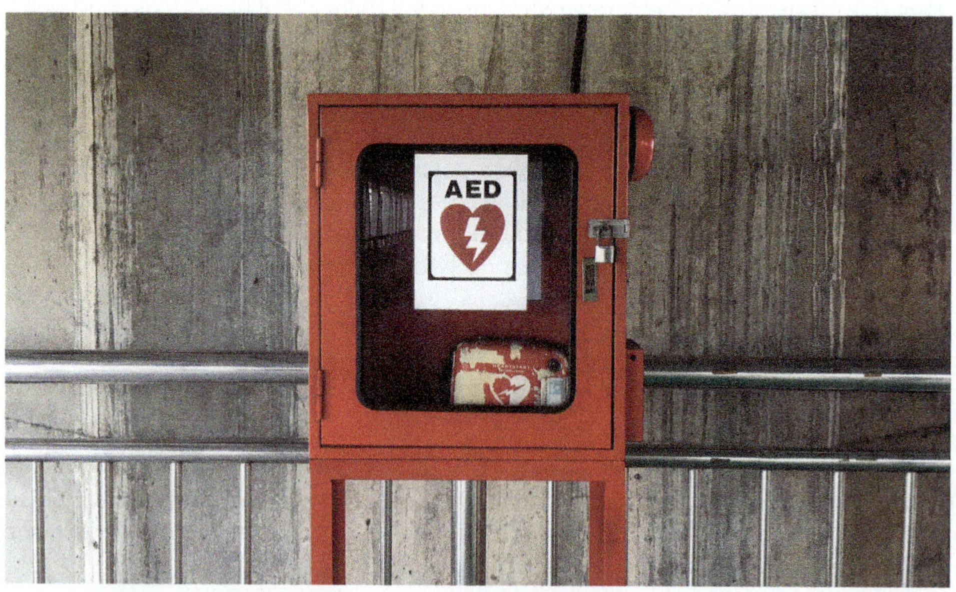

Figure 20 Many jobsites have an AED to help those experiencing sudden cardiac arrest. You should always know where AED devices are stored.
Source: PardiMer/Shutterstock

Your company will have an on-site medical professional or at least one member of the workforce who is trained and certified in first aid, including the use of an AED and CPR. **CPR (cardiopulmonary resuscitation)** is a lifesaving technique that is used when someone's heartbeat or breathing has stopped. The benefits of having personnel trained in first aid at jobsites include the following:

CPR (cardiopulmonary resuscitation): A lifesaving technique that is used when someone's heartbeat or breathing has stopped.

- Immediate and proper treatment of minor injuries. First aid may prevent minor injuries from developing into more serious conditions. OSHA does not require companies to record when simple first aid is used to address the injury or illness.
- Ability to determine if the injured person requires professional medical attention.
- Ability to help stabilize an injured person before professional medical care arrives.
- Help to eliminate or reduce medical expenses, lost work time, and sick pay.

If further medical treatment is needed, preparedness and a rapid communication process will mobilize the right people as soon as possible. Emergency numbers should be posted on the jobsite. As a leader, you should be very familiar with the construction safety plan and work with your Human Resources and safety leaders to ensure it stays updated. You need to locate your local off-site medical provider for emergency or nonemergency care prior to an incident occurring.

NOTE

The American Red Cross, Medic First Aid, and many other local organizations provide basic and advanced first aid courses at nominal costs. These courses include both first aid and CPR. The local area offices of these organizations can provide further details regarding the training available. CPR certifications must be renewed every two years.

4.4.2 Reporting and Investigating Incidents

As a frontline leader, you must know and follow your company's incident reporting processes. You also must ensure your team is following those policies. Your team should know from their training to immediately report any unusual event, incident, or injury. The policies and definitions around reporting should be included in safety manuals and construction safety plans.

If your employer is subject to the recordkeeping requirements of the OSH Act of 1970, it must maintain records of all recordable occupational injuries and illnesses using standard OSHA forms. In the event of an incident, the employer also is required to investigate the cause of the incident and determine how to avoid it in the future. According to OSHA regulations, the employer must investigate each work-related death, serious injury or illness, or incident having the potential to cause death or serious physical harm. The employer should document any findings from the investigation, as well as the action plan to prevent future occurrences. The company should complete these actions immediately, with photos or video if possible.

As a crew leader, you may be involved with an incident investigation. If required by company policy, you also will need to make a formal investigation and submit a report after an incident. An investigation looks for the causes of the incident by examining the circumstances under which it occurred and talking to the people involved. Investigations are perhaps the most useful tool in the prevention of future incidents. As shown in *Figure 21*, an incident investigation has four major parts.

NOTE

Three OSHA forms are used for injury and illness recordkeeping:

- OSHA Form 300, Log of Work-Related Injuries and Illnesses
- OSHA Form 300A, Summary of Work-Related Injuries and Illnesses
- OSHA Form 301, Injury and Illness Incident Report

All of these forms are available online at the OSHA website. OSHA has also launched their Injury Tracking Application (ITA) to allow for direct entry of OSHA Form 300A information. For more information, scan the QR code at the front of the module.

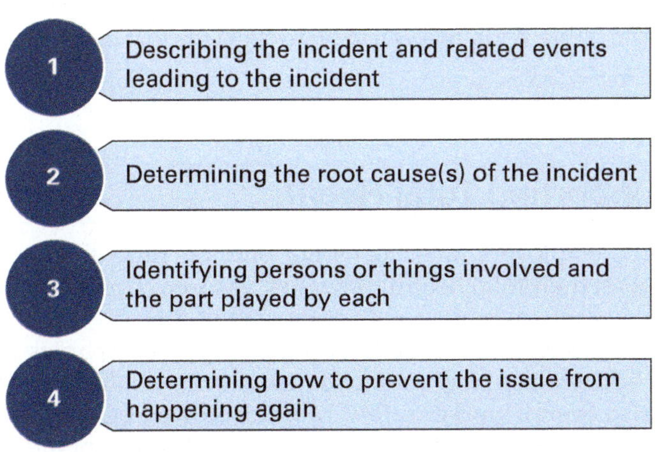

Figure 21 An incident investigation has four major parts.

Root cause: A core issue or fundamental reason for the occurrence of a problem.

A key step is determining the **root cause** of the incident. A root cause is a core issue or fundamental reason for the occurrence of a problem. Understanding the root cause is critical to avoiding similar incidents in the future. In many cases, the root cause was a flaw in the system that failed to recognize the unsafe condition or the potential for an unsafe act.

Root cause analysis (RCA): A method of problem-solving used for identifying the root causes of faults or problem.

Finding a root cause to a problem can be difficult. A **root cause analysis (RCA)** is a method of problem-solving used for identifying the root causes of faults or problems. A *Five Whys Analysis* is one type of root cause analysis that is helpful with incident investigations. The method is simple: When a problem occurs, you start with the broadest explanation for an incident and narrow it down to a more specific problem by asking "why" five times. The technique is even more effective when you include experienced team members who were actively part of the situation or have had similar experiences. It is especially useful when the root cause is not that obvious. *Figure 22* shows an example of how a Five Whys Analysis can apply to an incident investigation.

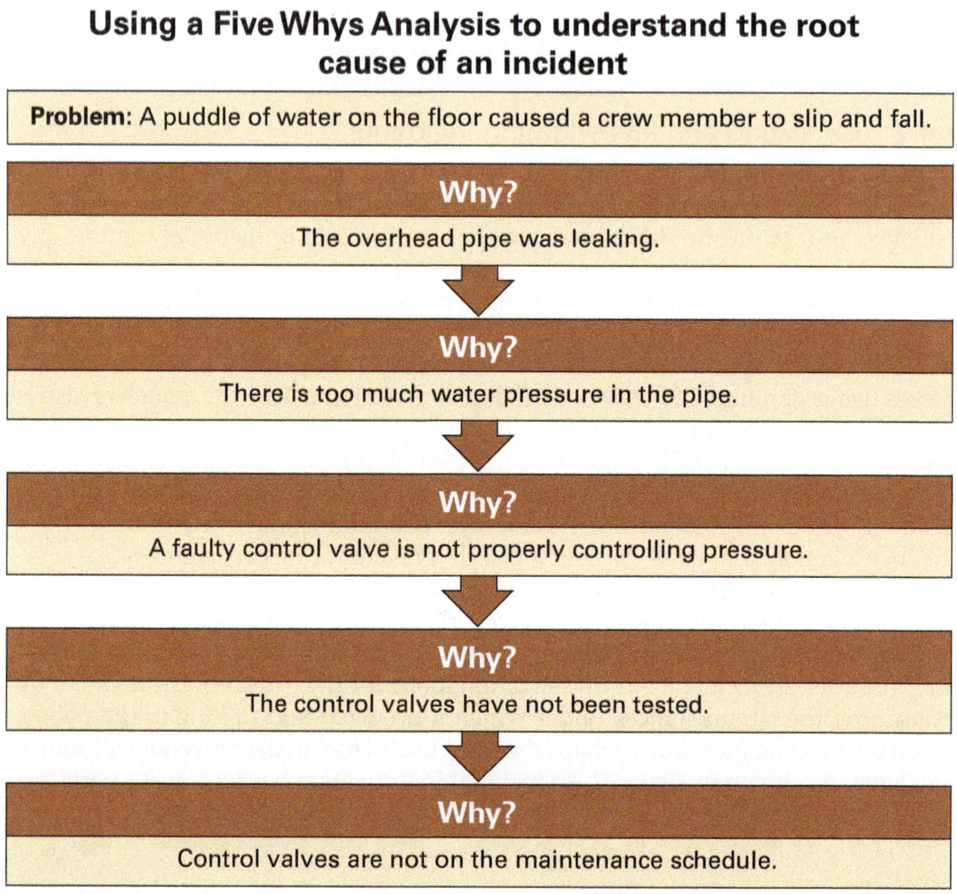

Figure 22 Using a Five Whys Analysis to understand the root cause of an incident.

4.5.0 Supporting Total Health

More than ever, employers must focus not only on their employees' physical well-being, but on their total health. A healthy work environment not only helps to ensure workers return home safely at the end of their shift, but also ensures they can have a healthy life outside of work. Many companies are adopting practices that address not only protection from work-related safety and health hazards, but also overall worker safety, health, and well-being.

Substance abuse: The inappropriate overuse of drugs and chemicals, whether they are legal or illegal.

Two key aspects of total health for your workforce include addressing **substance abuse** and mental health. As a frontline leader, you will not be able to address the root causes of these issues for your workers. You do, however, have the important responsibility to enforce company policies and procedures in these areas and work with your management or Human Resources Department to help employees in these situations.

4.5.1 Substance Abuse

Substance abuse is a serious safety issue in the workplace. Substance abuse is the inappropriate overuse of drugs and chemicals, whether they are legal or illegal. All substance abuse results in some form of mental, sensory, or physical **impairment**. Impairment is the point where someone's intake of substances affects their ability to perform appropriately. Alcohol can be abused by consuming to the point of intoxication. Legal prescription drugs can be abused by using the wrong dosage, using other people's medications, or self-medicating. Illegal drugs, such as cocaine, crystal meth, or marijuana in some states, can cause impairment.

Impairment: The point where someone's intake of substances affects their ability to perform appropriately.

Crew leaders must enforce company policies and procedures regarding substance abuse. There are legal consequences and safety implications associated with substance abuse. Suppose several crew members go out and smoke marijuana or drink during lunch. Then, they return to work to erect scaffolding for a concrete pour in the afternoon. If you can smell marijuana on the worker's clothing or alcohol on their breath, you must step in and take action. An impaired worker is more likely to cause an incident that could cause serious injury or death to themselves or others (*Figure 23*).

If you observe an employee showing impaired behavior for any reason, immediately contact your supervisor and/or Human Resources Department for assistance. You should work with management to deal with suspected drug and alcohol abuse and should not deal with these situations yourself. Reporting the concern helps you protect the business, and the employee's and other workers' safety.

It is often difficult to detect substance abuse because the effects can be subtle. The best way is to look for physical signs or sudden changes in behavior that are not typical of the employee. *Table 7* lists some examples of potential signs and behavioral changes that might indicate substance abuse.

Figure 23 An impaired worker is a dangerous one.
Source: vejaa/iStock/Getty Images

Did You Know?
Substance Abuse Does Not Pay

An employee who is involved in an incident while under the influence of drugs or alcohol may be denied workers' compensation insurance benefits.

TABLE 7 Potential Signs of Substance Abuse in a Worker

Physical Signs	Behavioral Changes
• Smelling of alcohol or drugs • Slurring of speech or an inability to communicate effectively • Wearing sunglasses indoors or on overcast days to hide dilated or constricted pupils, conditions which impair vision • Significant changes in personal appearance, cleanliness, or health	• Unscheduled absences • Failure to report to work on time • Significant changes in the quality of work • Unusual energy or tiredness • Sudden and irrational temper flare-ups • Sneaky behaviors, such as disappearing to wooded areas, storage areas, or other private locations • Attempting to borrow money from coworkers • Losing tools or company equipment

4.5.2 Mental Health

Mental health is a critically important, but often overlooked, part of overall construction safety and health. Construction workers have more risk factors than employees in other industries. These risk factors make workers more likely to experience mental health challenges, with a specific set of risks around suicide. Construction workers are six times more likely to die by suicide than they are to die from a fall.

With proper treatment, mental health conditions can be managed and overcome so those experiencing them can have full and productive lives. As with substance abuse, however, these conditions can lead to people being distracted, less productive than normal, and possibly unsafe and unable to perform their normal jobs. Because of this, mental health and suicide prevention need to be safety considerations.

As a leader, you cannot solve these challenges for your workers, but you can create a better environment. Use your toolbox talks to discuss mental health as part of overall total worker safety. In your safety huddles, check in on how your teammates are doing. Encourage your crew to talk to supervisors and colleagues when they feel stressed, anxious, or on the brink of burnout. Help ensure they are aware of any available counseling resources and that they are taking time off when they need it.

Even more so than substance abuse, it can be very difficult to identify a worker struggling with a mental health issue. *Table 8* lists some examples of potential signs and behavioral changes that might indicate someone is struggling with mental health issues.

If you observe an employee showing signs of mental health issues or have concerns for any reason, immediately contact your supervisor and/or Human Resources Department for assistance. You should work with management to deal with suspected mental health issues and should not deal with these situations yourself. Reporting the concern helps you protect the business, and the employee's and other workers' safety.

Did You Know?
Mental Health Is a Critical Part of Safety

Many organizations in the construction industry are focused on helping normalize conversations around mental health and make it less challenging to have these conversations. The National Institute for Occupational Safety and Health (NIOSH), for example, has a Total Worker Health® program focused on improving workforce well-being by protecting their safety and enhancing their health and productivity. Other focused nonprofits, like the Construction Industry Alliance for Suicide Prevention (CIASP), provide specific resources, like toolbox talks, that you can use to help with these conversations with your crew.

TABLE 8 Signs That May Indicate Someone Is Struggling with Mental Health Issues

Behavioral Changes	Performance Changes	Verbal Cues	Life Events
• Becoming withdrawn • Acting anxious or reckless • Misusing drugs and alcohol	• Missing work • Showing up late • Being less productive • Being unable to think clearly or solve problems • Having increased safety incidents	• Talking about feeling trapped • Talking about wanting to die • Saying they feel like a burden • Saying they feel like it would be better if they were gone	• Dealing with a divorce or breakup • Handling a death in the family • Working through a major illness or injury • Having financial issues

Source: Construction Industry Alliance for Suicide Prevention (ciasp.org)

4.0.0 Section Review

1. One of a crew leader's most important responsibilities to the employer is to ____.
   a. enforce company safety policies
   b. estimate material costs for a project
   c. make recommendations for setting up a crew
   d. provide input for fixed-price contracts

2. Which of these is *not* a function of OSHA?
   a. Issuing standards and rules for working conditions
   b. Researching occupational incidents and injuries
   c. Conducting workplace inspections
   d. Distributing company safety manuals

3. Which of the following should all employees receive upon being hired?
   a. A copy of the company's safety policies and procedures
   b. A new set of tools
   c. A leadership assessment to help map their career path
   d. First-aid and CPR training

4. A problem-solving method that is useful in investigating incidents and identifies the core issue of a problem is _____.
   a. problem-solving
   b. root cause analysis
   c. incident recordkeeping
   d. event playback

5. A focus on overall worker safety, health, and well-being, including mental health, is _____.
   a. home health
   b. complete worker safety
   c. occupational safety
   d. total health

5.0.0 Project Planning, Scheduling, and Cost and Resource Management

Objective

Demonstrate an understanding of construction project phases, project planning and scheduling, field reporting, and cost and resource management.

a. Identify construction project phases.
b. Summarize crew planning and scheduling responsibilities, including creating lookahead schedules.
c. Describe the frontline leader's role in field reporting.
d. Explain approaches to managing costs and controlling project resources, such as materials, tools, and equipment.

Performance Task

3. Develop and present a lookahead schedule.

This section describes the various phases of a construction project. It examines **estimating**, **planning**, **scheduling**, and **cost control** through managing material, people, tools, and other resources. All workers who participate in a job are responsible at some level for their assigned jobs. For example, the contractor's responsibility begins with obtaining the contract and it doesn't end until the client takes ownership of the project. The project manager is generally the person with overall responsibility for coordinating the project. Finally, the superintendent and frontline leadership, like a foreman, supervisor, or crew leader, are responsible for coordinating the work of one or more workers, one or more crews of workers within the company, or one or more crews of subcontractors.

Estimating: The process of calculating the cost of a project.

Planning: The process of laying out all details of a construction project before construction begins.

Scheduling: Involves defining which activities happen in what order and assigning dates and times to those tasks.

Cost control: The process that helps keep expenses under control by managing labor, material, and other costs to ensure that the project finishes on budget.

5.1.0 Construction Project Phases

A simple way to think about a construction project is in three phases: the *design and development phase*, the *preconstruction phase*, and the *construction phase*. *Figure 24* shows the flow of a typical project from beginning to end through these phases.

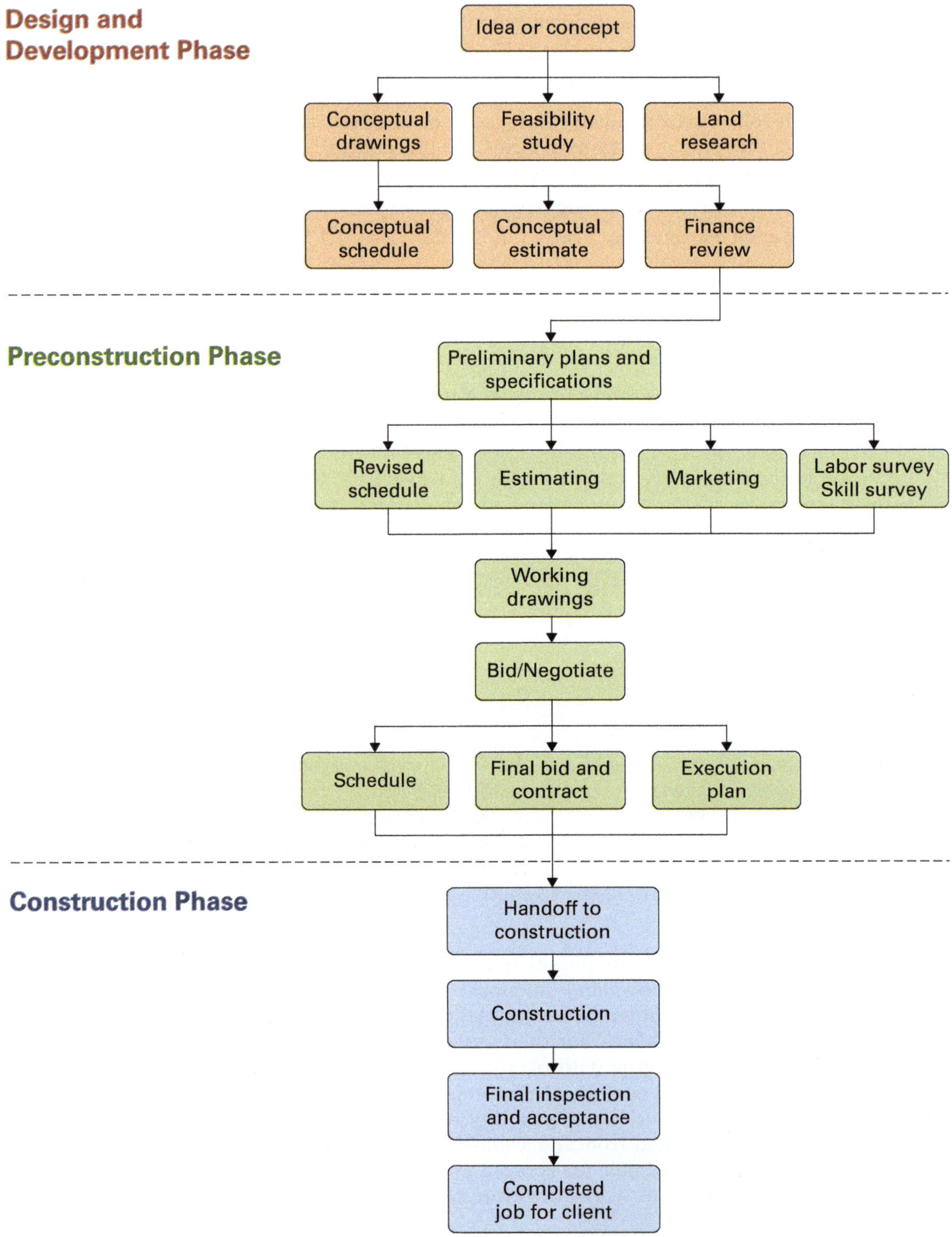

Figure 24 Construction project phases and flow.

5.1.1 Design and Development Phase

The design and development phase is the first stage of a project (*Figure 25*). It starts when an owner decides to build a new facility or add to an existing facility. This process involves land research and feasibility studies to ensure that the project has merit. A **feasibility study** is an analysis that considers all of a project's relevant factors—including budget, technical elements, legal aspects, and scheduling—to determine the likelihood that the project can be completed successfully.

Feasibility study: An analysis that considers all of a project's relevant factors—including budget, technical elements, legal aspects, and scheduling—to determine the likelihood that the project can be completed successfully.

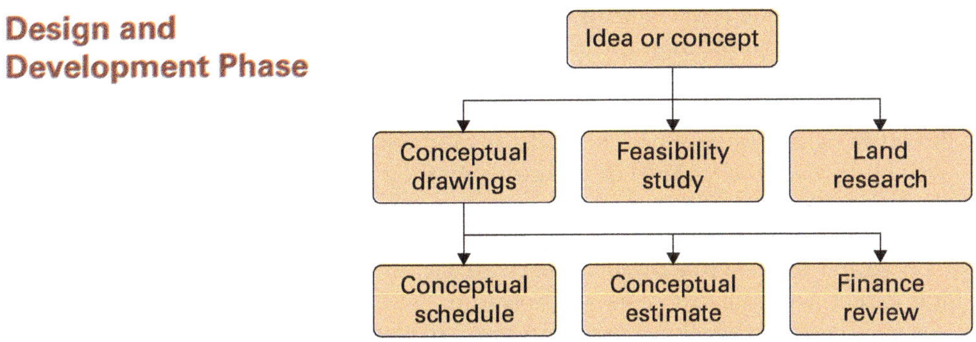

Figure 25 The design and development phase is the first stage of a project.

Architects or engineers develop the conceptual drawings that define the project graphically. They then provide the owner with sketches of room layouts and elevations and make suggestions about what construction materials to use.

During the design and development phase, architects, engineers, and/or the owner develop an estimate for the proposed project and establish a preliminary budget. Once that budget is established, the owner will determine how to pay for, or finance, the project. The owner will also complete a review to determine whether the project costs will exceed its market value and that the project provides a reasonable profit during its existence. If needed, the development team begins preliminary reviews for zoning, building restrictions, landscape requirements, and environmental impact studies.

5.1.2 Preconstruction Phase

If all steps in the design phase are completed successfully, the project moves to the preconstruction phase (*Figure 26*). The preconstruction stage includes planning that occurs before the start of construction. The preconstruction planning process does not always involve you, as a crew leader, but you should understand it to do your job well.

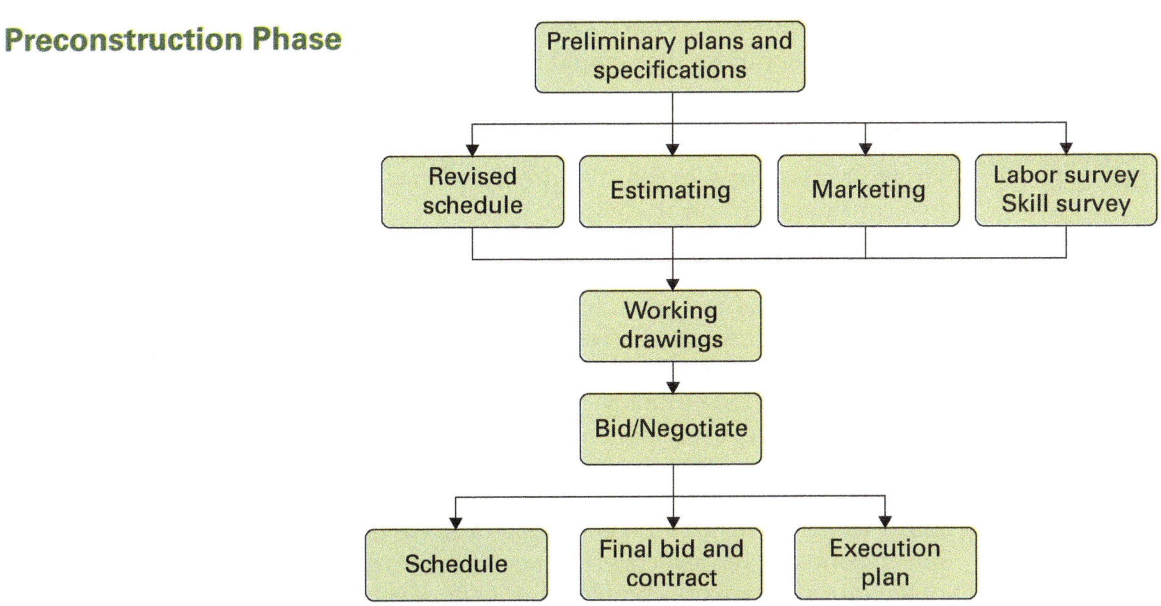

Figure 26 The preconstruction phase is the second stage of a project.

When the architects and engineers begin to develop the project drawings and specifications, they consult with other design professionals such as structural, mechanical, and electrical engineers. They perform the calculations, make a detailed technical analysis, and check details of the project for accuracy.

The design professionals create preliminary drawings and specifications. They use these drawings and specifications to communicate the necessary information to the contractors, subcontractors, suppliers, and workers who will contribute to the project.

A key next step is developing the proposal, bid, or negotiated price for the job. This is when the estimator, the project manager, and the field superintendent develop a preliminary plan for completing the work. They apply their experience and knowledge from previous projects to develop the plan. The process involves determining what methods, personnel, tools, and equipment the work will require and what level of productivity they can achieve.

Part of this process is determining how much the project will cost. Before building a project, an estimate must be prepared. Estimating is the process of calculating the cost of a project. There are two types of costs to consider, including direct and indirect costs as shown in *Table 9*. Direct costs, also known as *general conditions*, are those that planners can clearly assign to a budget. Indirect costs are overhead costs shared by all projects. Planners generally calculate these costs as an overhead percentage to labor and material costs.

TABLE 9 Examples of Direct and Indirect Costs

Direct Costs	Indirect Costs
• Materials	• Office rent
• Labor	• Utilities
• Tools	• Telecommunications
• Equipment	• Accounting
	• Office supplies and signage

Did You Know?

Profit Powers Businesses

Profit is the fuel that powers a business. It allows the business to invest in new equipment and facilities, to provide training, and to maintain a reserve fund for times when business is slow. In large companies, profitability attracts investors who provide the capital necessary for the business to grow. For these reasons, contractors can't afford to consistently lose money on projects. If they can't operate profitably, they are forced out of business. As a frontline leader, you can help your company remain profitable by managing budget, schedule, quality, and safety as outlined in the drawings, specifications, and project schedule.

Planners must also consider other factors when estimating project costs, including the following:

- Timing of all phases of work
- Types of materials to be installed and their availability
- Equipment and tools required and their availability
- Personnel requirements and availability
- Site and local conditions, such as soil types, accessibility, or available staging areas
- Climate conditions that should be anticipated during the project
- Relationships with the other contractors and their representatives on the job

On a simple job, crews can handle these items almost automatically. However, larger or more complex jobs require the planner to give these factors more formal consideration and study.

The bid price includes the estimated cost of the project as well as the **profit**. Profit refers to the amount of money that the contractor will make after paying all the direct and indirect costs. If the direct and indirect costs exceed the estimate for the job, the difference between the actual and estimated costs must come out of the company's profit. This reduces what the contractor makes on the job.

Many companies employ professional cost estimators to determine the cost of a project. Estimators will focus on **production** and **productivity rate**. Production is a quantity or amount of construction materials or work put in place. It is the quantity of materials that will be installed on a job, such as 1,000 linear feet of waste pipe installed for a task. On the other hand, the productivity rate depends on the level of efficiency of the work. The productivity rate relates to the quantity of work completed or materials put in place by the crew over a certain period of time. For example, productivity could be the amount of waste pipe installed each day by one worker or a crew.

Profit: The amount of money that the contractor will make after paying all the direct and indirect costs.

Production: The amount of construction materials put in place. It is the quantity of materials installed on a job.

Productivity rate: A measure of how much work gets done over a period of time, also called *output per labor hour*.

Production levels are set during the estimating stage. The estimator determines the total amount of materials to be placed based on the drawings, plans, and specifications. After the job is complete, supervision can assess the actual amount of materials installed and can compare the actual production to the estimated production.

During the estimating stage, the estimator uses company records of productivity to determine how much time and labor it will take to place a certain quantity of materials. From this information, the estimator calculates the productivity necessary to complete the job on time.

For example, say you need a crew of two people to paint 8,100 ft^2 in nine days. To calculate the required average productivity, divide 8,100 ft^2 by 9 days. The result is 900 ft^2/day. As the work progresses, you can compare the daily production of any crew of two painters doing similar work with this average (*Figure 27*).

NOTE

Many software programs and mobile apps are available to simplify the cost-estimating process. Estimating programs are typically set up to include a takeoff form and a form for estimating labor by category. Most of these programs include a database that contains current prices for labor and materials, so they automatically price the job and produce a bid. The programs can also generate purchase orders for materials on the job. Many of them are tailored to specific industries, such as construction or manufacturing, as well as specific trades within the industries, such as plumbing, electrical, HVACR, and insulating contractors.

Production
A quantity or amount of construction materials or work put in place

Productivity
A rate amount of work done by a worker or crew in a given timeframe

Production
Paint 8,100 ft^2

Productivity
One crew member can paint 450 ft^2/day
Two crew members can paint 900 ft^2/day

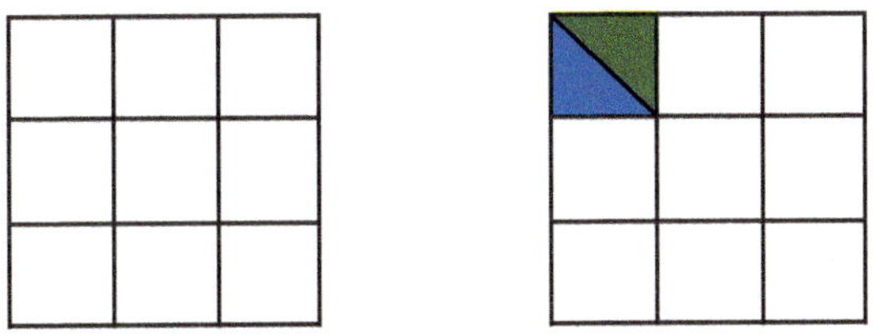

Figure 27 Production and productivity.

During the preconstruction phase, the owners hold many meetings to review and refine estimates, adjust plans to conform to regulations, and secure a construction loan. If the project is one that requires a marketing program, such as a condominium, office building, or shopping center, the marketing may start during preconstruction. In such cases, the selling of the building space often starts before actual construction begins.

Next, the design team produces a complete set of drawings, specifications, and bid documents. After this, the owner will select the method to obtain contractors. The owner may choose to negotiate with several contractors or select one through competitive bidding. Everyone concerned must also consider safety as part of the planning process. A safety crew leader may walk through the site as part of the pre-bid process.

Contracts for construction projects can take many forms. The most common types of contracts are *lump sum* and *time and materials* contracts.

In a *lump sum contract*, also called a *fixed price* or *firm-fixed price contract*, the owner generally provides detailed drawings and specifications, which the contractor uses to calculate the cost of materials and labor. The contractor also adds a percentage to these costs representing company overhead expenses, such as office rent, insurance, and accounting/payroll costs. At the end, the contractor adds a profit factor.

When submitting the bid, the contractor will state very specifically the conditions and assumptions on which the company based the bid. The **scope**, or *scope of work*, is the list of the commitments and activities that all contractors, subcontractors, and suppliers are expected to do during a project.

Scope: The list of the commitments and activities that all contractors, subcontractors, and suppliers are expected to do during a project. Also called *scope of work*.

In a *time and materials contract*, also known as a *T&M contract*, the owner pays back the contractor for labor, materials, and other costs encountered in the performance of the contract. Typically, the contractor and owner agree in advance on hourly or daily labor rates for different categories of worker. These rates include an amount representing the contractor's overhead expense. The owner also reimburses the contractor for the cost of materials and equipment used on the job. Since the buyer reimburses the contractor for the cost of materials and pays an hourly wage, any unexpected delays, roadblocks, and other changes to the scope of work are covered. T&M contracts thus work best for projects in which the scope of work is not well-defined.

NOTE

You may also hear about other types of contracts, such as cost-plus and guaranteed maximum price. Some types of contracts, such as unit-rate and hybrid, combine several aspects of these other contract types. Scan the QR code in the front of this module to access additional resources on construction contract types.

On a time and materials contract project, you play an important role by understanding the scope of the work and making sure that everything is tracked so that your company is reimbursed for the work. You must identify changes or problems that increase the scope or amount of planned labor or materials. You also must ensure you and your team are following your company policies and getting proper authorization before starting any work.

After the client awards the contract, planners create a final schedule and an execution plan that defines the actual work methods and resources needed to perform the work.

5.1.3 Construction Phase

After the contract is signed, the construction phase can start (*Figure 28*). The designated contractor enlists the help of mechanical, electrical, elevator, and other specialty subcontractors to complete the construction phase. The contractor may perform one or more parts of the construction and rely on subcontractors for the remainder of the work. However, the general contractor is responsible for managing all the trades necessary to complete the project.

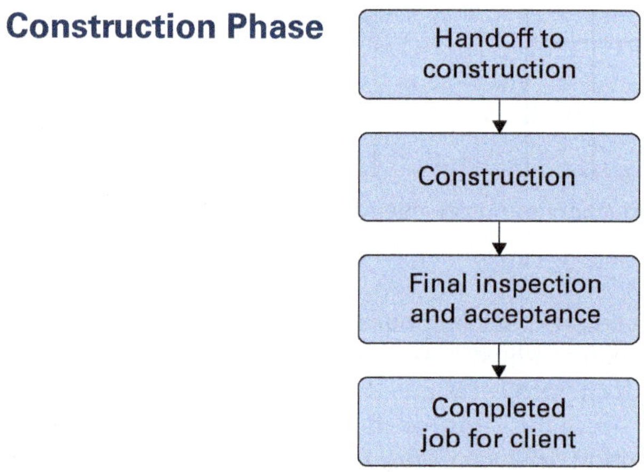

Figure 28 The construction phase is the third stage of a project.

The initial set of drawings for a construction project reflects the completed project as conceived by the architect and engineers. During construction, changes may be necessary because of factors unforeseen during the design phase. For example, when electricians must reroute cabling or conduit, or the installed equipment location is different than shown on the original drawing, such changes must be marked on the drawings. Without this record, technicians called to perform maintenance or modify the equipment later will have trouble locating all the cabling and equipment.

Project supervision must document any changes made during construction or installation on the drawings as the changes occur. Before digital drawings, architects would note changes on hard-copy drawings using a colored pen or pencil. These marked-up versions are commonly called **redline drawings**. With new digital drawings created using building information modeling tools, architects and engineers can revise and share the latest drawing versions almost instantly to mobile phones and tablets. After the drawings have been revised to reflect any redline changes, the final drawings are called **as-built drawings**. As-built drawings are a revised set of drawings submitted by a contractor upon completion of a project or job. These drawings, sometimes called *record drawings*, reflect all changes made during construction and show the exact specifications of the completed work.

Redline drawings: Marked-up versions of construction drawings documenting any changes made during construction or installation.

As-built drawings: A revised set of drawings that reflect all changes made during the construction process and show the exact specifications of the completed work.

The main part of your role as a frontline leader happens during the construction phase, including daily work planning and scheduling, cost control, and resource management, including material delivery and storage, tool control, and equipment control. All of these tasks drive production and productivity, as well as the overall profitability of the project.

As construction nears completion, the architect/engineer, owner, and government agencies start their final inspections and acceptance of the project. If the general contractor has managed the project, the subcontractors have performed their work, and the architects/engineers have regularly inspected the project to ensure it satisfies the local code, then the inspection process can finish up in a timely manner. This results in a satisfied client and a profitable project for all.

On the other hand, if the inspection reveals faulty workmanship, poor design, incorrect use of materials, or violation of codes, then the inspection and acceptance will take longer. It may result in a dissatisfied client or an unprofitable project.

Construction **project delivery methods** provide different ways to organize the services of development, planning, and construction in order to execute a project.

NOTE

Scan the QR code in the front of this module to access additional resources on construction project delivery methods.

Project delivery methods: Different ways to organize the services of development, planning, and construction in order to execute a project.

5.2.0 Planning and Scheduling

In the previous sections, you learned about the phases of a construction project, including the design and development phase, the preconstruction phase, and the construction phase. As noted, the main part of your role as a frontline leader happens during the construction phase.

During construction, you may be directly involved in daily work planning and scheduling. You also may be involved in cost control and resource management, including material delivery and storage, tool control, and equipment control.

5.2.1 Understanding Overall Schedules

Construction planning and scheduling refers to the process through which you map out what will take place during a project, what resources you need, when key activities will take place, which personnel will be involved, and anything else pertinent to the project.

Construction planning refers to the process of laying out all details of a construction project beforehand. Construction scheduling involves defining which activities happen in what order and assigning dates and times to those tasks. Construction schedules are linked to resources such as materials, labor, and equipment to help create the most efficiency in your project. Construction scheduling is deeply connected to planning.

Planning and scheduling are closely related and are both very important to a successful job. Planning identifies the required activities and how and in what order to complete them. Scheduling involves establishing start and finish times/dates for each activity.

A schedule for a project typically shows the following:

- Operations listed in sequential order
- Units of construction
- Duration, or length of time, of an activity
- Estimated dates to start and complete each activity
- Quantity of materials to be installed

Figure 29 shows an example of a high-level project schedule. As noted, the development of a schedule for a project starts as early as the design and development phase. As the project moves through each phase and the work is better defined, the schedule also becomes more clearly defined.

When a crew is close to completing specific work—say, within three weeks of working on Foundation 1—a **Level 4 schedule** is created. A Level 4 schedule offers additional detail to provide working schedules to the crews doing the

NOTE

Scan the QR code in the front of this module to access additional resources on schedule levels.

Level 4 schedule: A schedule that offers additional detail to provide working schedules to the crews doing the work, usually within three to six weeks of completing specific work.

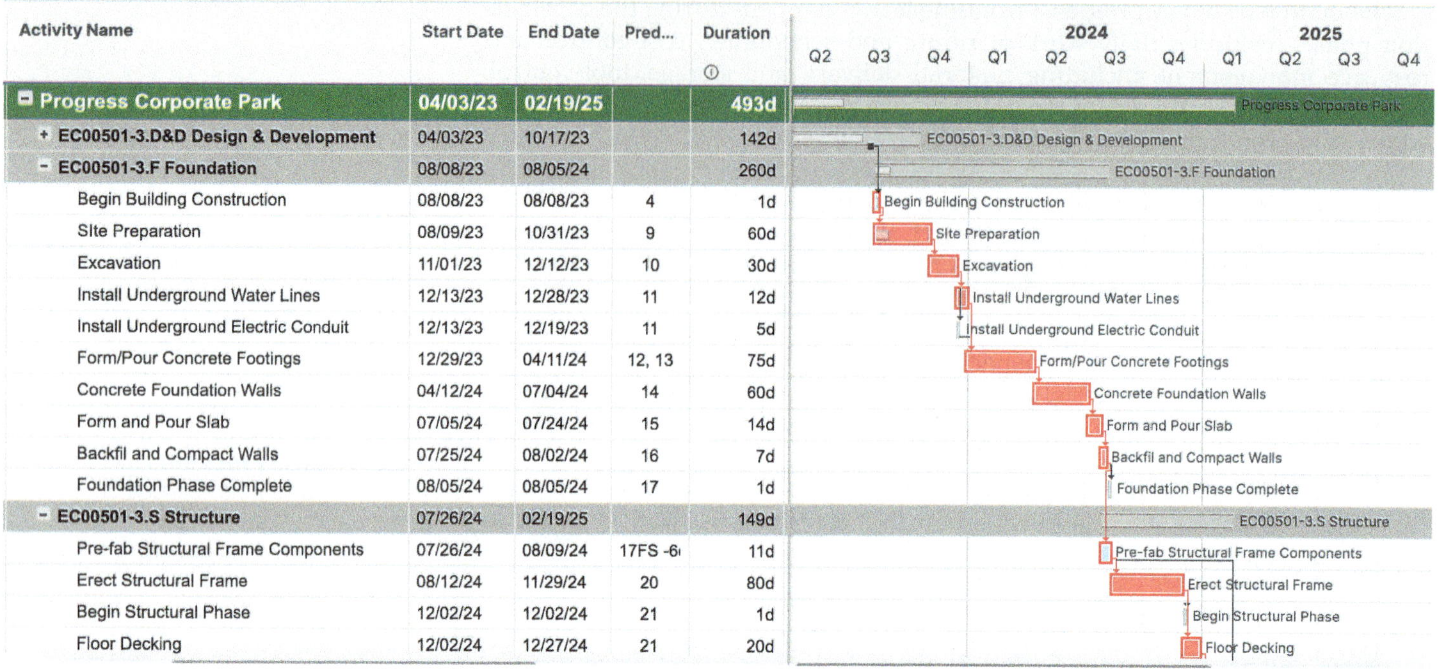

Activity Name	Start Date	End Date	Pred...	Duration
Progress Corporate Park	04/03/23	02/19/25		493d
+ EC00501-3.D&D Design & Development	04/03/23	10/17/23		142d
- EC00501-3.F Foundation	08/08/23	08/05/24		260d
Begin Building Construction	08/08/23	08/08/23	4	1d
SIte Preparation	08/09/23	10/31/23	9	60d
Excavation	11/01/23	12/12/23	10	30d
Install Underground Water Lines	12/13/23	12/28/23	11	12d
Install Underground Electric Conduit	12/13/23	12/19/23	11	5d
Form/Pour Concrete Footings	12/29/23	04/11/24	12, 13	75d
Concrete Foundation Walls	04/12/24	07/04/24	14	60d
Form and Pour Slab	07/05/24	07/24/24	15	14d
Backfil and Compact Walls	07/25/24	08/02/24	16	7d
Foundation Phase Complete	08/05/24	08/05/24	17	1d
- EC00501-3.S Structure	07/26/24	02/19/25		149d
Pre-fab Structural Frame Components	07/26/24	08/09/24	17FS -6	11d
Erect Structural Frame	08/12/24	11/29/24	20	80d
Begin Structural Phase	12/02/24	12/02/24	21	1d
Floor Decking	12/02/24	12/27/24	21	20d

Figure 29 Example of a construction project schedule.

work (*Figure 30*). A Level 4 schedule allows a supervisor to create a plan, schedule handoffs from crew to crew, and estimate all of the resources that must be in place for the work to happen.

In the schedule in *Figure 30*, ironworkers must fabricate and set the rebar and carpenters must build the forms before the handoff is made to place the concrete. Each of these activities is a critical path activity. A **critical path** is the sequence of dependent tasks that form the longest duration. In *Figure 29* and *Figure 30*, the critical path in the schedule is shown in red. In other words, until one activity is complete, workers cannot start the other tasks, and the project will likely be delayed by the amount of time the previous activity was delayed.

Critical path: In manufacturing, construction, and other types of creative processes, the required sequence of tasks that directly controls the ultimate completion date of the project.

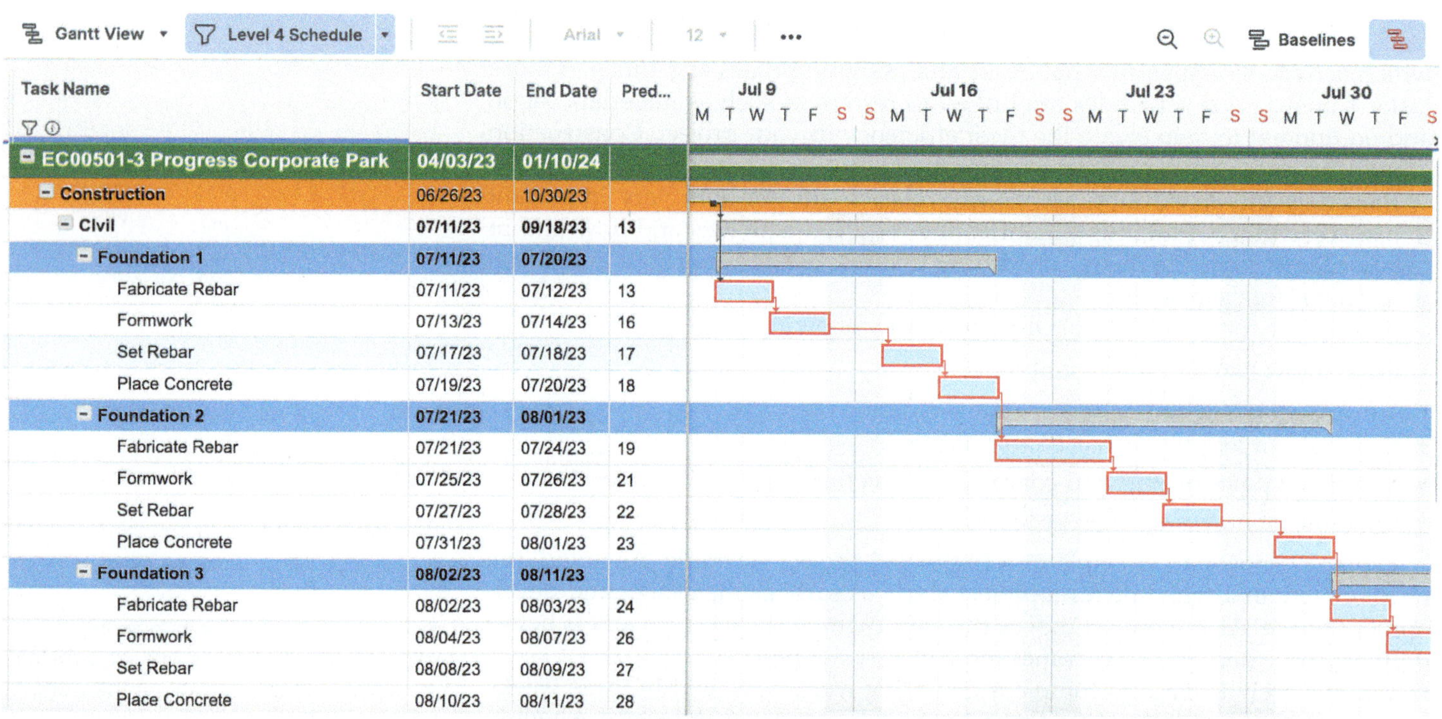

Task Name	Start Date	End Date	Pred...
EC00501-3 Progress Corporate Park	04/03/23	01/10/24	
Construction	06/26/23	10/30/23	
Civil	07/11/23	09/18/23	13
Foundation 1	07/11/23	07/20/23	
Fabricate Rebar	07/11/23	07/12/23	13
Formwork	07/13/23	07/14/23	16
Set Rebar	07/17/23	07/18/23	17
Place Concrete	07/19/23	07/20/23	18
Foundation 2	07/21/23	08/01/23	
Fabricate Rebar	07/21/23	07/24/23	19
Formwork	07/25/23	07/26/23	21
Set Rebar	07/27/23	07/28/23	22
Place Concrete	07/31/23	08/01/23	23
Foundation 3	08/02/23	08/11/23	
Fabricate Rebar	08/02/23	08/03/23	24
Formwork	08/04/23	08/07/23	26
Set Rebar	08/08/23	08/09/23	27
Place Concrete	08/10/23	08/11/23	28

Figure 30 A Level 4 schedule provides enough details for project foremen to forecast handoffs and needed resources.

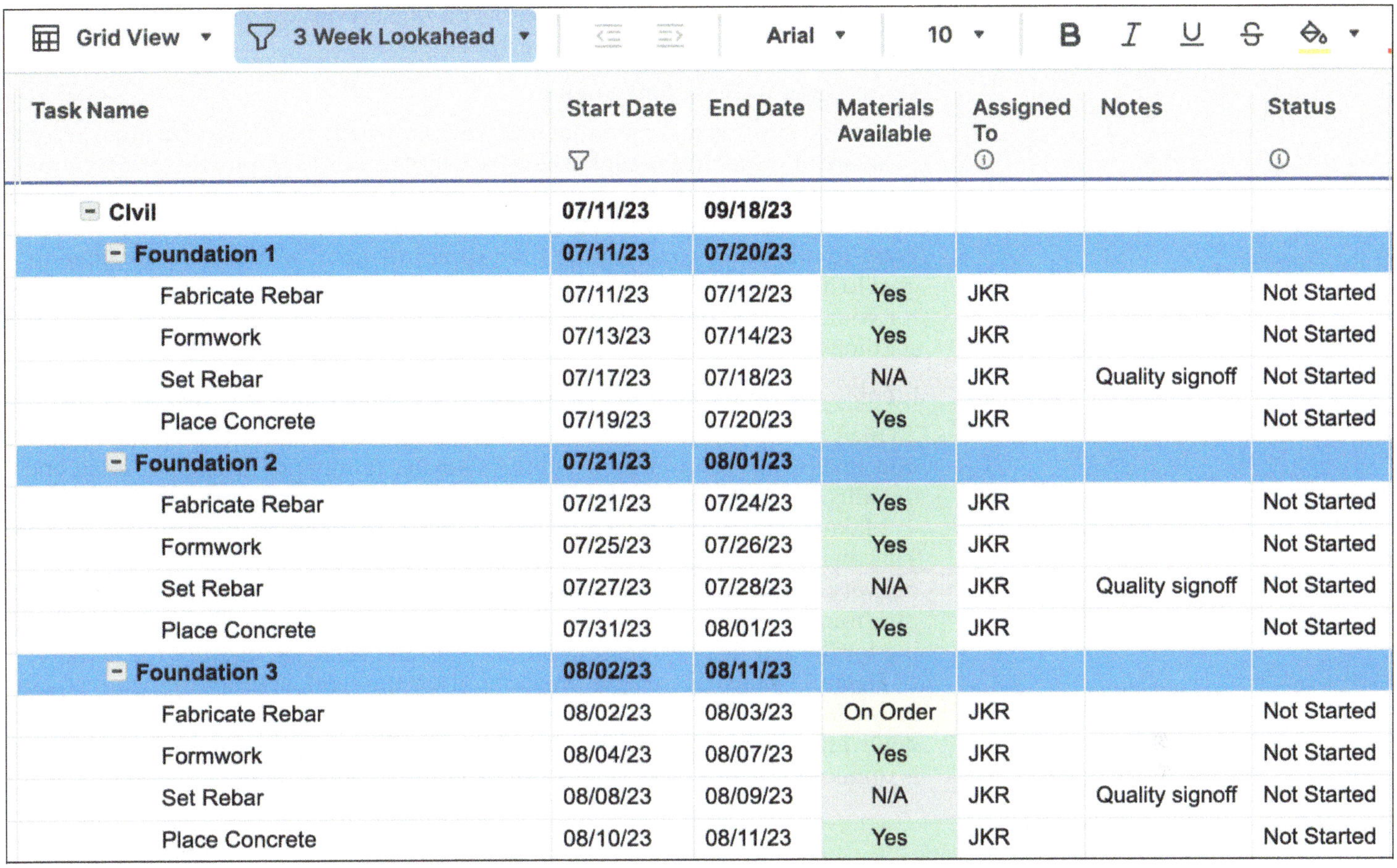

Task Name	Start Date	End Date	Materials Available	Assigned To	Notes	Status
Civil	**07/11/23**	**09/18/23**				
Foundation 1	**07/11/23**	**07/20/23**				
Fabricate Rebar	07/11/23	07/12/23	Yes	JKR		Not Started
Formwork	07/13/23	07/14/23	Yes	JKR		Not Started
Set Rebar	07/17/23	07/18/23	N/A	JKR	Quality signoff	Not Started
Place Concrete	07/19/23	07/20/23	Yes	JKR		Not Started
Foundation 2	**07/21/23**	**08/01/23**				
Fabricate Rebar	07/21/23	07/24/23	Yes	JKR		Not Started
Formwork	07/25/23	07/26/23	Yes	JKR		Not Started
Set Rebar	07/27/23	07/28/23	N/A	JKR	Quality signoff	Not Started
Place Concrete	07/31/23	08/01/23	Yes	JKR		Not Started
Foundation 3	**08/02/23**	**08/11/23**				
Fabricate Rebar	08/02/23	08/03/23	On Order	JKR		Not Started
Formwork	08/04/23	08/07/23	Yes	JKR		Not Started
Set Rebar	08/08/23	08/09/23	N/A	JKR	Quality signoff	Not Started
Place Concrete	08/10/23	08/11/23	Yes	JKR		Not Started

Figure 31 A step-by-step lookahead in a Level 4 schedule provide details for frontline leaders to plan handoffs.

For work to happen, you will need resources such as materials, available crew, available construction equipment, completed permits, and more. Developing a **lookahead schedule** helps ensure that all resources are available for the project when needed. This is often included in a Level 4 schedule. A lookahead schedule is useful for planning and scheduling work that is happening in the short term, usually within one to three weeks (*Figure 31*).

Lookahead schedule: For planning and scheduling work that is happening in the short term, usually within one to three weeks; used for anticipating material, labor, tool, and other resource requirements, as well as identifying potential schedule conflicts or other problems. Also called *short-term schedule*.

5.2.2 Creating a Short-Term or Lookahead Schedule

As a frontline leader of a crew, you must be able to create, read and interpret a lookahead schedule. You can use this information to plan work more effectively, set realistic goals, and compare the starts and completions of tasks to those on the schedule. As the leader of the team executing the work, you are a key expert to help produce a plan at this level of detail. *Table 10* provides a checklist for planning and scheduling.

TABLE 10 Crew Leader Planning and Scheduling Checklist

Did You…?	Yes/No
Determine the best method for performing the job.	
Identify the responsibilities of each person on the work crew.	
Determine the duration, or length in time, and sequence of each activity.	
Identify the tools and equipment needed to complete the job.	
Ensure that the required materials are at the worksite when needed.	
Make sure that heavy construction and other equipment is available when required.	
Work with other contractors in such a way as to avoid interruptions and delays.	

The following outlines the process you must complete to develop a plan and a schedule and to use the schedule to manage work.

Understand or Establish a Goal

A goal is a specific outcome that one works toward. The goals you must accomplish are often set in the higher-level schedules. For example, the project superintendent of a residential construction project could establish a goal to have a house dried-in or ready for the application of roofing and siding by a certain date. To meet that goal, you and the superintendent would need to agree to a goal to have the framing completed by a given date. You then establish subgoals for your crew to complete each element of the framing (floors, walls, roof) by a set time.

Identify the Required Work Activities and Divide Them into Tasks

To plan effectively, you must be able to break a work activity assignment down into smaller tasks. Large jobs include a greater number of tasks than small ones, but all jobs can be broken down into manageable components.

Make a list of all the activities that the plan requires to complete the job. This includes individual work tasks and special tasks, such as inspections or the delivery of materials. Next, start to divide work activities into tasks. As you do this, remember that almost every work activity has three general parts: preparing for the activity, performing the activity, and cleaning up after the activity.

Figure 32 shows a suitable breakdown for the work activity to install square vinyl floor tiles in a cafeteria. You could create even more detail by breaking down any one of the tasks into subtasks. In this case, however, that much detail is unnecessary and wastes your time and the project's money.

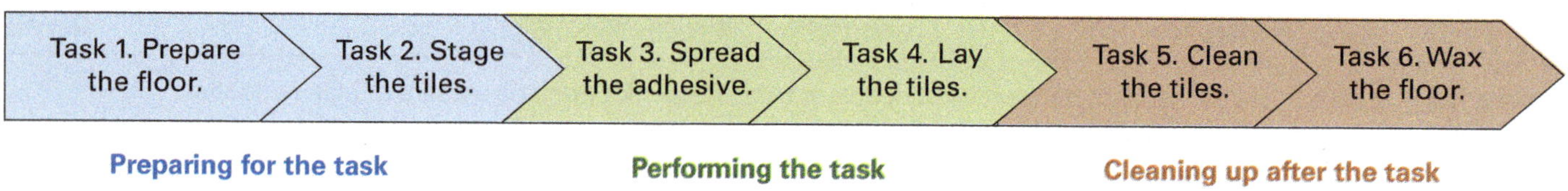

Figure 32 Example task breakdown of the work activity to install square vinyl floor tiles in a cafeteria.

When breaking down an activity into tasks, make each task *identifiable* and *definable*. A task is identifiable when you know exactly the types and amounts of resources it requires. A task is definable if it has a specific duration. The activity breakdown does not need to be overly detailed or complex unless the job has never been done before or must be performed with strictest efficiency.

At this point, you should just be concerned with generating a list, not with determining how to accomplish the activities, who will perform them, how long they will take, or the necessary sequence to complete them.

Plan Resources Needed for Each Task

Once a job has been broken down into activities and tasks, you must identify and account for all of the resources you need to complete them. Resource planning includes considerations for safety, materials, site-specific issues, equipment, tools, and labor (*Table 11*). Each of these is discussed in more detail in the sections that follow.

Organize the Work into a Logical Sequence

The next step in planning is to use the list of activities and resources needed to organize the work activities into a logical sequence. When doing this, keep in mind that certain steps can't happen before the completion of others. For example, concrete placement cannot happen until after rebar is set. Some construction work activities, such as installing 12" footing forms, are done so often that they require little planning. However, other jobs, such as placing a new type of mechanical equipment, require substantial planning.

Assign Responsibility for Each Task

After you have identified the tasks, you must determine who will complete each activity. This requires that you are aware of the experience, skills, and abilities of

TABLE 11 Key Resource Considerations During Planning

Resource Planning Area	Key Responsibilities
Safety Planning	Using the company safety manual as a guide, you must assess the safety issues associated with the job and take necessary measures to minimize any risk to the crew.
Materials Planning	You may be required to estimate quantities of materials during construction. You also have a key role in the receipt, storage, and control of the materials after they reach the jobsite.
Site Planning	Many planning elements are involved in sitework, including access roads, parking, site security, material and equipment staging and storage, sedimentation control, stormwater runoff, and more.
Equipment Planning	You may need to plan the types of equipment needed, the use of the equipment, and the length of time it will be on the site. You must work with the main office to make certain that the equipment reaches the jobsite on time. You must reserve time for equipment maintenance to prevent equipment failure.
Tool Planning	To plan tool usage, you must determine the tools required, tell your crew who will provide the tools, ensure that the crew is qualified to use the tools safely and effectively, and establish proper tool controls.
Labor Planning	In many companies, the project manager or job superintendent determines the size and makeup of the crew. Then, supervision expects you to accomplish the goals and objectives with the crew provided. If you are responsible for planning for labor, you must identify the skills needed to perform the work, determine the number of people having the required skills, and decide who will be on the crew.

everyone in your team. You must put this knowledge to work by matching the crew's skills and abilities to specific tasks. Your job is to draw from their expertise to get the job done well and in a safe and timely manner.

Assign a Duration to Each Task

Next, assign a duration or length of time that it will take to complete each activity and determine the start date and time for each. Then place each activity into a schedule format. This step is important because it helps you compare the task time estimates to the scheduled completion date or time.

Make note of any activities that cannot start until previous work has been completed and ensure those do not start before the other work is complete. You also will need to ensure you have not assigned too much work to any one person or team.

Some of the information to support short-term scheduling comes from the estimate or cost breakdown. *Figure 33* shows how you can use estimate or cost breakdown information to set short-term production goals for a team.

Scenario: A carpentry crew on a retaining wall project is about to form and pour catch basins and put up wall forms. The crew will work on both the basins and the wall forms at the same time. The crew has put in several catch basins, so the crew leader is sure that they can perform the work within the estimate. However, the crew leader is concerned about their production of the wall forms and wants to set a clear target for the amount of work to be done each day to hit the schedule.

The scheduling process in this scenario could be handled as follows:

1. Review the estimate or cost breakdown.	2. Consider available workers.	3. Determine the job goal.	4. Analyze how much work the existing workers can complete.	5. Communicate production goal with the crew.
• Production factor for wall forms = 16 worker-hours (w-h) per 100 ft^2 • Work to be done by measurement = 800 ft^2 • Total time: (800 ft^2 × 16 w-h/100 ft^2) = 128 w-h	The carpentry crew includes: • One carpentry crew leader • Four carpenters • One laborer	The crew leader determines the goal for the job should be set at 128 w-h (from the cost breakdown).	If the crew remains the same (six workers), the work should be completed in about 21 crew-hours (128 w-h ÷ 6 workers/crew = 21.3 crew-hours).	• Crew leader shares the production goal of completing 800 ft^2 in 21 crew-hours with the crew. • Crew leader encourages them to work together to meet the goal of getting the forms erected within the estimated time.

Figure 33 Crew leaders can use lookaheads or short-term scheduling to set production goals.

In this example, the crew leader used the short-term schedule to translate production into work hours and to schedule work so that the crew can accomplish it within the estimate. A crew can improve production when they know the amount of work to accomplish and the time available to complete the work, and can provide input when setting goals. In addition, setting production targets provides the motivation to exceed schedule targets.

Communicate the Schedule and Assigned Responsibilities

After the schedule is complete, you must then assign work responsibilities to the crew. Each person in the crew should know what they are responsible for accomplishing on the job.

Use the Schedule to Monitor and Control Work

NOTE

Most scheduling apps and programs include this feature that automatically helps you focus on the lookahead period for a job.

Schedules only have value if they are actively used as tools to monitor and control work. A lookahead or short-term schedule provides you with the information you need to plan for work and track progress. Before starting a job, actively review the schedule to:

- Determine the materials, tools, equipment, and labor needed to complete the job.
- Determine when the various resources are needed.
- Follow up to ensure that the resources are available on the jobsite when needed.

You should also verify the availability of needed resources three to four working days before the start of the job. This should occur even earlier for larger jobs. Advance preparation will help avoid situations that could potentially delay starting the job or cause it to fall behind schedule. As work continues, use your schedule to monitor whether actual production begins to slip behind estimated production and develop plans to recover the schedule as needed.

Follow Up and Confirm Work Is Completed Well

After you delegate tasks to the appropriate crew members, you must follow up to confirm that the crew completed the tasks correctly and efficiently. To do this effectively, you must be present on the jobsite to make sure all the resources are available to complete the work, to confirm that team members are working on their assigned activities, to answer any questions, and to help to resolve any problems that occur while the work is being done.

5.3.0 Crew Leader Field Reporting Responsibilities

A field-reporting system consists of a series of forms that are completed by you and other leaders on the project to share updates on labor, materials, tools, equipment, and more.

As a frontline leader, you may not be directly involved in updating overall schedules. However, the daily field or progress reports you submit will be used by the company to keep the schedule up to date.

The following sections provide some examples of field reporting. Each company has its own forms and methods for obtaining information. You should always follow your company's policies and procedures for field reporting.

5.3.1 Completing Daily Reports

Job diary: A written record that a supervisor maintains periodically (usually daily) of the events, communications, observations, and decisions made during the course of a project.

You will be involved with many activities on a day-to-day basis. As you work each day, you will want to keep a **job diary** to track events such as job changes, interruptions, and visits. A job diary, also called a *daily report, site diary, construction log,* or *site journal,* is a notebook or digital application in which you record activities or events that take place on the jobsite that may be important later (*Figure 34*).

> July 8, 2023
>
> Weather: Hot and Humid
>
> Project: Opal Garage Annex
>
> - The paving contractor crew arrived late (10 am).
> - The owner representative inspected the footing foundation at approximately 1 pm.
> - The concrete slump test did pass. Two trucks had to be ordered to return to plant, causing a delay.
> - David Lopez had an incident on the second floor. I sent him to the on-site medical personnel for medical first aid treatment. The cause of the incident is being investigated.

Figure 34 Sample page from a job diary.

For this comparison process to be of use, the information you and others submit from the field must be accurate. In the event of a legal or contractual conflict with the client over costs, job diaries can be helpful as evidence in legal proceedings or in reaching a settlement. *Table 12* outlines an example scenario where your action of accurate field reporting will help benefit your company.

TABLE 12 Accurate Field Reporting Benefits Your Company

Scenario	Field Reporting Action
You are running a crew of five concrete finishers for a specialty contractor. When you and your crew show up to finish a slab, the general contractor (GC) informs you, "We're a day behind on setting the forms, so I need you and your crew to stand down until tomorrow."	Follow your company's field reporting policy to report the change. For example, you might first call your office to let them know about the delay. Then, you would record it in your job diary. A six-man crew for one day represents 48 worker-hours. If your company charges $30 an hour, that's a potential loss of $1,440, which the company would want to recover from the general contractor. If there is a dispute, your entry in the job diary could result in a favorable decision for your employer.

Today, many companies use mobile applications that allow you to complete daily reporting on your mobile phone or tablet while you are on-site. You can quickly enter information throughout the day and attach photos and other documents to add more information. They generally also make it easy to send to your supervisor via email or text message (*Figure 35*). Your supervisor will be notified when the report is complete.

Whether print or digital, when making entries in a job diary, make sure that the information is accurate, factual, complete, consistent, organized, and up to date. Do not include any emotional details; keep the report focused on facts. This is especially true if documenting personnel problems. If the company

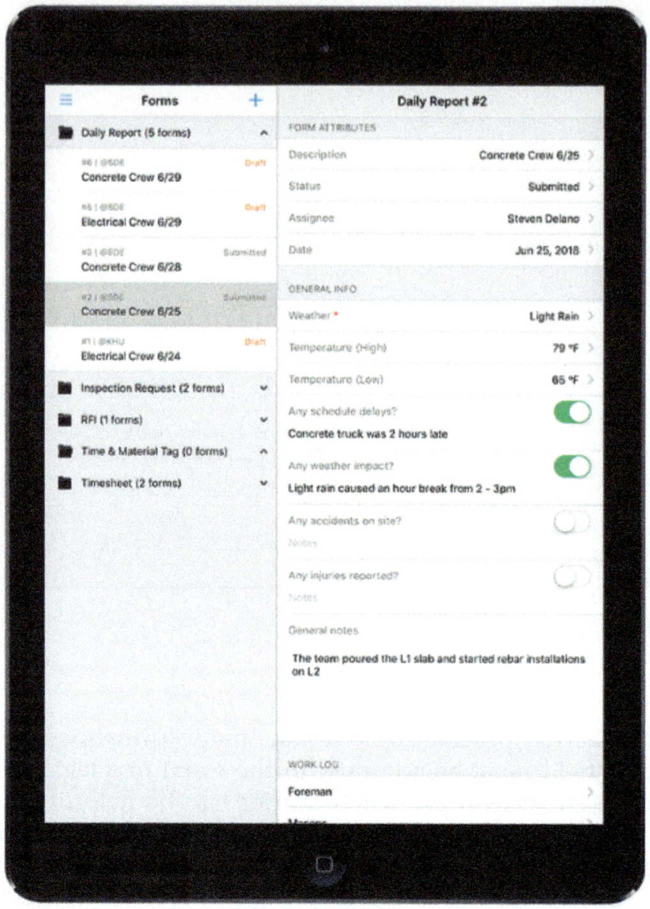

Figure 35 Mobile apps make it easier to complete daily reporting on-site.
Source: Fieldwire, https://www.fieldwire.com/

requires maintaining a job diary, follow company policy in determining which events and what details you should record. The information needs to be detailed enough so that, if needed, you could determine what happened on any day years later. If you are unsure what to include, it is better to have more information than too little. Check in with your supervisor to confirm if your daily reporting is including the right information.

5.3.2 Tracking Employee Costs Through Timesheets

Work breakdown structure (WBS): A method of organizing the entire scope of work in a logical manner based on how the work will be executed.

When the project budget is set up, many companies create a work breakdown for each project. A **work breakdown structure (WBS)** organizes the entire scope of work in a logical manner based on how the work will be executed. Within the WBS, planners assign each major task a unique charge number or code (*Figure 36*). Anyone working on that task charges that number or code on their timesheet, so that project managers can readily track cost performance.

As a frontline leader, you not only need to ensure everyone is completing their timesheets, but also that they are coding those timesheets to the proper codes in the WBS. Doing so ensures you and your company are tracking and managing labor costs against specific work.

For example, as shown in *Figure 36*, if all timesheets are correctly coded, you and the company can determine the labor cost to install each of the three types of pipe: carbon steel piping, stainless steel piping, and the high-density polyethylene (HDPE) piping. You can also calculate how much it costs to install the piping for the entire project.

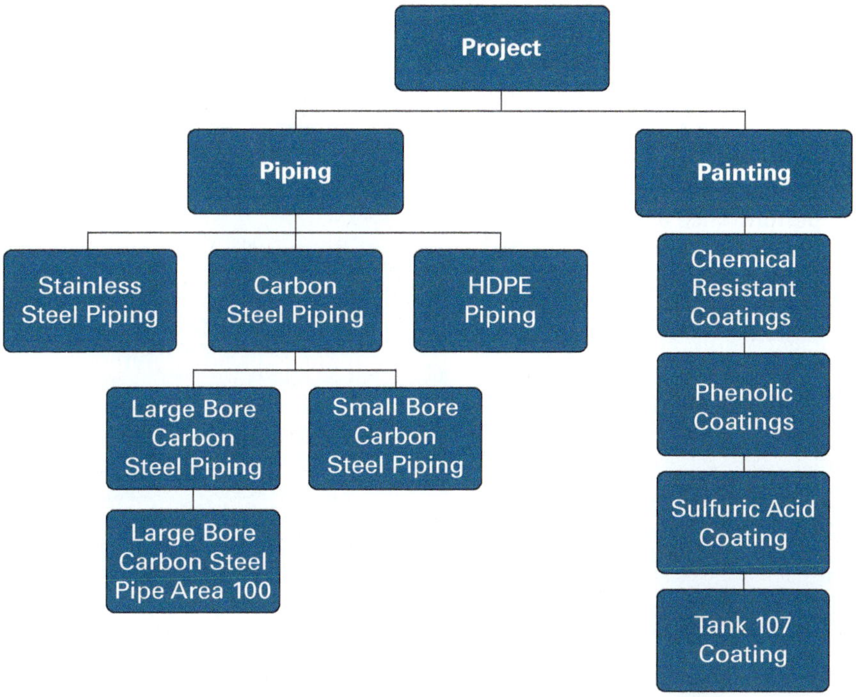

Level	Work	Account Code
Level 0	Piping	B
Level 1	Stainless Steel Piping	B-SS
Level 1	Carbon Steel Piping	B-CS
Level 1	HDPE Piping	B-HP
Level 2	Large Bore Carbon Steel Piping	B-CS-L
Level 2	Small Bore Carbon Steel Piping	B-CS-S
Level 3	Large Bore Carbon Steel Pipe Area 100	B-CS-L-100
Level 0	Painting	L
Level 1	Chemical Resistant Coatings	L-CR
Level 2	Phenolic Coating	L-CR-P
Level 3	Sulfuric Acid Tanks Coating	L-CR-P-ST
Level 4	Tank 107 Coating	L-CR-P-ST-107

Figure 36 An example of a work breakdown structure (WBS) for piping and painting on a project.

5.4.0 Frontline Leader Cost and Resource Management Responsibilities

Being aware of costs and controlling them is the responsibility of every employee on the job. As a frontline leader, it is your job to ensure that employees uphold this responsibility. Cost control is the process by which you keep expenses under control by managing labor, material, and other costs to ensure that the project finishes on budget.

5.4.1 Managing Costs

As you have learned, profit refers to the amount of money that the contractor will make after paying all the direct and indirect costs. If the costs exceed the estimate for the job, the difference between the actual and estimated costs must come out of the company's profit. This reduces what the contractor makes on the job.

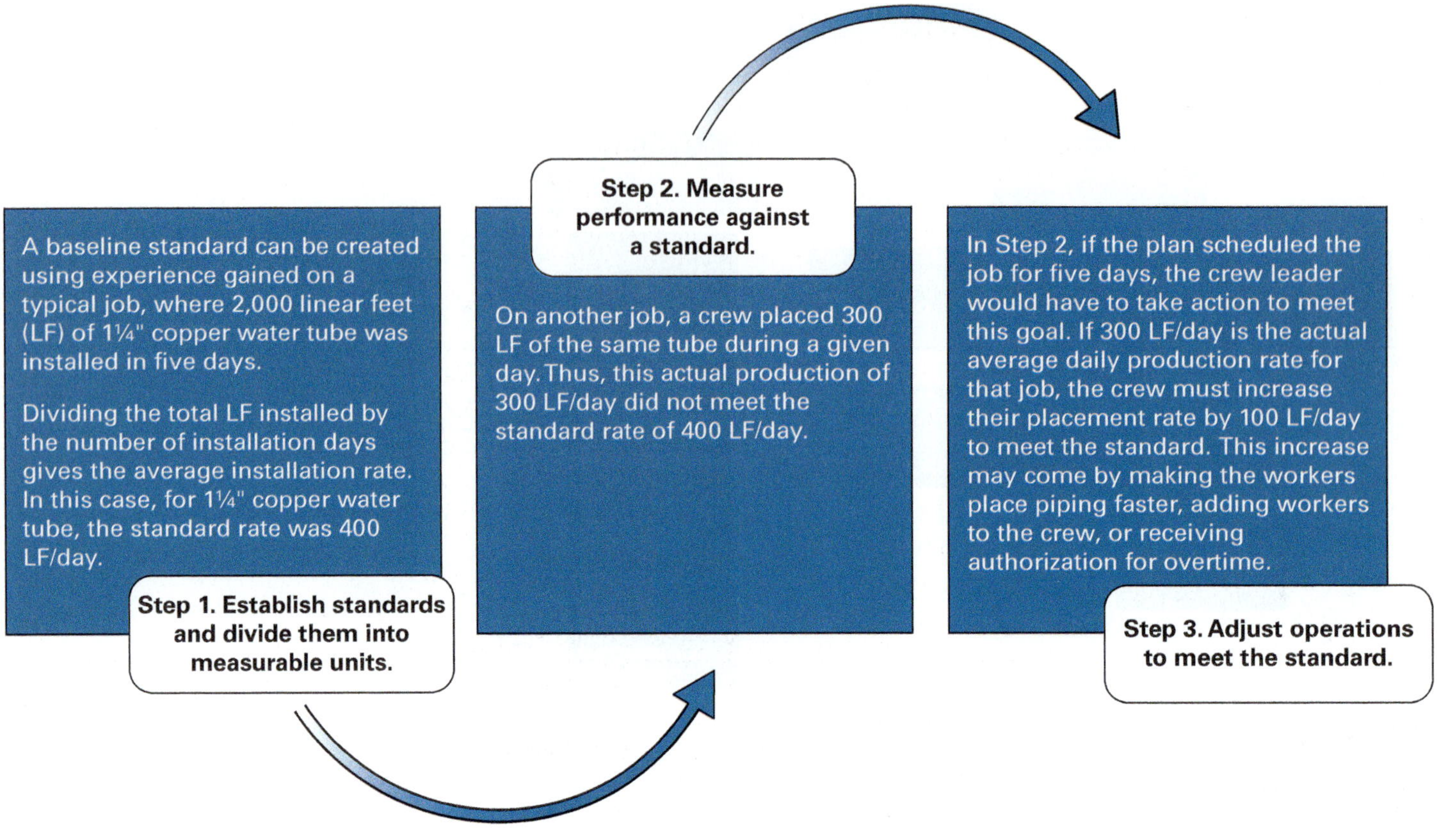

Figure 37 An effective cost control process monitors production and productivity to help manage costs.

NOTE

Before starting any action, you should check with company supervision to see that the action proposed is acceptable and within the company's policies and procedures.

Early identification of cost changes is very important and a key part of your role. As shown in *Figure 37*, an effective cost control process to monitor production and productivity includes several key steps. Identifying changes provides enough notice to correct the causes of overruns and manage those costs. You must learn to anticipate changes from plans and specifications based on experience and take measures to prevent them from occurring. If a situation cannot be fully corrected, early notification allows time for communicating and for managing expectations. No one likes surprises when it comes to increased project costs.

Table 13 lists some of the factors that can make actual costs exceed estimated costs, along with some potential actions you can take to avoid these situations.

A key factor in managing costs is controlling waste. Waste in construction can add up to loss of critical and costly materials and may result in job delays. You need to ensure that every crew member knows how to complete the work assigned and use materials efficiently. One example of waste is a carpenter who saws off a piece of lumber from a full-sized piece, when he could have found the length needed in the lumber scrap pile.

Doing rework is another form of waste (*Figure 38*). For example, less-experienced or less-trained crew members may make mistakes that cause work to have to be done over. You can help avoid these errors by providing on-the-spot training and by checking on newer, less experienced workers more often. Rework also can result from a conflict in craft scheduling that requires removal and reinstallation of finished work. Rework also increases safety risks on a job. Rework often involves demolition, time pressure, and frustration from the team, all of which create a higher likelihood of an incident.

Some issues that cause rework, such as mistakes on drawings, may be outside of your control. However, you can manage this risk by ensuring that you always have the most current set of drawings and examining the drawings carefully before you start work. Finding errors before the work begins increases safety and saves time, labor, materials, and more, which saves costs overall.

TABLE 13 Factors That Drive Cost Changes

Factor	Proactive Action
Client-related changes	Clients may want to change the scope of a project after it has started. You must be able to assess the potential impact of such changes and, if necessary, confer with the employer to determine the course of action. If contractor-related losses are occurring, you and the superintendent will need to work together to get the costs back in line.
Late delivery of materials, tools, and/or equipment	Review and update your lookahead schedules on a regular basis. Plan ahead to ensure that job resources will be available when needed.
Inclement weather	Work with the superintendent and other field leaders to have alternate plans ready.
Higher-than-expected wages or unexpected overtime due to low manpower	Plan ahead to ensure that labor resources will be available when needed. Ensure you are accurately recording daily work using your field reporting systems and job diary. Work with your supervisor or the superintendent to identify other staffing approaches.
Contractor or owner delays due to work availability	Ensure you are accurately recording daily work using your field reporting systems and job diary. Work with the superintendent and other field leaders to have alternate plans ready.
Unmotivated workers	Apply the leadership and team motivation strategies discussed earlier in this course. Remember, different things motivate different people in different ways. This means there is no single best approach to motivating all of your crew members.
Incidents	Continue to lead by example, reinforce the critical nature of safety, and follow the guidelines discussed earlier in the course. Work with the superintendent and other field leaders to help evaluate and improve the safety and health program, if needed.
Owner-related changes	Owners may want to change the scope of a project after it has started. You must be able to assess the potential impact of such changes and, if necessary, confer with the employer to determine the course of action. If contractor-related losses are occurring, you and the superintendent will need to work together to get the costs back in line.
Waste due to rework, material usage, and more	Train less-experienced workers on how to complete work. Review and update your lookahead schedules on a regular basis. Ensure crew members know how to use materials efficiently. Ensure you have the latest drawings and reference them daily.

Figure 38 Rework is another form of waste on a construction site.
Source: Vladimir Cetinski/iStock/Getty Images

5.4.2 Ensuring Crew Productivity

People typically represent more than half the cost of a project and thus have an enormous impact on profitability. For that reason, it is essential for you to manage your crew and their work environment in a way that maximizes their productivity.

One way to manage costs and increase productivity is to make sure you have the right skills on your crew to execute the level of tasks that you have to complete. *Crew mix* is a concept that describes the number of experienced or journey-level workers relative to the number of apprentices or less-experienced workers in a crew. For safety, quality, and productivity reasons, you do not want a worker trying to complete a task for which they are not skilled. However, you also do not want to have higher-wage crew members doing work that can be done by lower-wage crew members. For example, paying a journeyman electrician to pull wire costs more than paying an electrician helper who is skilled enough to do that work under the supervision of a journeyman electrician. As a frontline supervisor, you may not have control of your specific crew mix, but you can work to align their skills levels most appropriate to the specific work. Doing this well can reduce the overall project costs.

Other practices can help to ensure productivity:

- Keeping the crew adequately staffed at all times to avoid job delays
- Ensuring that all workers have the required resources when needed
- Ensuring that everyone knows where to go and what to do after each task is completed
- Making reassignments as needed
- Ensuring that all workers have completed their work properly

The time on the job should be for business, not for taking care of personal problems. Anything not work related should be handled after hours, away from the jobsite, if possible. Even the planning of after-work activities, arranging social functions, or running personal errands should occur after work or during breaks. Limited and necessary exceptions to these rules should be permitted (for example, making or meeting medical appointments), and these should be spelled out in company or crew policies.

Given the nature of construction work, frequent breaks can be needed to ensure worker safety. However, if crew members are often taking unscheduled breaks or goofing off, you will need to confront the behavior and coach these workers. Document any discussions in your job diary. If needed, you should redirect repeated violations to the attention of higher management as guided by company policy.

You can minimize or avoid delays caused by other crews or contractors by carefully tracking the project schedule. In doing so, you can anticipate delays that will affect your work and either take action to prevent the delay or redirect the crew to another task.

The availability and location of materials, equipment, and tools can have a serious effect on a crew's productivity. In the sections that follow, you will learn more about how to manage these resources to minimize such problems.

5.4.3 Handling Material Delivery and Storage

NOTE

Scan the QR code in the front of this module to access additional resources on the steps involved in the development of an estimate, including creating a quantity takeoff sheet.

Your specific responsibilities for material delivery and storage will depend on the policies and procedures of the company. In general, you will be responsible for ensuring on-time delivery of materials, verifying the materials were delivered, and storing the materials (*Figure 39*). You also must prevent the waste and theft of materials.

You will also be involved in planning materials for tasks such as job-built formwork and scaffolding or acquiring materials if you run out. If you run out of a specific material, be sure to consult the appropriate supervisor on how to proceed, because most companies have specific purchasing policies and procedures.

It is essential that the materials required for each day's work be on the jobsite when needed. You should confirm in advance the placement of orders for all

Figure 39 The receipt, storage, and control of materials after they reach the jobsite is an important part of a frontline leader's work.
Source: July Alcantara/E+/Getty Images

Source: metamorworks/Shutterstock

Did You Know?

Just-in-Time (JIT)

Just-in-time (JIT) delivery is a strategy in which materials arrive at the jobsite when needed. This means that a crew may install the materials right off the truck. This method reduces the need for on-site storage and staging areas. It also reduces the risk of loss or damage while moving products about the site. Other modern material management methods include the use of radiofrequency identification (RFID) tags that make it easy to inventory and locate material in crowded staging areas. Using an RFID inventory control system requires special scanners.

materials and that they will arrive on schedule. A week or so before the delivery date, follow up to make sure there will be no delayed deliveries or items on backorder. If other people are responsible for providing the materials for a job, you must follow up to make sure that the materials are available when needed. Otherwise, delays can occur as crew members wait for the delivery of materials.

You may be responsible for the receipt of materials delivered to the worksite. When this happens, you should require a copy of the shipping invoice or similar document and check each item on the invoice against the actual materials to verify delivery of the correct amounts.

You should also check the condition of the materials to verify that nothing is defective before signing the invoice. This can be difficult and time-consuming because it requires the crew to open cartons and examine their contents. However, this step cannot be overlooked, because a signed invoice indicates that the recipient accepted all listed materials and in an undamaged state. If the crew leader signs for the materials without checking them, and then finds damage, no one will be able to prove that the materials came to the site in that condition (*Figure 40*).

Figure 40 When receiving materials, check the condition of the materials to verify that nothing is defective before signing the invoice.
Source: kali9/E+/Getty Images

After checking and signing the shipping invoice, you should give the original or a copy to the superintendent or project manager. The company then will file the invoice for future reference because it serves as the only record the company has to check bills received from the supplier company.

Another essential element of materials control is selecting where to store materials on a jobsite. *Table 14* outlines three factors to consider when selecting an appropriate material storage location.

TABLE 14 Considerations When Selecting a Material Storage Location

Factor	Description
Access	If possible, store materials near where the crew will use them, so the crew can access them easily. Not having to carry the materials long distances will greatly reduce time, effort, and installation costs.
Damage Protection	Store materials in a secure area to avoid damage. Be sure that the storage area suits the materials. For instance, temperature-sensitive materials, such as chemicals or paints, should be stored in climate-controlled areas to prevent waste.
Security	Theft or vandalism of construction materials increases direct and indirect costs. These include replacement materials, lost production time while replacing needed materials, and the costs of additional security precautions. In addition, a contractor's insurance costs may increase due to theft and vandalism.

The best way to avoid theft and vandalism is to have a secure jobsite. At the end of each workday, store unused materials and tools in a secure location, such as a locked construction trailer. If the jobsite is fenced or the building has lockable access, the materials can be stored within. Many sites have security cameras and/or intrusion alarms to help minimize theft and vandalism.

5.4.4 Planning and Controlling Tool Usage

Among companies, various policies govern who provides hand and power tools to employees. Some companies provide all the tools, while others furnish only the larger power tools. You should be familiar with and enforce any company policies related to tools.

As a crew leader, you are responsible for planning tool usage for a job. This task includes the following:

- Determining the tools required
- Informing the workers who will provide the tools (company or worker)
- Making sure the workers are qualified to use the tools safely and effectively
- Determining what controls to establish for tools

Tool control involves controlling the issue, use, and maintenance of all tools provided by the company. It also involves storing the tools in a location that provides easy access to your crew (*Figure 41*).

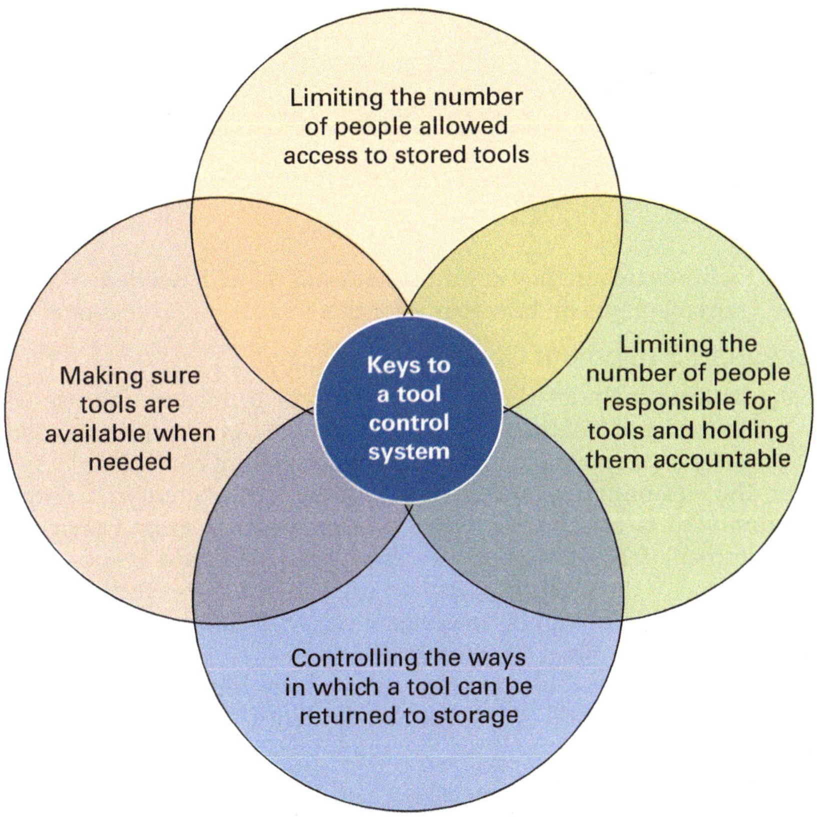

Figure 41 Studies show there are four keys to an effective tool control system.

Using the proper tools correctly saves time and energy. In addition, proper tool use reduces the chance of damage to the tool during use, as well as injury to the user and to nearby workers. This is especially true of edge tools and those with any form of fixed or rotating blade. A dull tool is far more dangerous than a sharp one.

Tools must be adequately maintained and properly stored. Making sure that tools are cleaned, dried, and lubricated prevents rust and ensures that the tools are in the proper working order. This applies to tools supplied by the company as well as tools that belong to the workers.

If a tool is damaged, it is essential to repair or replace it promptly. Otherwise, an incident or injury could result. Users should take care of company-issued tools as if they were their own. Workers should not abuse tools simply because they are not their property.

One of the major sources of low productivity on a job is the time spent searching for a tool. To prevent this from occurring, supervision should establish a storage location for company-issued tools and equipment. You should make sure that crew members return all company-issued tools and equipment to this designated location after use. Similarly, workers should organize their personal toolboxes so that they can readily find the appropriate tools and return their

NOTE

Regardless of whether a worker or the company owns a tool, OSHA holds companies responsible for the consequences of using a tool on a jobsite. The company is accountable if an employee receives an injury from a defective tool. Therefore, the crew leader needs to be aware of any defects in the tools the crew members are using.

Did You Know?

Connected Tools Are Controlled Tools

Tool companies are using technology to help you track tools and equipment on a jobsite. These smart tool systems have a software application that allows you to add tools enabled with Bluetooth® technology to the tracking system. You can also attach powered tags to large equipment or materials and add those to the tracking system. As crew members sign out tools, the assignment is tracked in the system, so you can track which tools are in use, by whom, and where they are on the jobsite at any point. You can also track materials and equipment in the same way. By helping to ensure all of your equipment, tools, and materials are accounted for and easy to find, these systems help companies control costs by saving time and replacement costs.

Source: DEWALT Industrial Tool Co.

tools to their toolboxes when they are finished using them. If needed, coach less-experienced team members on how to master this important professional skill.

5.4.5 Ensuring Equipment Control

Equipment planning includes the types of equipment needed, the use of the equipment, and the length of time it will be on the site. As a frontline leader of a crew, you may not be responsible for long-term equipment control.

However, the equipment required for a specific job is often your responsibility. The first step is to schedule when a crew member must transport the required equipment from the shop or rental yard. You are responsible for informing the shop of the location of its use and confirming a worker returns it to the shop when finished with it. You must work with the main office to make certain that the equipment reaches the jobsite on time.

Coordinating the use of the equipment is also very important. Some equipment operates in combination with other equipment (*Figure 42*). For example,

Figure 42 Coordinating equipment requires knowing which equipment operates in combination with other equipment.
Source: RyanOverman/iStock/Getty Images

dump trucks are generally required when loaders and excavators are used. You should also coordinate equipment with other contractors on the job. Sharing equipment can save time and money and avoid duplication of effort.

It is common for equipment to lie idle at a jobsite because of a lack of proper planning for the job and the equipment arriving early. For example, if wire-pulling equipment arrives at a jobsite before the conduit is in place, this equipment will be out of service while awaiting the conduit installation. In addition to the wasted rental cost, the equipment could also be damaged, lost, or stolen while awaiting use.

You need to control equipment use, ensure that the crew operates the equipment according to its instructions, and make sure they are adhering to time and cost guidelines. You must also assign or coordinate responsibility for maintaining and repairing equipment as indicated by the applicable preventive maintenance schedule. Delaying maintenance and repairs can lead to costly equipment failures.

You must also ensure that crew equipment operators are properly trained. This includes ensuring that the equipment operators have the necessary credentials to operate the equipment, including applicable licenses.

You are also responsible for the proper operation of all other equipment resources, including cars and trucks. Reckless or unsafe operation of vehicles will likely result in damaged equipment, leading to disciplinary action for the workers, fines and repair costs levied by the vehicle vendors, and a loss of confidence in your ability to control the actions of your crew.

You should also arrange to secure all equipment at the close of each day's work to prevent theft. If continued use of the equipment is necessary for the job, you should make sure to lock it in a safe place. If it is not, return it promptly.

5.0.0 Section Review

1. Which of these activities occurs during the design and development phase of a project?
   a. Architect/engineer sketches are prepared, and a preliminary budget is developed.
   b. Final inspections are completed.
   c. Detailed project drawings and specifications are prepared.
   d. Contracts for the project are awarded.

2. Which of these activities is *not* part of creating a lookahead schedule?
   a. Identifying the required work activities and dividing them into tasks
   b. Planning resources needed for each task
   c. Organizing the work into a logical sequence
   d. Assigning cost estimates to each task

3. A job diary should typically record any of the following *except* _____.
   a. job interruptions and visits
   b. personal details and emotions
   c. changes needed to project drawings
   d. the actual time each job task took to complete

4. Which of the following is a *correct* statement regarding project cost?
   a. Cost is handled by the Accounting Department and isn't a concern of the crew leader.
   b. The difference between the estimated cost and the actual cost affects a company's profit.
   c. Wasted material is factored into the estimate and is never a concern.
   d. The contractor's overhead costs aren't included in the cost estimate.

Module 46100 Review Questions

1. Who is generally the on-site field supervisor who is responsible for one or more crew leaders?
   a. Project Manager
   b. Lead Engineer
   c. Superintendent
   d. Pre-Construction Manager

2. Accepting responsibility for job completion and being personally answerable for the results is known as _____.
   a. responsibility
   b. delegation
   c. authority
   d. accountability

3. Which of the following is *not* true of situational leadership?
   a. It involves adjusting the way you lead to the specific situation and needs of the group.
   b. It assumes all situations do not require the same leadership.
   c. It allows you to adapt your leadership style to the situation.
   d. It has four styles, including directing, coaching, supporting, and delegating.

4. When you communicate with your team verbally, be sure to _____.
   a. assess your audience before you speak
   b. use technical language as much as possible
   c. speak quickly to save time
   d. assume everyone has the same basic understanding of the topic

5. The *first* step in problem solving should be _____.
   a. choosing a solution to the problem
   b. evaluating the solution to the problem
   c. defining the problem
   d. testing the solution to the problem

6. Which of these is a way to lower the motivation and morale of your employees and team?
   a. Focusing on task completion only, not professional growth
   b. Giving recognition and rewards
   c. Setting clear expectations and goals
   d. Providing opportunities for training and upskilling

7. During an employee's first week of work on your crew, you should be sure to set aside time to _____.
   a. eat lunch together
   b. do a task verification
   c. quiz the employee on company policies
   d. evaluate social engagement

8. Which of these steps is important when delegating work to an individual or a team?
   a. Letting individuals set their own standards of measurement
   b. Checking in only after the work is complete
   c. Ensuring crew members understand what to do and have the skill to do it
   d. Allowing the team to set their own deadlines and get their own resources

9. Which common jobsite documents can help a crew leader overcome a language barrier with a worker?
   a. OSHA incident records
   b. Contracts and schedules
   c. Plans and drawings
   d. Permits

10. As a leader, you can help to manage performance issues by _____.
    a. ignoring the problem that needs resolving
    b. gathering information and preparing well for the employee conversation
    c. discussing the performance issues publicly with the rest of the team
    d. having the conversation when the employee is unprepared

11. The negative impact of incidents may include all of the following *except* _____.
    a. lower workers' compensation costs
    b. damage to the overall morale of everyone on a project or jobsite
    c. increased project costs to the company
    d. project delays

12. OSHA was established in 1970 to _____.
    a. collect dues for membership
    b. ensure safe and healthful working conditions for workers
    c. manage workers' compensation claims
    d. conduct medical research

13. Which of the following statements regarding safety training sessions is *correct*?
    a. The project manager usually holds them.
    b. They are held only for new employees.
    c. They should be conducted frequently by a frontline leader.
    d. They are required only after an accident has occurred.

14. A key step in an incident investigation is to determine the incident's _____.
    a. root cause
    b. secondary driver
    c. cost impacts
    d. design flaws

15. If you observe an employee showing impaired behavior or signs of mental health issues, you should work with _____.
    a. OSHA
    b. the safety inspector
    c. the owner or client
    d. your supervisor and/or Human Resources

16. What is completed during the design phase to ensure a project can be delivered successfully?
    a. Feasibility study
    b. As-built drawings
    c. Installation work packages
    d. Project schedule

17. A contract in which the client pays the contractor for their labor, materials, equipment, and other costs they incur during the performance of the contract is known as a _____.
    a. fixed-price contract
    b. time-spent contract
    c. time and materials contract
    d. cost-reimbursable contract

18. Which of the following would be tracked in a job diary?
    a. Job changes, interruptions, and visits
    b. The crew's time off
    c. Training needs and career development
    d. Personal information provided by an employee

19. Which of the following is a source of data in the field-reporting system?
    a. The cost-estimate for the project
    b. The daily job diary
    c. The Level 1 schedule
    d. The lookahead schedule

20. To prevent job delays due to late delivery of materials, the crew leader should _____.
    a. demand a discount from the supplier to compensate for the delay
    b. tell a crew member to go look for the delivery truck
    c. refuse the late delivery and re-order the materials from another supplier
    d. check with the supplier in advance of the scheduled delivery

Answers to odd-numbered review questions are found in the Module Review Answer Key at the back of this book.

Answers to Section Review Questions

Answer	Section	Objective
Section One		
1. b	1.1.2	1a
2. b	1.2.2	1b
Section Two		
1. d	2.1.2	2a
2. b	2.2.3	2b
3. a	2.3.0	2c
Section Three		
1. b	3.1.1	3a
2. c	3.2.0	3b
3. c	3.3.3	3c
Section Four		
1. a	4.0.0	4a
2. d	4.2.0	4b
3. a	4.3.1	4c
4. b	4.4.2	4d
5. d	4.5.0	4e
Section Five		
1. a	5.1.1	5a
2. d	5.2.2	5b
3. b	5.3.1	5c
4. b	5.4.1	5d

MODULE 02403

Water Pressure Booster and Recirculation Systems

Source: Courtesy of Grundfos

Objectives

Successful completion of this module prepares you to do the following:

1. Describe the characteristics of a water pressure booster system and identify its components.
   a. Describe the types of water pressure booster systems.
   b. Describe the components of water pressure booster systems.
   c. Describe the design of water pressure booster systems.
   d. Describe how to install and maintain water pressure booster systems.
2. Describe the characteristics of a recirculation system and identify its components.
   a. Describe the types of recirculation systems.
   b. Describe the components of recirculation systems.
   c. Describe how to install and maintain recirculation systems.

Performance Tasks

Under supervision, you should be able to do the following:

1. Size the basic components of a water pressure booster system.
2. Size the basic components of a recirculation system.

Overview

Water pressure booster systems boost pressure up to the level required by a building's plumbing system. The components of booster systems include storage tanks, pumps, controls, and water hammer arresters. Recirculation systems are special installations that constantly circulate hot water throughout a building's water supply piping. The location of the water heater in the building determines the method used to circulate water back to the heater. Large recirculation systems often require more than one water heater or storage tank.

CODE NOTE

Codes vary among jurisdictions. Because of the variations in code, consult the applicable code whenever regulations are in question. Referring to an incorrect set of codes can cause as much trouble as failing to reference codes altogether. Obtain, review, and familiarize yourself with your local adopted code.

NCCER Industry-Recognized Credentials

If you are training through an NCCER-accredited sponsor, you may be eligible for credentials from NCCER. The ID number for this module is 02403. Note that this module may have been used in other NCCER curricula and may apply to other level completions. Contact NCCER at 1.888.622.3720 or go to **www.nccer.org** for more information.

You can also show off your industry-recognized credentials online with NCCER's digital credentials. Transform your knowledge, skills, and achievements into credentials that you can share across social media platforms, send to your network, and add to your resume. For more information, visit **www.nccer.org**.

Digital Resources for Plumbing

SCAN ME

Scan this code using the camera on your phone or mobile device to view the digital resources related to this craft.

1.0.0 Water Pressure Booster Systems

Performance Task

1. Size the basic components of a water pressure booster system.

Objective

Describe the characteristics of a water pressure booster system and identify its components.

a. Describe the types of water pressure booster systems.
b. Describe the components of water pressure booster systems.
c. Describe the design of water pressure booster systems.
d. Describe how to install and maintain water pressure booster systems.

Water pressure booster system: A plumbing installation that increases water pressure in the fresh water supply system.

Water supply systems sometimes need help to provide potable water at the right pressure. Many rural houses rely on deep wells for potable water, but the water pressure alone may not be enough to raise the water to the houses. Tall city office buildings can connect to the city water supply, but the city water pressure may not be enough to raise water to the top floor. Plumbers install **water pressure booster systems** to solve problems like these. Water pressure booster systems do exactly what their name suggests: They boost water pressure up to the level required by the plumbing system. Booster systems are designed to meet the specific needs of individual buildings. As a plumber, you will need to be able to determine when there is inadequate water pressure to maintain proper flow. This will help you decide if a booster system is needed.

1.1.0 Identifying Types of Water Pressure Booster Systems

Elevated-tank system: The oldest and simplest type of water pressure booster system, featuring a water tank on the roof operated by the weight of the water column.

Hydropneumatic tank system: A water pressure booster system that uses air pressure to circulate the water.

Constant-speed system: A water pressure booster system in which water circulates continuously through the system and pumps replacement water from the main.

Variable-capacity system: A type of water pressure booster system that uses two or more pumps to provide water in response to peaks and lows in demand.

There are four main types of water pressure booster systems:

- **Elevated-tank system**
- **Hydropneumatic tank system**
- **Constant-speed system**
- **Variable-capacity system**

Each system was designed as an improvement over the previous one. Earlier systems are considered outdated because of new technology and standards.

1.1.1 Elevated-Tank Systems

The elevated-tank system is the oldest and simplest type of water pressure booster system. In an elevated-tank system, a water tank is installed on the top floor of a building. A small pump fills the tank. The weight of the water column provides the required pressure to the fixtures below (*Figure 1*). Elevated tanks are also called gravity tanks.

Elevated-tank systems are no longer installed. Many years ago, they were well-suited for smaller buildings not connected to a city service. However, they are not effective in tall buildings or compatible with modern plumbing systems. In addition, the weight of a large water tank on a tall building can cause severe structural problems.

1.1.2 Hydropneumatic Tank Systems

The next major innovation in water pressure booster systems was the hydropneumatic tank system (*Figure 2*). These systems are used in smaller installations, such as in a floor of a building. Air pressure forces water from a tank to various parts of a building. These systems have several drawbacks. The constant contact between air and water can cause excessive corrosion, which can cause unsanitary conditions in the system. Over time, air diffuses into the

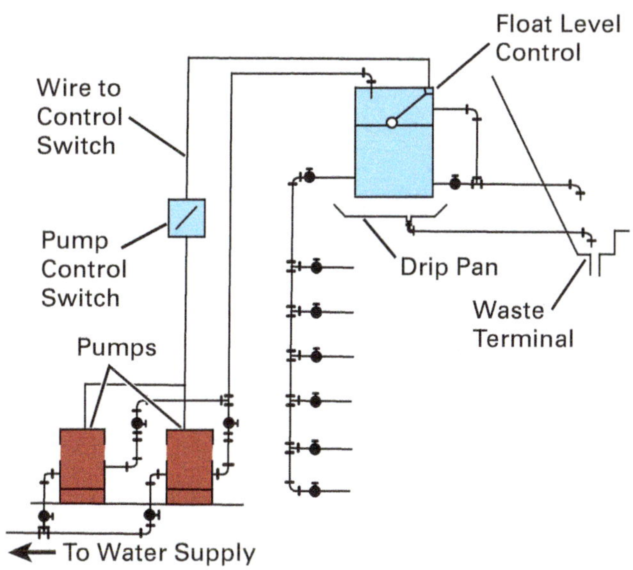

Figure 1 Schematic of an elevated-tank system.

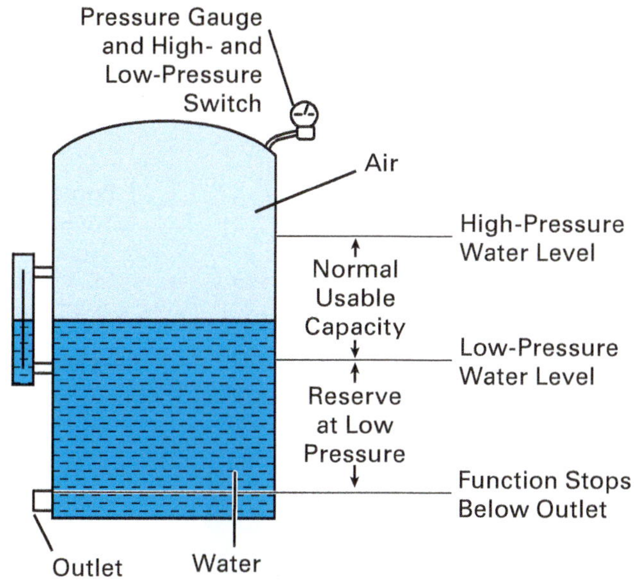

Figure 2 Hydropneumatic tank.

water, which causes the tank to become waterlogged and reduces the system pressure. Water in the plumbing system is also at risk from contamination by compressor oil. These problems can plague older systems, but newer and more sophisticated systems are less likely to have them.

Install the largest hydropneumatic tank possible for the installation's needs. Select a pump that will be able to charge the tank fully. The tank should be able to maintain a minimum of a 10 percent water seal at full air volume. The system pressure will be highest when the tank is full, and it will be lowest when the tank is nearly empty. Set the pump to activate when the tank reaches a specified low pressure.

Consult the tank specifications for this information. The goal is to design a system that operates economically and efficiently. Select a variation that makes sense. Hydropneumatic tanks should only be used in nonpotable systems.

1.1.3 Constant-Speed Systems

In a constant-speed system (*Figure 3*), water is pumped continuously through the plumbing installation. When water is used, the system pumps replacement water from the main. This system does not use air to create the water pressure, which eliminates the risk of corrosion that can affect hydropneumatic tank systems. While some constant-speed systems use storage tanks, others do not. Instead, these systems use a flow-through design with transducers monitoring pressure and controlling pump speed.

Constant-speed systems have some drawbacks, however. They require large pumps to maintain pressure during peak demand, but during periods of low demand, the pumps still do the same amount of work. This wastes a lot of electricity. Modern energy-use guidelines limit the installation of constant-speed systems, so review the local code before installing one of these systems.

1.1.4 Variable-Capacity Systems

The variable-capacity system was invented to correct the shortcomings of the constant-speed system. Variable-capacity systems use more than one pump (see *Figure 4*). The pumps engage one at a time or all together in response to pressure drops in the system. A plumber programs the pumps to activate at specific times, depending on the anticipated demand. This approach is more energy-efficient than having one pump operate all the time, which makes this system more efficient than constant-speed systems.

CAUTION

Because of contamination concerns, some jurisdictions prohibit hydropneumatic tank systems. Others may require special testing and certification to ensure the tank's structural soundness. Check your local code for restrictions on these systems in your area.

CAUTION

Hydropneumatic tanks are under high pressure. When working near a tank, avoid striking the tank with any hard or sharp object (for example, a hammer, chisel, or other tool). A rupture in the tank could cause severe damage and personal injury. Maintenance of hydropneumatic tank systems should be performed only by qualified technicians. Some jurisdictions require the installation of pressure-relief valves on hydropneumatic tanks.

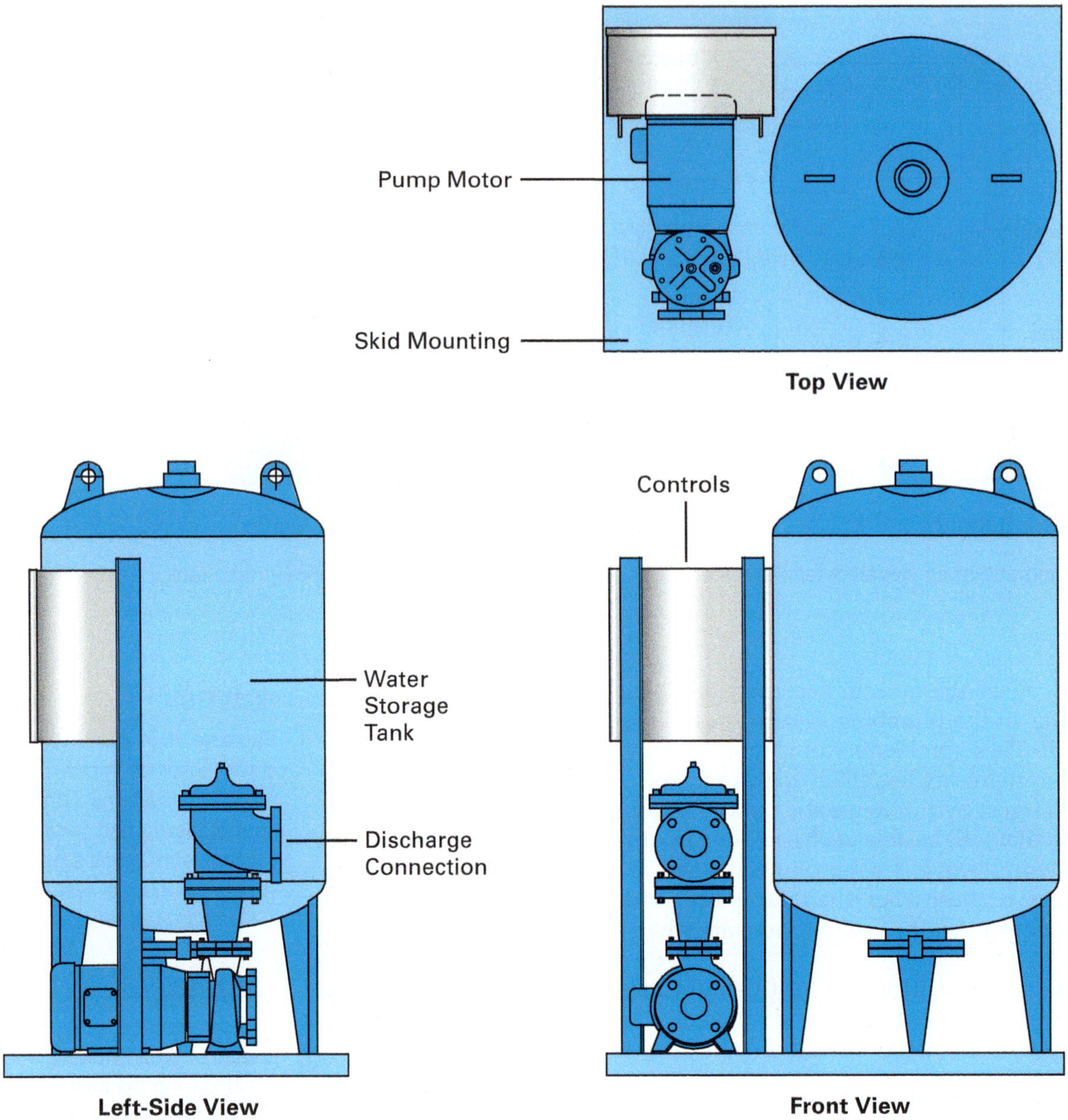

Figure 3 Constant-speed system pump and storage tank.

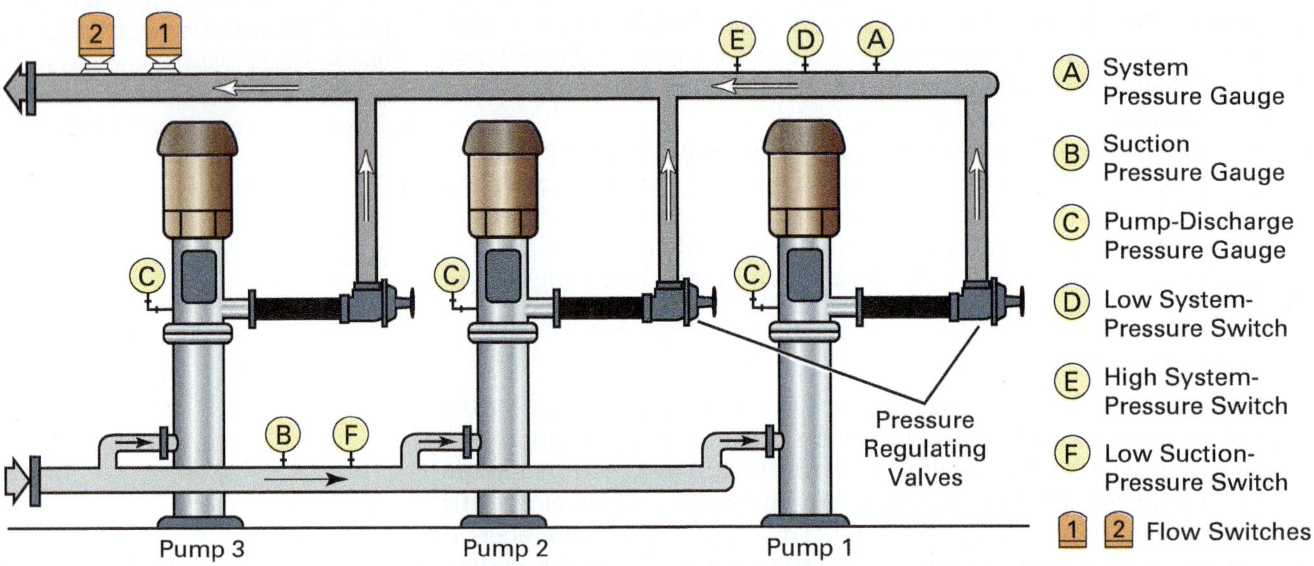

Figure 4 Variable-capacity system.

Before a variable-capacity system can be installed, you must establish the water-use cycle for an installation. This can take the form of a chart showing use over time. Then, select pumps that will provide enough water pressure to meet the projected demand at all times. Using the chart in *Figure 5*, for example, a plumber might choose to install two pumps. One would provide 33 percent of the system's capacity, and the other would provide 67 percent. During low-flow periods, the small pump operates. As demand increases, the second pump engages, and the first pump shuts down. To meet the peak demand, both pumps operate together (see *Figure 6*).

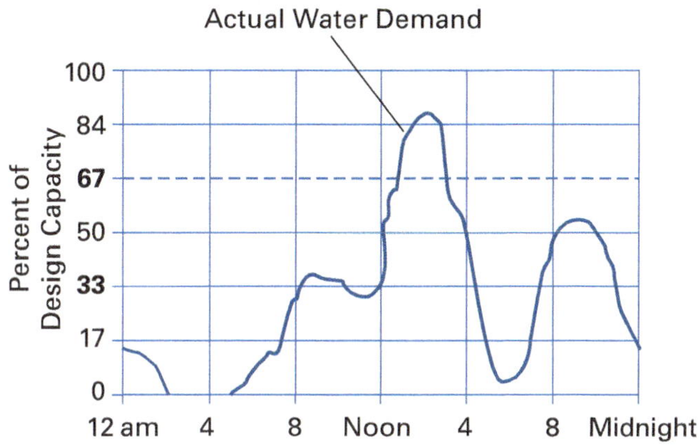

Figure 5 Water-use chart.

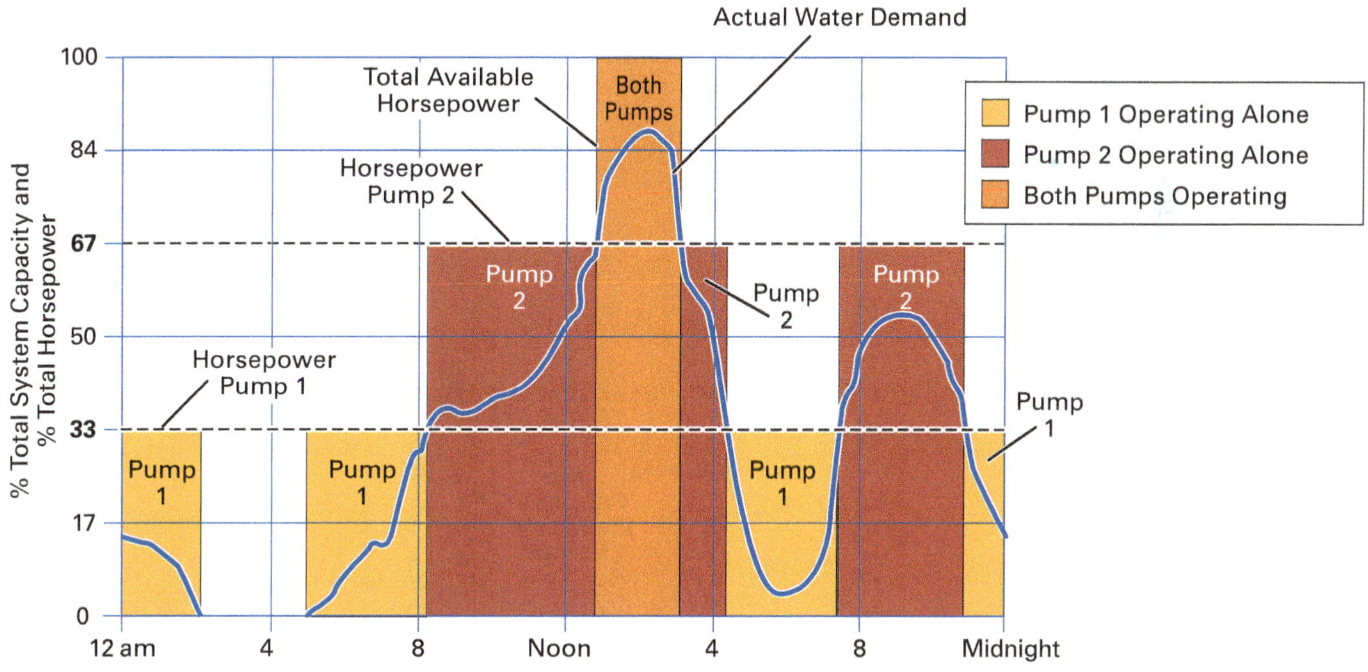

Figure 6 Variable-capacity system designed to meet water demand.

Some variable-capacity systems use a **pressurized air-bladder storage tank** to replenish water in the system. This is a tank in which water pressure is kept constant by compressed air. Pressure builds up in the tank until it reaches the prescribed charge. A flexible bladder or diaphragm keeps the air and water apart. The system's pumps operate only to fill the tank. A pressure switch shuts off the system during periods of no flow (see *Figure 7*).

Pressurized air-bladder storage tank: A tank used in a water pressure booster system that uses compressed air to maintain water pressure.

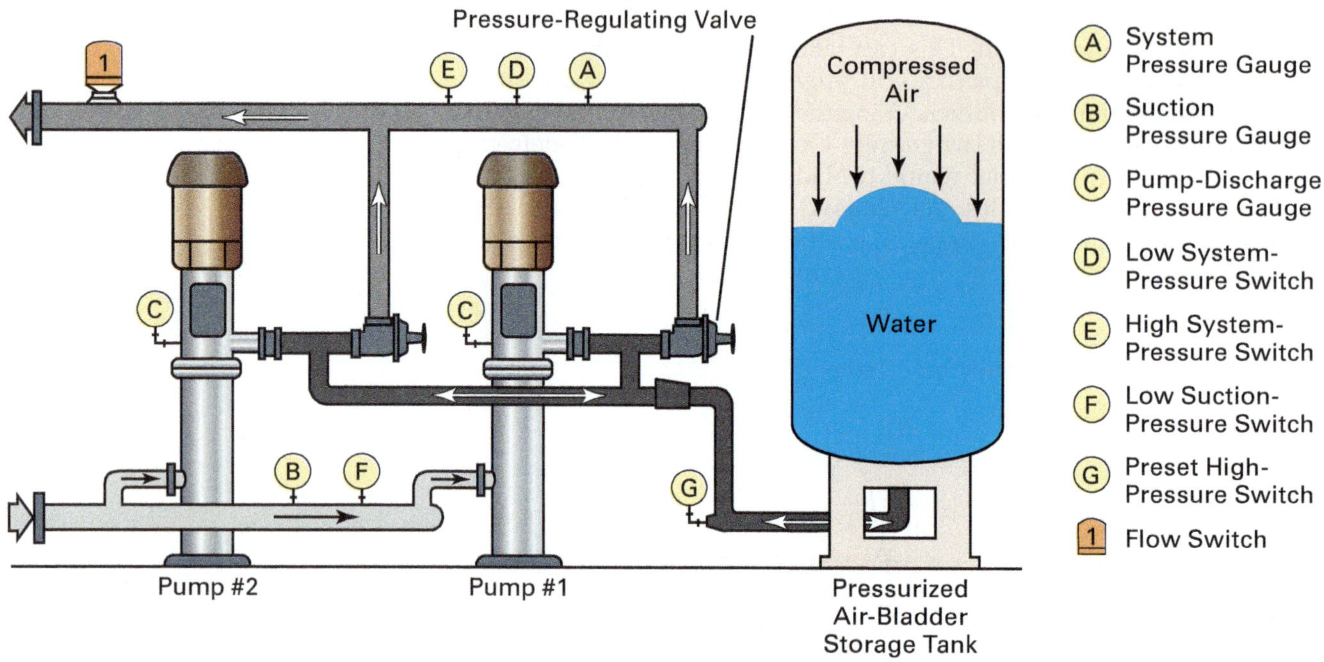

Figure 7 Variable-capacity booster system using a pressurized air-bladder storage tank.

Pressurized tanks have several advantages. The bladder eliminates the threat of corrosion, and because the pumps do not have to cycle on and off as often, they last longer. A pressurized tank also eliminates the need for constant pump cycling during periods of no flow caused by minor leaks. A properly designed system with a pressurized tank is highly energy-efficient. *Figure 8* illustrates a variable-capacity system.

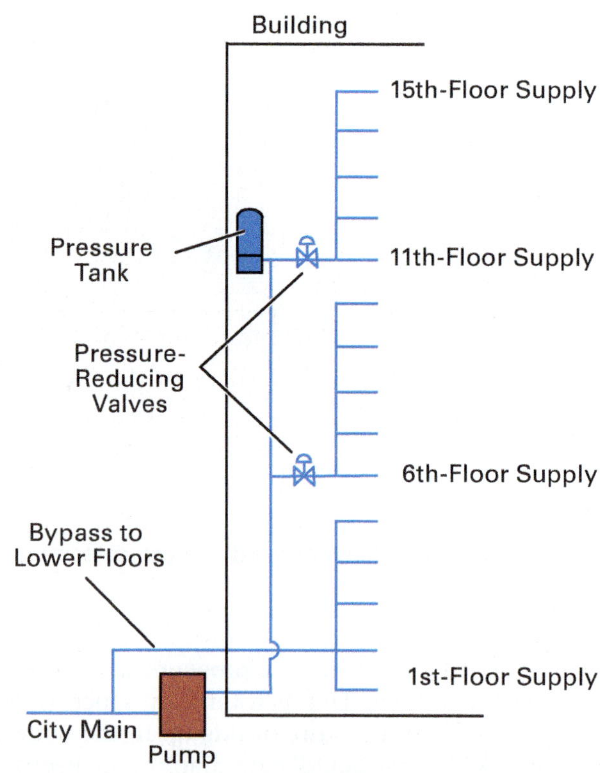

Figure 8 Variable-capacity booster system with an elevated pressurized tank in a multi-story structure.

1.2.0 Identifying Components of Water Pressure Booster Systems

When designing a water pressure booster system to meet the specific needs of a building or structure, you must consider the intended use, predicted capacity, and energy requirements. These factors will help determine what components are needed. Components of a water pressure booster system include the following:

- Storage tanks
- Pumps
- Pump motors
- Controls
- Water hammer arresters

Ensure that the components are suitable for the type of water system installation. Refer to the local code for specific requirements governing booster system components.

1.2.1 Storage Tanks

Booster system tanks are manufactured from materials such as galvanized steel, polyethylene, fiberglass-reinforced plastic, and even steel with a glass or cement lining. Tanks range in size from 2 gallons to more than 5,000 gallons. Tank installations may require flexible hoses, manual or automatic float valves, manual shutoff valves, a sight gauge, and a drain. The tank can be purchased from the manufacturer as a kit, or it can be built on site. Ensure that the tank materials and accessories are permitted by the local code.

1.2.2 Pumps

Centrifugal pumps (*Figure 9*) are the most widely used type of water pump in plumbing systems because of their mechanical simplicity and low cost. They are most efficient when pumping large amounts of water. Centrifugal pumps can be classified according to how many suction cavities, or intakes, the impeller has. A **single-suction impeller** has a single suction cavity on one side of the impeller. A **double-suction impeller** has two cavities that are on opposite sides of the impeller. Both single- and double-suction impeller pumps can be used in booster systems. Single-suction impellers work well in low-flow conditions, while double-suction impellers are designed for high-flow conditions.

NOTE

Tanks located on roofs must be supported in accordance with the International Building Code. This includes using masonry, reinforced concrete, steel, or heavy timber construction to support roof tanks with a capacity of more than 500 gallons. The material supporting the tank must be fire-resistance rated for Type IA construction. Consult the code for other requirements.

NOTE

Many local codes require the use of stainless-steel pumps in water pressure booster systems that provide potable water.

Single-suction impeller: A type of impeller used in centrifugal pumps that have a single-suction cavity. Single-suction impellers are used in low-flow water pressure booster systems.

Double-suction impeller: A pump impeller with two cavities, one each on opposite sides of the impeller. It is used to channel water in high-flow conditions in water pressure booster systems.

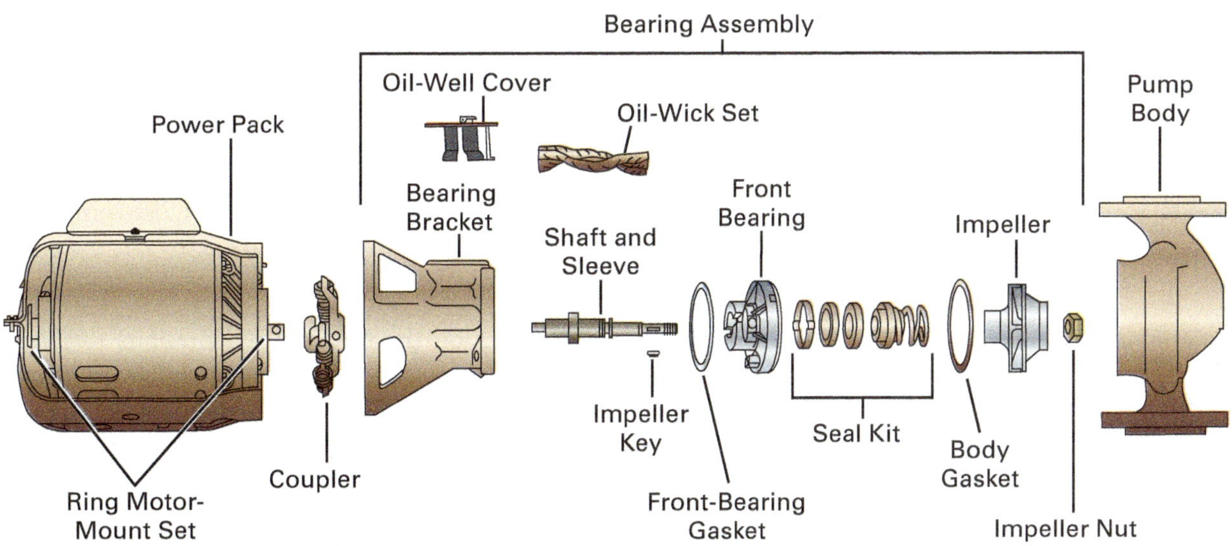

Figure 9 Centrifugal water booster pump.

Volute: In centrifugal pumps, a geometrically curved outlet path.

Single-suction pumps can suffer from a high side-load thrust, which is an unbalanced load caused by water exiting from only one side of the impeller. Side-load thrust can deform the motor shaft over time. Pump seals and impellers will wear out if the condition is allowed to persist. Avoid excess wear of pump seals and impellers by installing a **volute** in the pump body (*Figure 10*). A volute is a geometrically curved outlet path that directs water from the impeller into the discharge pipe more efficiently than the standard pump-outlet design. Volutes improve pump efficiency and reduce wear.

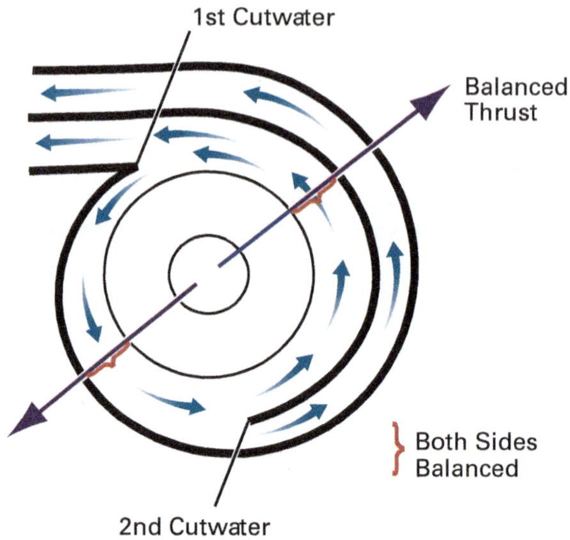

Figure 10 Booster-pump volute.

Vertical turbine pump: A pump used in water pressure booster systems where high-pressure and low-flow conditions exist.

Stage: In a vertical turbine pump, the arrangement of a single impeller and its water passage.

Vertical turbine pumps (*Figure 11*) are also used in water pressure booster systems. They consist of a series of centrifugal impellers stacked one above each other in **stages**. A stage is a single impeller and its water passage. Water is drawn into the first impeller from the bottom. The water is pumped from one stage to the next. The pressure increases at each stage. The pressurized water is finally discharged after it passes through the last impeller. Vertical turbine pumps work well in high-pressure and low-flow conditions, such as in high-rise buildings. They do not experience side-load thrust problems.

1.2.3 Pump Motors

Many booster systems are designed to provide low flow at high pressure. Select a pump that offers the most efficient combination of water flow and pump speed. Note that doubling the pump capacity results in a fourfold increase in head. In other words, a pump that operates at 3,500 revolutions per minute (rpm) can pump twice the capacity of a pump that operates at 1,750 rpm, but the 3,500-rpm pump generates four times the head of the 1,750-rpm pump. Faster pumps operate more efficiently than slower ones.

Vibration isolators: Devices such as rubber pads and flexible line connectors that reduce the effects of pump vibration.

Faster pumps often operate with more noise than slower pumps; however, each pump requires a certain rpm to operate at its design specifications. Install **vibration isolators** to cut down on the transfer of vibrations to the structure. These include rubber pads beneath the pump base and flexible connectors in the inlet and discharge lines.

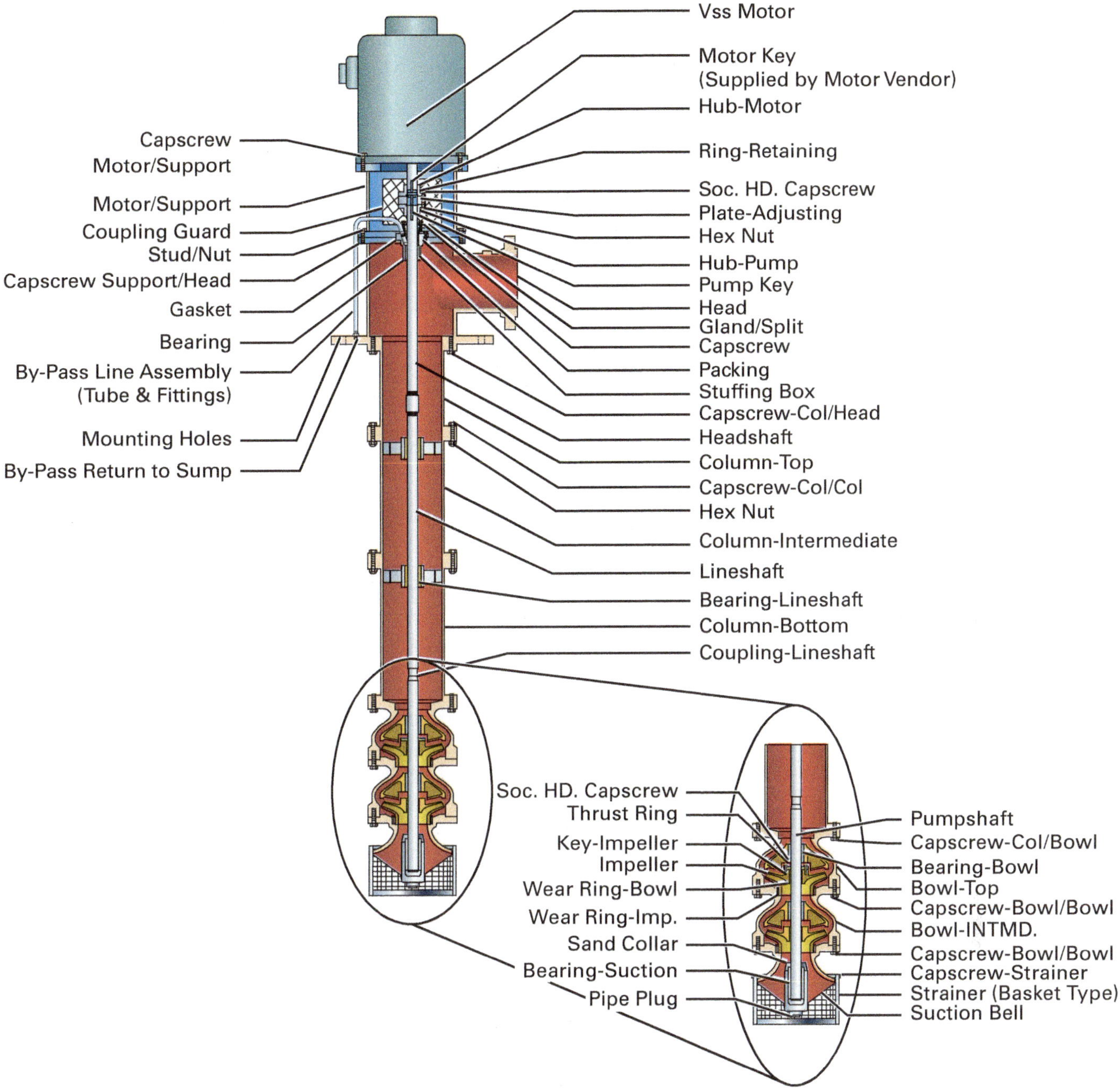

Figure 11 Vertical turbine pump.

1.2.4 Controls

Booster system controls include sensors for water pressure, temperature, and flow. A packaged booster system comes with its own controls, or you can use other controls. Use a pressure switch for pumps installed in residential water systems serviced by a well. Install flow sensors in a variable-capacity system to determine which pumps should be functioning. Temperature sensors, also called thermal sensors, shut the pumps down during no-flow periods. A temperature sensor detects the heat caused by a spinning impeller after flow stops, and then the sensor shuts off the pump until flow starts again.

1.2.5 Water Hammer Arresters

Water hammer can occur in water pressure booster systems when the flow of water is suddenly stopped. It occurs more often in higher-pressure systems, and

Water hammer arrester: A device that absorbs the pressure surge that occurs when water flow is suddenly stopped in a system, known as water hammer.

it can shorten the service life of the water supply system. Install **water hammer arresters** (*Figure 12*) to absorb the pressure surge and protect the water supply system from damage. They will also help quiet the booster system's operation.

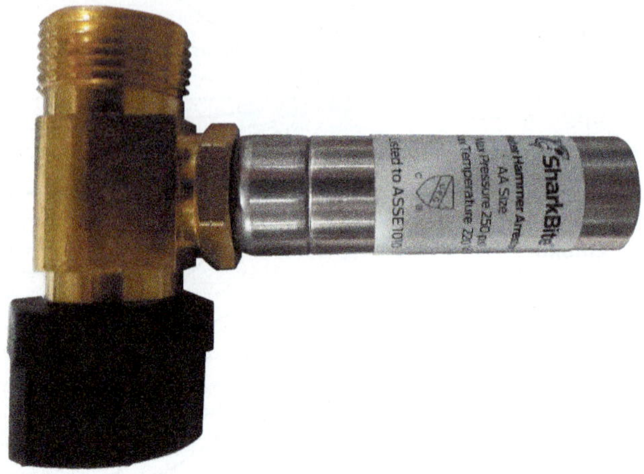

Figure 12 Water hammer arrester.

1.3.0 Designing Water Pressure Booster Systems

Intermittent demand: Demand for water caused by fixtures that are used no more than about five minutes at a time.

To work effectively and efficiently, a water pressure booster system must be selected and designed to meet the needs of the installation. For example, if the system is intended for a building with **intermittent demand**, such as a hotel or an office building, consider a variable-capacity system. This system is well-suited for a wide variation between peak and low demand. A constant-supply system might be appropriate for a hospital, where water is required at all times. The booster system's performance depends on correct estimates of several factors, including the following:

- Energy efficiency
- Type of pump
- Flow rates
- Anticipated water demand

The system should be designed so it achieves the best balance of all these considerations. Each consideration is discussed in more detail below.

1.3.1 Energy Efficiency

Inefficient booster systems cost more to operate, which is an important consideration with rising energy costs. Inefficient booster systems waste water and electricity. The difference in operational costs between a 10-horsepower (hp) pump motor and a 50-hp pump motor over a one-year period could amount to hundreds of dollars. Be sure to calculate the pumping needs of a system carefully.

Thermal purge valve: A valve that allows heated water to drain from the recirculation system when no-flow conditions exist.

Some water pressure booster systems have a **thermal purge valve** installed. These valves drain heated water out of the system when a no-flow condition exists, a procedure that wastes large amounts of water. Avoid installing thermal purge valves where operating cost is a concern.

1.3.2 Pump Selection

Selecting the right type of pump involves considering factors such as cost, ease of maintenance, reliability, and operational efficiency (*Table 1*). Double-suction centrifugal pumps and vertical turbine pumps are very reliable, but they are also the most expensive. The cost of pump replacement parts is often less than the cost of the labor needed to install them. If replacement parts for a particular pump are not available locally, the cost of shipping replacement parts from a distant location is an additional consideration.

Table 1 Guidelines for Selecting an Appropriate Booster Pump

	Single-Suction Centrifugal				
	Close-Mounted	Frame-Mounted	In-Line	Double-Suction Centrifugal	Vertical Turbine
Efficiency	Especially good in low-capacity range			Good in high-capacity range	Good in all ranges
Reliability	Good	Good	Good	Best (except at shutoff)	Best
Suction-Lift Ability	Good	Good	Good	Best	Best*
Bearings to Maintain	Motor	Motor + 2	Motor	Motor + 2	Motor
Number of Shaft Seals	1	1	1	2	1
Cost to Repair Seal	High	Moderate	High	Highest	Low
Hours to Repair Seal	2 to 4	1 to 3	2 to 4	4 to 8	1 to 2
Ability to Operate at Shutoff	Poor	Poor	Poor	Poor	Good
Pump Curve	Flat	Flat	Flat	Flat**	Steep*
Motor Availability	Fair	Good	Poor***	Good	Fair
Space Required	Least	Medium	Least	Most	Least
Original Price	Low	Moderate	Moderate	High	High

Notes:
*Can also tank-mount pumps to eliminate suction lift.
**Multistage pumps can be selected further to left or right on curve.
***If specially designed shaft is used.

1.3.3 Flow and Demand

Engineers are responsible for calculating flow and demand in a water pressure booster system (*Table 2*). **Average flow** is considered to be the flow rate under average conditions. This rate cannot be used to determine maximum operating capability of a system, but it can be important when calculating off-peak loads. **Maximum flow** is the total flow possible if all outlets are opened at the same time. Because it is highly unlikely that this would happen, this flow factor is unrealistic. The most important factor in sizing water pressure booster systems is **maximum probable flow**. This term refers to the flow under peak demand conditions.

Additional factors that deserve consideration are **continuous demand** and intermittent demand. Outlets such as hose bibbs, irrigation pumps, and other relatively constant-flowing devices are considered continuous-demand items. When

Average flow: The flow rate in a water pressure booster system under average operating conditions, used to calculate off-peak loads.

Maximum flow: The theoretical total flow in a water pressure booster system if all outlets are simultaneously opened.

Maximum probable flow: The flow in a water pressure booster system during periods of peak demand.

Continuous demand: Demand for water caused by outlets, pumps, and other devices with a relatively constant flow.

Table 2 Sample Table for Calculating Booster System Flow and Demand

Design Data			
System Capacity ________ GPM			
Pressure Required at Discharge Header ________ PSIG			
Pressure Drop Through Package (include discharge valve) ________ PSIG			
Minimum Suction Pressure ________ PSIG			
Pump Data	**Pump No. 1**	**Pump No. 2**	**Pump No. 3 (optional)**
Gallons per Minute	________	________	________
Pump TDH (feet)	________	________	________
Header size ________ inch			
Discharge Valve:			
☐☐PRV or Check Valve	Size ________	________	________
Series ________	Size ________	________	________
Motor HP ________			
Motor RPM ________	Voltage ________	Hertz ________	Phase ________
Configuration	**Pump Orientation**	**Header Material**	
	☐Vertical	☐Copper	
	☐Horizontal	☐Galvanized Steel	

these devices are used, the water flow is constant, and makeup water must be at hand to replace it. Intermittent demand, as its name implies, refers to fixtures such as sinks, lavatories, showers, water closets, and similar devices with typical usage time of about five minutes or less. Continuous and intermittent demand within a system significantly affect the maximum probable-demand factor.

1.4.0 Installing and Maintaining Water Pressure Booster Systems

There are several elements involved in the process of installing and maintaining water pressure booster systems, including the system's piping materials, water tanks, motors, and pumps. Plumbers must be aware of the particular requirements for each of these components when installing a water pressure booster system. These components must also be taken into consideration when performing service or maintenance work on the system. The following sections provide guidance for installing and servicing each of these components. Refer to the manufacturer's instructions and the local applicable code before installing or servicing water pressure booster systems.

1.4.1 Installing and Maintaining Piping Materials

Ensure that the pipes and connections are suitable for use in a potable water system. Refer to your local code. Pipes must not corrode or degrade as a result of contact with potable water. Local codes also specify limits on lead content in pipes. Do not use any of the following joints or connections in a water supply:

- Cement or concrete
- Nonapproved fittings
- Solvent joints between different types of plastic pipe
- Saddle-type fittings

Permitted pipe materials include acrylonitrile-butadiene-styrene (ABS), polybutylene (PB), cross-linked polyethylene (PEX), polyvinyl chloride (PVC), and chlorinated polyvinyl chloride (CPVC) plastic; asbestos-cement; brass; gray and ductile iron; copper; galvanized-steel pipe; and stainless-steel pipe types 304/304L and 316/316L. Consult your local code for other approved materials. When connecting pipes made of different materials, use proper joining methods, in accordance with the local applicable code.

1.4.2 Installing and Maintaining Water Tanks

Depending on the design, the plumber may be responsible for locating the water tank. Be sure to provide access for tank maintenance and tank removal. When designing a booster system for a multi-story building, remember that installing the tank higher allows a greater head. This not only increases the service pressure but reduces the pumping requirement.

Provide adequate support for the tank. Ensure that all underground tanks are covered to prevent unauthorized access, contamination, and infestation. Do

Table 3 Size of Drainpipes for Water Tanks

Tank Capacity (Gallons)	Drainpipe (inches)
Up to 750	1
751 to 1,500	1½
1,501 to 3,000	2
3,001 to 5,000	2½
5,001 to 7,500	3
More than 7,500	4

Source: International Code Council (ICC), from the 2021 International Plumbing Code (IPC), Table 606(5)(7), Size of Drain Pipes for Water Tanks

not locate tanks or manholes for potable water pressure tanks underneath soil- or waste-piping, or any other possible source of contamination. Install a valved drainpipe at the lowest point of each tank; refer to *Table 3* for drainpipe sizing requirements. Provide a pressure-relief valve at the top of the tank (pressurized air-bladder storage tanks may not require a relief valve). On hydropneumatic systems, install a pressure-relief valve on the supply pipe. Ensure that the relief valve is equal to the tank's rating.

Leaks in a pressure tank can cause excessive pump cycling and low water pressure. Test for leaks by applying a soap-and-water solution to the exterior of the tank above the waterline. Bubbles will appear where air is escaping from the tank. Repair the leak using methods approved for potable water systems.

1.4.3 Installing and Maintaining Motors and Pumps

Install a low-pressure cutoff on booster system pumps to prevent back siphonage caused by negative pressure. Prevent tank overflow by installing ball cocks on inlet piping. Follow manufacturers' instructions when installing pumps. Ensure the pump is primed before testing, if required.

Knowing how to diagnose pump problems quickly and accurately will not only save you time, but will also save the owner money. Whether testing a new pump or repairing an existing one, the most common problems include the following:

- Pump will not start.
- Pump motor has overheated.
- Pump cycles excessively.
- Pump will not turn off.
- Pump delivers little or no water.

Review the troubleshooting tips that follow for each type of problem before answering a service call. The more experience you have in the field, the more familiar you will become with pump problems and solutions. Ask other experienced plumbers for advice. Some problems may be more common in your area due to climate and local codes.

Pump Not Starting

If a pump motor will not start, examine the pump's electrical system. If there is a blown fuse or if the motor is short-circuiting, contact an electrician to secure or replace the wiring. This work should be done only by a qualified electrician. Mechanical problems may also keep a pump motor from starting. Examine the booster system controls. Reset improperly adjusted temperature, pressure, and flow switches. Ensure that the tubing to the pressure switch is not plugged. Test the impeller by first turning the pump off, then attempting to turn the impeller with a screwdriver. If the impeller does not rotate, it may be blocked. Remove the pump casing, and inspect the impeller for blockage.

Pump Motor Overheating

Pump motors have overload protectors to trip the motor off in the event of overheating. If a motor overheats, use a voltmeter to test the line voltage. If the voltage is below the required level, contact an electrician to ensure that the wiring from the electrical main to the motor is the correct size. The electrician should also ensure that the motor is wired according to the manufacturer's instructions. Only those qualified to correct electrical problems should attempt to do so.

High air temperature and inadequate venting can also cause overheating. Ensure that the pump is adequately vented and that the surrounding air temperature is within the pump's specified limits. Another cause of overheating is prolonged operation at low water pressure. If this is the case, install a globe valve on the water discharge line, and throttle it to provide the desired pressure.

Excessive Pump Cycling

Excessive cycling may be caused by leaks in various locations in the system. Seals are the most frequently required repairs for domestic water pumps. Inspect all seals for leaks, and replace them if necessary. To test for leaks on the discharge side, shut off all fixtures in the system. Inspect the discharge piping while listening for the sound of running water, which will indicate a leak. Check all valves, especially ball cocks, for leaks, and repair or replace any valves that are leaking. Use a pressure gauge to find a leak on the suction side of a shallow well system. To identify a leak on the suction side of a deep well system, use the following steps:

Step 1 Attach a pressure gauge to the pump.

Step 2 Close the discharge line valve.

Step 3 Using a compressor or hand pump, apply a pressure of about 30 pounds per square inch (psi) to the system.

Step 4 Once this pressure has been reached, stop pumping and read the pressure gauge. If the pressure drops, there is a leak somewhere in the suction side. Inspect all pipes and valves.

Pump Not Turning Off

If the pump will not turn off, check the pressure switch. It may be set too high for the actual operating conditions, or it may have "drifted" to an incorrect setting. An electrician can inspect the pressure switch for defects caused by arcing in the electrical system. Clear the pressure-switch tube. Ensure that the pump is correctly primed. Inspect the pump ejector, and remove any blockage.

WARNING!

You must follow safe procedures when immersing pump motors in water. When doing this type of work, you are at risk of electric shock, and even electrocution. Before working on a pump motor, ensure that it is turned off and that power to the pump motor is disconnected. Dry the pump motor thoroughly before operating. Wear appropriate personal protective equipment, and follow proper installation and maintenance procedures.

Pump Delivering Little or No Water

Electrical and mechanical problems can cause a pump to deliver little or no water. If you haven't been trained to deal with electrical problems, consult with a qualified electrician. To avoid air lock, always make sure you prime the pump prior to starting. To clear an air lock, turn off the pump for about one minute, and then restart it. Repeat this several times until the air is cleared from the pump. Ensure that the piping to the pump is correctly sized and free of leaks. Inspect the air volume control for leaks, and repair or replace the control, if required. Check the setting of pressure-regulating valves against the manufacturer's instructions. Ensure that the pump ejector is correctly sized. Remove any blockages in the intake, impeller, and ejector. Inspect the pump for worn parts, and install appropriate replacements.

Fresh Water Supply in America's First Modern Hotels

Boston's Tremont Hotel opened in 1829. It has been called America's first modern hotel. The four-story building featured metal bathtubs in the basement and indoor toilets on the ground floor—an unheard-of luxury in those days. The water for these fixtures came from a metal water tank on the roof. A steam engine pumped the water from ground level to the tank. Five years later, the Tremont's designer, Isaiah Rogers, was commissioned to build the six-story Astor House in New York City. This hotel had 300 guest rooms. Rogers equipped 17 of the guest rooms on the upper floors with water closets and baths. The Tremont Hotel and Astor House were the first large modern buildings in the United States equipped from the outset with their own plumbing systems.

1.0.0 Section Review

1. Because their pumps operate at the same speed during periods of low and high demand, modern energy-use guidelines limit the installation of _____.
   a. elevated-tank systems
   b. variable-capacity systems
   c. underground-tank systems
   d. constant-speed systems

2. Doubling the capacity of a pump results in a _____.
   a. fourfold increase in head
   b. twofold increase in head
   c. fourfold decrease in head
   d. twofold decrease in head

3. Average flow can be used to determine a system's _____.
   a. operating pressure range
   b. pump sizing
   c. maximum operating capability
   d. off-peak loads

4. Install a valved drainpipe at a water tank's _____.
   a. highest point
   b. lowest point
   c. center line
   d. maximum fill line

2.0.0 Recirculation Systems and Components

Objective

Describe the characteristics of a recirculation system and identify its components.

a. Describe the types of recirculation systems.
b. Describe the components of recirculation systems.
c. Describe how to install and maintain recirculation systems.

Performance Task

2. Size the basic components of a recirculation system.

When you take a shower at home, you may have to let the water run for a little while before it gets hot. However, in a hotel, the water is usually hot as soon as you turn on the tap. This is because the hotel uses a **recirculation system**. This is a special installation that constantly circulates hot water throughout a building's water supply piping. Recirculation systems can make hot water more readily available to any fixture in the system.

Recirculation system: A plumbing installation that circulates hot water within a building, providing customers with hot water on demand.

Recirculation systems work best in larger buildings, such as hotels, apartments, schools, factories, and restaurants. They are less effective in buildings where the farthest hot water tap is less than 100' from the water heater. Most recirculation systems are designed by plumbing engineers. However, plumbers need to know about the different types of recirculation systems, their components and layouts, and the basics of installation.

NOTE

Hot water systems with piping under 2" require thermal expansion protection.

Upfeed system: A type of recirculation system that supplies hot water as it travels up the system.

Downfeed system: A recirculation system that supplies hot water as it travels down the system.

Combined upfeed and downfeed system: A recirculation system in which hot water is supplied in the upward and downward flow through the system.

2.1.0 Identifying Types of Recirculation Systems

There are three types of recirculation systems. In an **upfeed system**, hot water is supplied to fixtures as the water travels up the system (*Figure 13*). In a **downfeed system**, hot water is supplied as the water travels down the system (*Figure 14*). In a **combined upfeed and downfeed system**, hot water is supplied on both the upward flow and downward flow (*Figure 15*). If required by the design, each of these systems may include hot water storage tanks. In all three types of recirculation systems, the water heater and storage tank can be located at either the top or the bottom of the installation. Plumbers need to know how to install and maintain all types of recirculation systems.

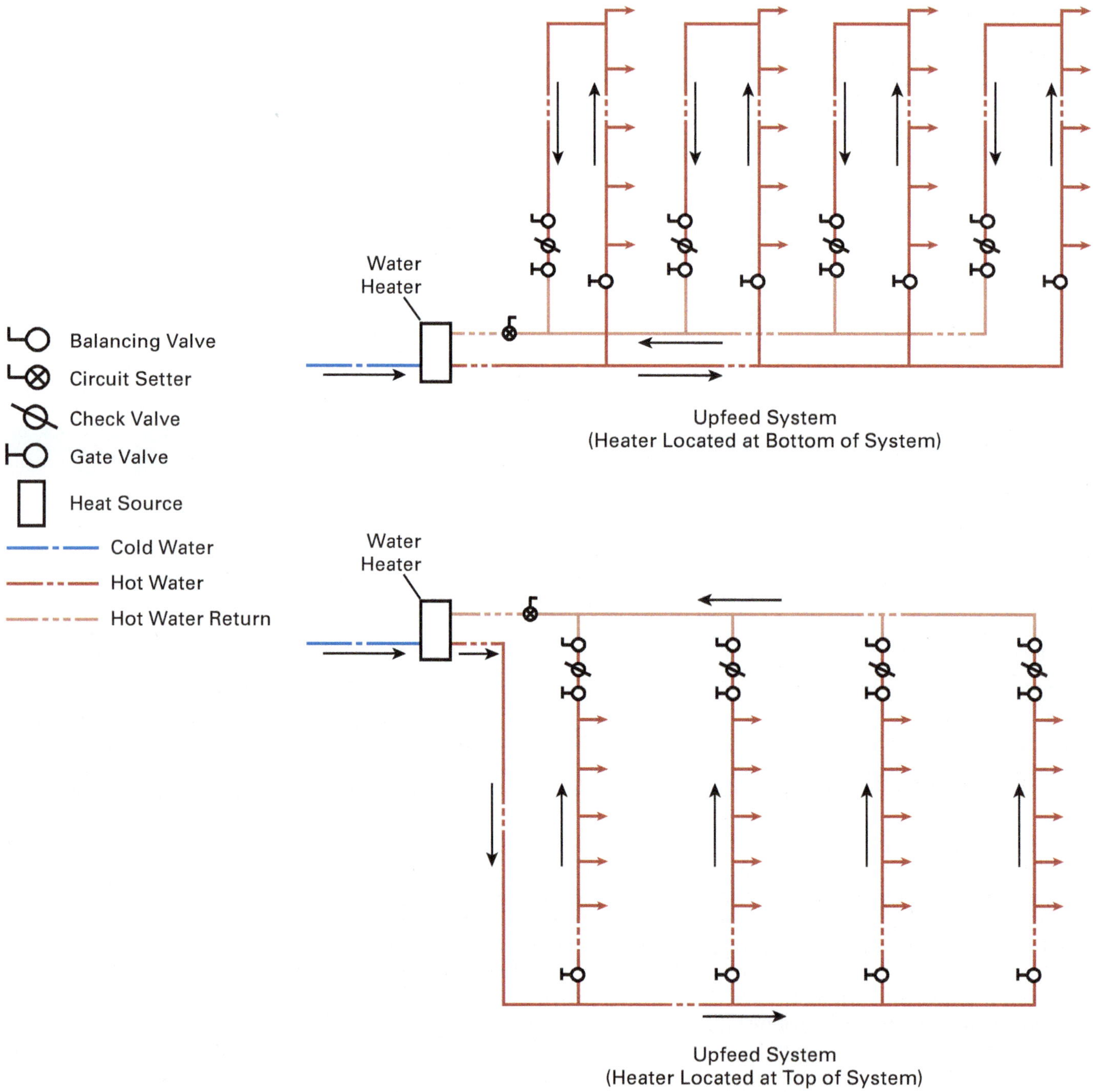

Figure 13 Upfeed recirculation system.

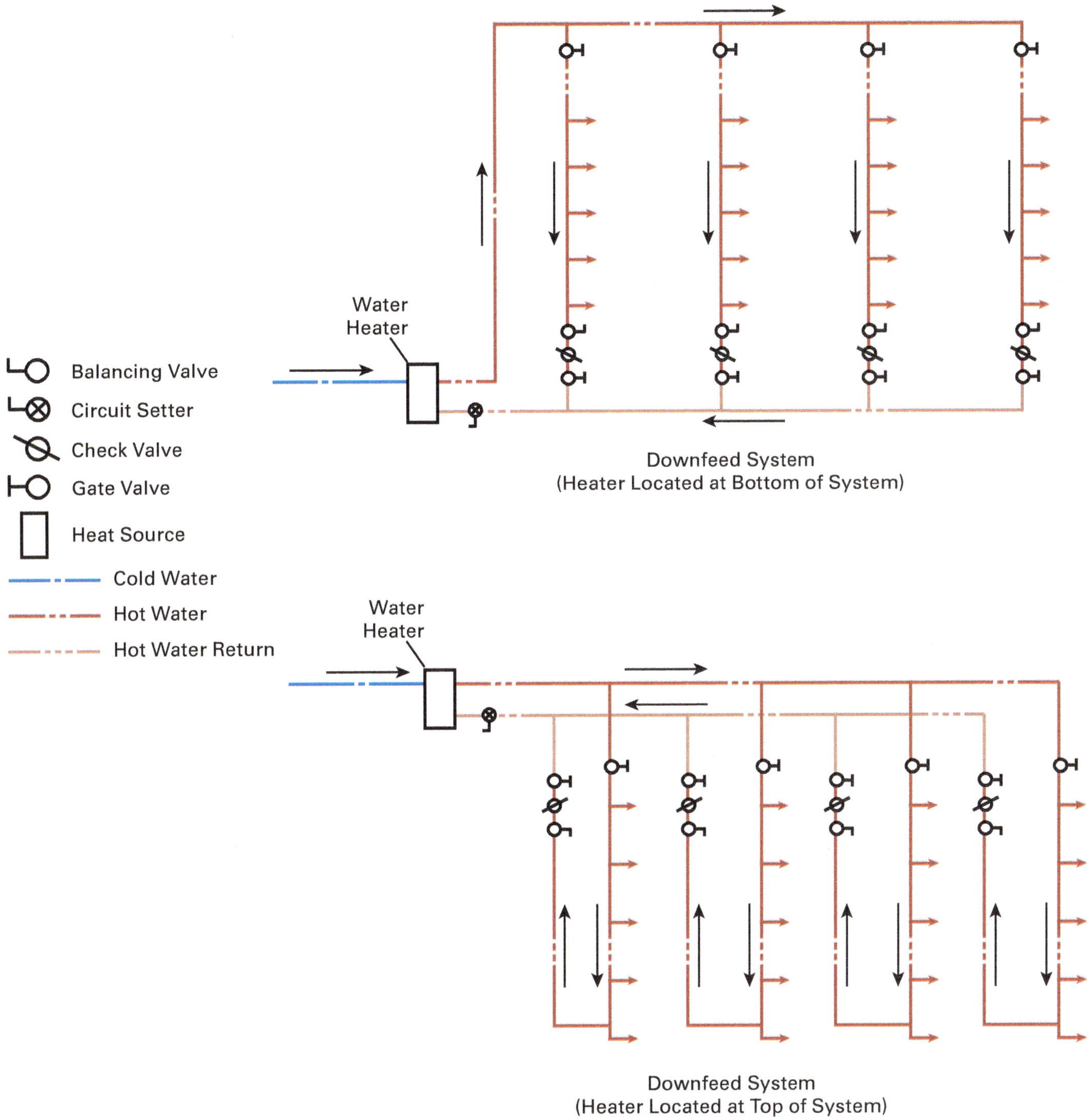

Figure 14 Downfeed recirculation system.

When selecting a recirculation system, there are several things to consider. These include whether it is an upfeed or a downfeed system, and whether the water heater is positioned at the top or the bottom of the building. The various ways these systems can be configured require more or less piping. For example, as shown in *Figure 13*, an upfeed system with the water heater on top requires less piping than an upfeed system with the water heater on the bottom. Conversely, in a downfeed system (shown in *Figure 14*), less piping is required when the water heater is located at the bottom, and more piping is required when the water heater is located at the top. Regardless of which system you select, be aware that the potential for leaks increases according to the amount of pipe used, not according to the location of the water heater.

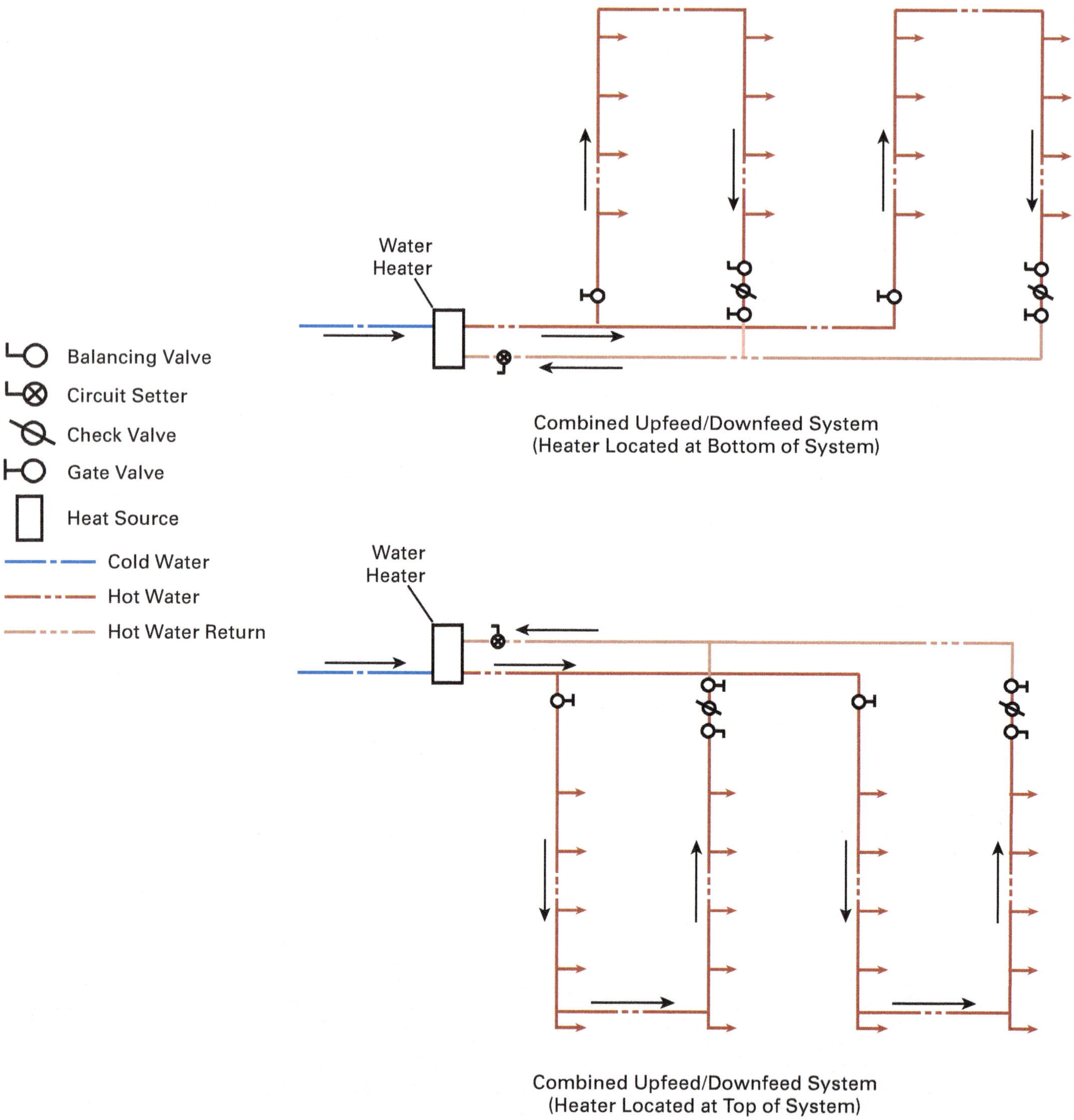

Figure 15 Combined upfeed and downfeed recirculation system.

Forced circulation system: A recirculation system that uses a pump to circulate hot water through the system.

Gravity return system: A recirculation system in which the force of gravity circulates hot water throughout the system. These systems are no longer permitted by code.

Systems that have the heater and storage tank at the top usually require a pump to circulate the water back to the heater. These are called **forced circulation systems**. Today, most hot water systems use forced circulation. A smaller number of systems use the force of gravity to return water to the heater. These are referred to as **gravity return systems** (*Figure 16*) and are no longer permitted by code. Gravity return systems can be found mostly in smaller and older buildings.

2.1.1 Laying Out a Forced Circulation System

Forced circulation can be used in both upfeed and downfeed systems. This design is often used in newer and taller buildings. Forced circulation can also be used in buildings with long horizontal runs of pipe, such as schools. The layout is essentially the same as that of a gravity return system. The main difference

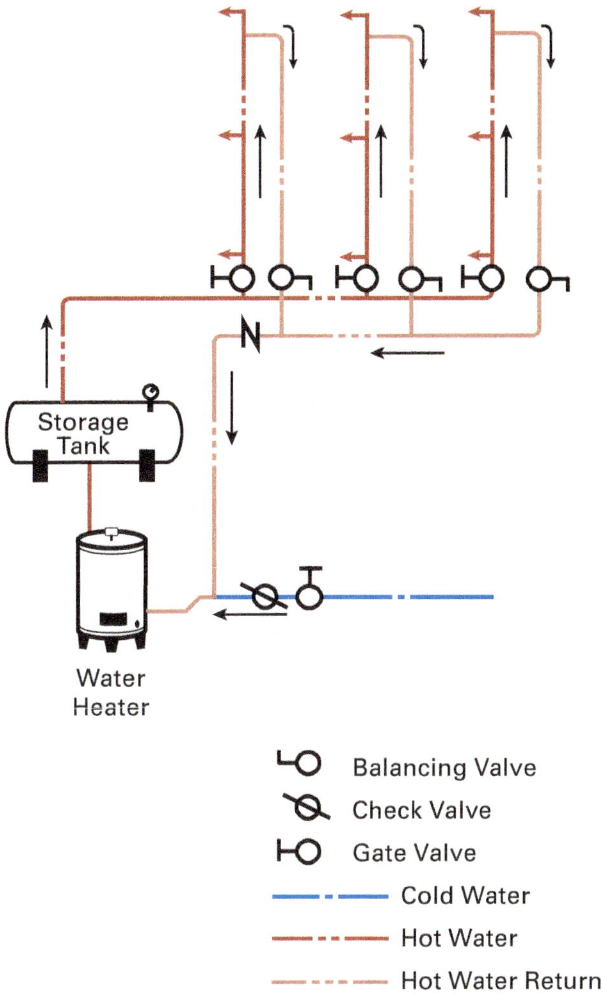

Figure 16 Gravity return system.

is the addition of a pump to circulate the hot water (*Figure 17*). Because forced circulation systems do not rely on the force of gravity to move the water, select a pump that is rated for the system's operating conditions. Install the pump on the return line before the water heater. Provide a gate valve on the inlet and outlet sides of the pump, and install a check valve on the inlet pipe after the gate valve.

2.1.2 Multiple Heaters and Storage Tanks

Large recirculation systems often require more than one heater or storage tank. The number of units is determined by the plumbing engineer based on anticipated hot water demand. Designers can combine heaters and storage tanks in several ways. Basic guidelines for designing systems with multiple heaters and storage tanks are as follows:

- Do not allow hot water to circulate through a cold heater.
- Arrange piping and valves so that the heaters and tanks can function independently or together.
- Use as few valves as possible.
- On long pipe runs from heaters to storage tanks, ensure that the pipe runs have the proper degree of slope.
- Connect two water heaters to the discharge pipe using 45-degree wyes in the direction of flow.

These guidelines are general recommendations only. Consult the water heater or boiler system manufacturer's piping diagrams for specific installation instructions.

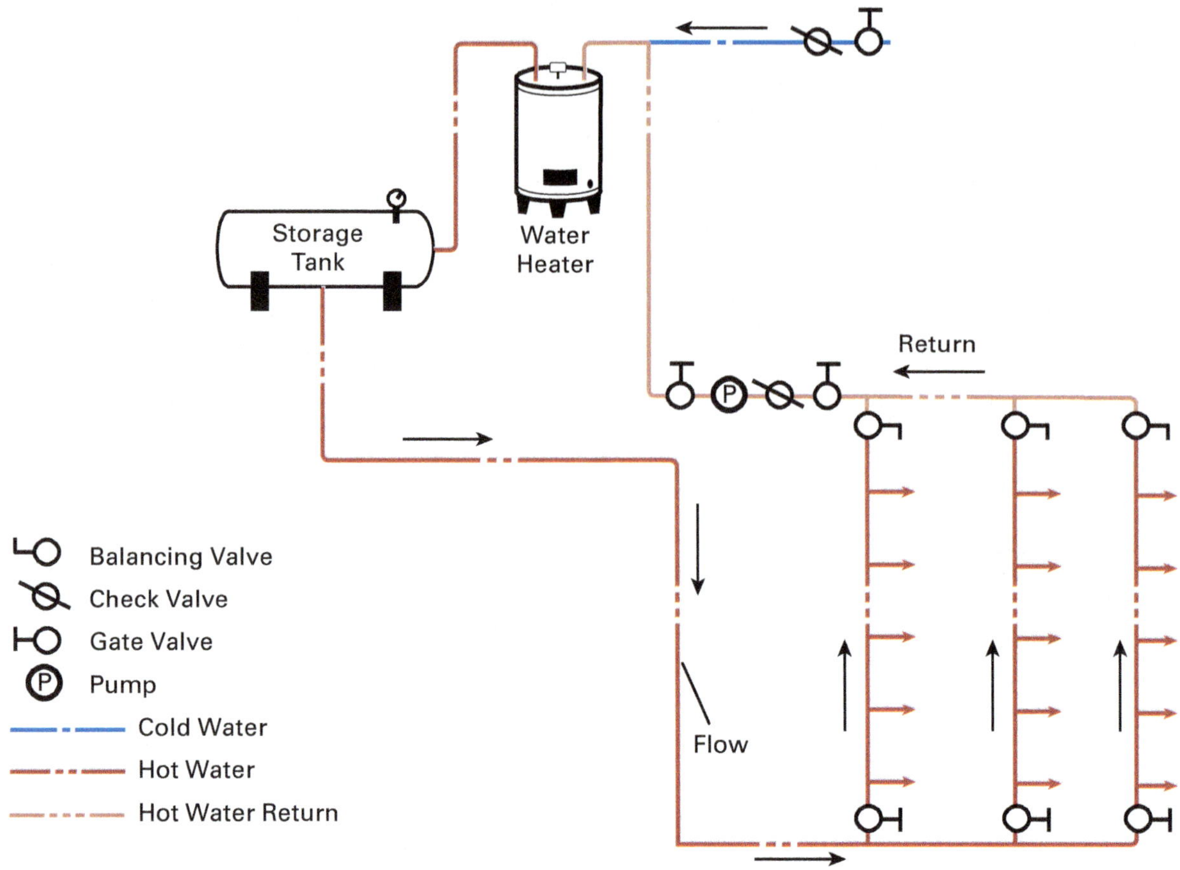

Figure 17 Forced circulation in an upfeed system.

Hot Water in Ancient Rome

Long before the development of recirculation systems, ancient Rome used hot water in its public and private baths. Public baths served as community centers where people gathered to hear news, talk with friends, and relax. The bathhouses were huge, ornate structures. The indoor baths of the Emperor Diocletian, for example, could seat 3,000 people. Most bathhouses had separate baths with cold, warm, and hot water. Water for the baths was heated by an underground furnace that also heated the air in the baths. The walls were covered in mosaic tiles. Water poured out from silver spigots shaped like fish and mythical creatures.

To supply the baths, Roman engineers accomplished some of the most complex engineering feats of their day. Water was channeled to the city's baths through 220 miles of gravity-powered aqueducts and tunnels. At its height, this network provided about 300 gallons of water for each citizen per day. By 400 AD, Rome had 11 public baths, 1,352 public fountains and cisterns, and 856 private baths.

2.2.0 Identifying Components of Recirculation Systems

In addition to water heaters, hot water recirculation systems use storage tanks, pumps, valves, and temperature controls to manage the flow of hot water. Standard pipes and fittings carry the water. Plumbing engineers choose the components that suit the needs of each installation. In this section, you will learn about the components of a recirculation system. Consult your local code for guidelines that are specific to your area.

WARNING!

Do not install valves on temperature and pressure (T/P) relief lines. Valves on a relief line could cause the line to block, which could damage the water heater or even cause it to explode. A malfunctioning water heater can result in water damage to property. An explosion can cause injury, and even death.

2.2.1 Storage Tanks

Some systems use automatic storage tank water heaters (*Figure 18*), which heat and store water at a controlled temperature and deliver it on demand. Other systems require a separate storage tank. Storage tanks can be used with most electric and gas-fired water heaters. Do not install storage tanks in systems that use indirect (instantaneous) water heaters. Storage tanks must be pressurized; install pressure-relief valves, as required. The size of the storage tanks depends on the size of the facility. Engineers will estimate the size of the tank. Plumbers should consult the construction drawing to determine the tank size and refer to the local code for specific requirements.

Figure 18 Hot water storage tank.
Source: n.tati.m/Shutterstock

2.2.2 Pumps

Forced circulation systems use centrifugal pumps to circulate water through the hot water lines. Pumps increase the efficiency of a recirculation system. They can be designed to operate one of two ways. One option is to have the pump on constantly, which allows the water to circulate constantly through the system. The other option is to operate the pump by a temperature control, which allows the pump to switch on when the water cools below a preset minimum temperature. The choice depends on the type of installation and the demand for hot water. Some codes require that pumps be shut off when the system is not operating.

Use stainless steel or bronze pumps to handle potable water. These metals do not rust, which would contaminate water and may clog pipes. Consult local codes for additional requirements for recirculation pumps.

WARNING!

The maximum operating pressure of a pump is listed on its nameplate. Do not exceed this pressure, or the system could fail. This could not only damage the system, but could also cause injury or death.

Figure 19 Expansion joint.
Source: Itsanan/Shutterstock

2.2.3 Pipes

Piping materials used in hot water recirculation systems are generally the same as those that are used with cold water. Plumbing engineers specify their size and type. Codes require that hot water pipes withstand a given pressure at high temperatures. Select only valves and joints that are made from compatible materials, and ensure that the lead content of the pipe material does not exceed local standards. Pipes should be properly insulated to help make the system more energy-efficient. Follow the building plans closely when installing pipe. Undersized or oversized pipes will cause the system to malfunction.

High water temperatures will cause pipes and joints to expand. Install **expansion joints** (*Figure 19*) in hot water piping as directed by the building plans. These will allow the pipes to expand and flex when hot water flows through them. Expansion joints can be installed several ways. Consult your local code. The building plan may call for expansion loops made from runs of pipe; specifications will be provided in the plans.

Expansion joint: A mechanical device installed on a pipe that allows the pipe to expand and flex as hot water flows through it.

2.2.4 Valves

You may already be familiar with a type of **tempering mixing valve** that is used in single-handle faucets. Hot water recirculation systems also use tempering

Tempering mixing valve: A valve that adds cold water to the hot water flow to control water temperature.

valves (*Figure 20*). They control water temperature by adding cold water to the hot water flow. This ensures that the water in the system will remain at the right temperature. Install tempering mixing valves in systems that use indirect heaters. Recall that closed systems will also need an expansion tank.

One use for tempering mixing valves is in installations that require dual hot water systems. For example, a commercial kitchen might need two systems to provide different hot water temperatures for dishwashers and hand sinks.

(A)

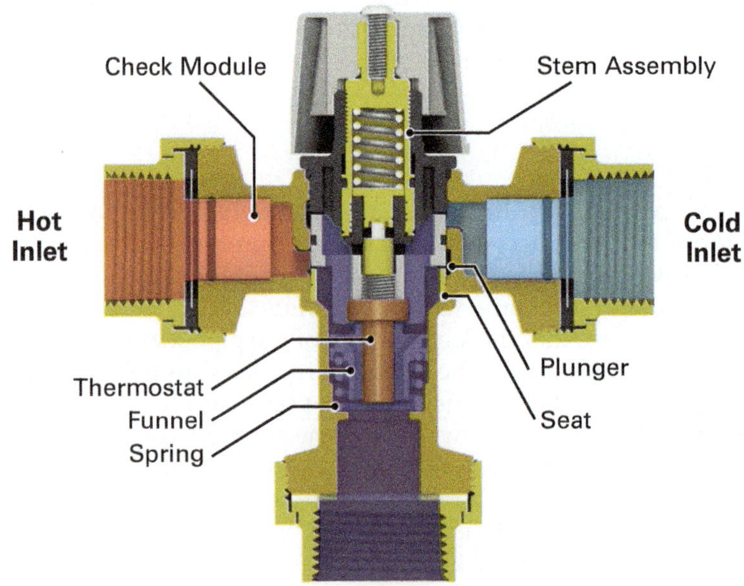

(B)

Figure 20 Tempering mixing valve.
Source: Courtesy of Watts Water Technologies, Inc.

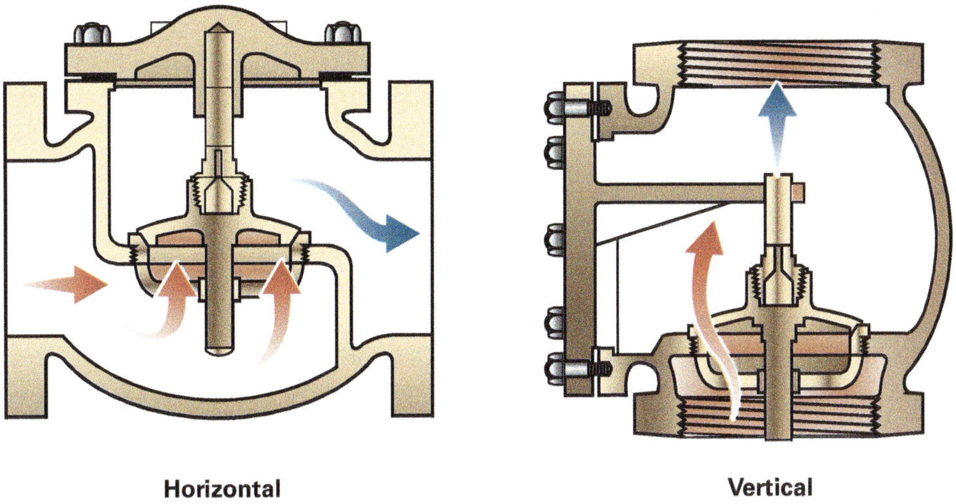

Figure 21 Check valves.

Check valves are automatic valves that permit liquids to flow in only one direction. They protect against backflow caused by back pressure and back siphonage. Check valves are widely used in a variety of plumbing installations. They are available in horizontal and vertical styles (*Figure 21*).

Install either a manual or an automatic air-relief valve at the highest point in the system to allow air in the pipes to be purged. An adjustable balancing valve is installed at the base of each water line to set the most efficient flow (refer to *Figures 13, 14,* and *15*). **Balancing valves** compensate for pipe friction and other irregularities, ensuring a smooth and steady flow. On the return line near the water heater, install a special type of balancing valve with a dial that allows for fine adjustments to the water flow, such as the Bell & Gossett **Circuit Setter™**, to permit balancing of the entire circuit.

2.2.5 Aquastats

An aquastat (*Figure 22*) is a thermostat that regulates the temperature of hot water in a boiler. It can be used to control the water heater's operation. Install aquastats on or near the water heater or at high points in the water lines. **High-limit aquastats** are a special kind of aquastat. They turn off the boiler when the water inside reaches maximum temperature. These aquastats act as safety devices. Electronic aquastats are also now available and provide added energy efficiency.

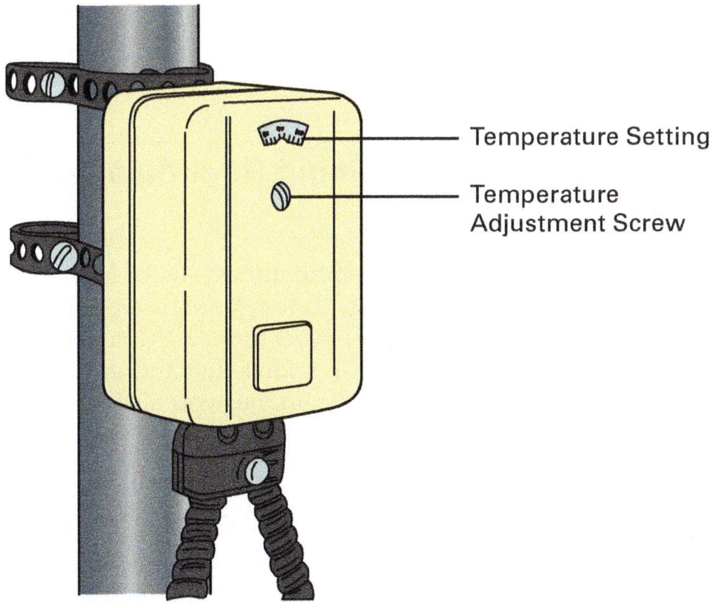

Figure 22 Aquastat.

Cold-Weather Tip

A recent estimate shows that homes can waste anywhere from 27 to more than 100 gallons of water a day during the wait for running water to get hot. That means an efficient hot water recirculation system can save between 9,800 and 38,000 gallons per year in just one home. Depending on the installation, a recirculation system could represent significant cost savings to the customer and reduced wastewater in the community.

Balancing valves: Valves in a recirculation system that ensure steady water flow.

Circuit Setter™: A type of balancing valve used to regulate the flow on the return of an entire circuit, with a dial permitting fine adjustments.

High-limit aquastat: A safety device that deactivates a hot water boiler when the water reaches the maximum temperature.

2.2.6 Expansion Tanks

Expansion tank: A tank attached to the supply pipes in a recirculation system that allows water to cool and slow down as it flows.

Some designs require the installation of **expansion tanks** (*Figure 23*) to help with water hammer and hydraulic issues created when pumps turn on and off or change speed. Expansion tanks are attached to the hot water supply pipes. The tanks shown in *Figure 23* use a flexible diaphragm, or bladder, that separates system water from pressurized air. Pressurized expansion tanks often come precharged with air, but the pressure can be adjusted through a charge valve, if required.

Install expansion tanks above the system where the hottest water collects. Review the building plans to determine this location. Use pipes or a combination of pipes and a tank to allow the desired expansion. Check your local code for expansion tank requirements for your jurisdiction.

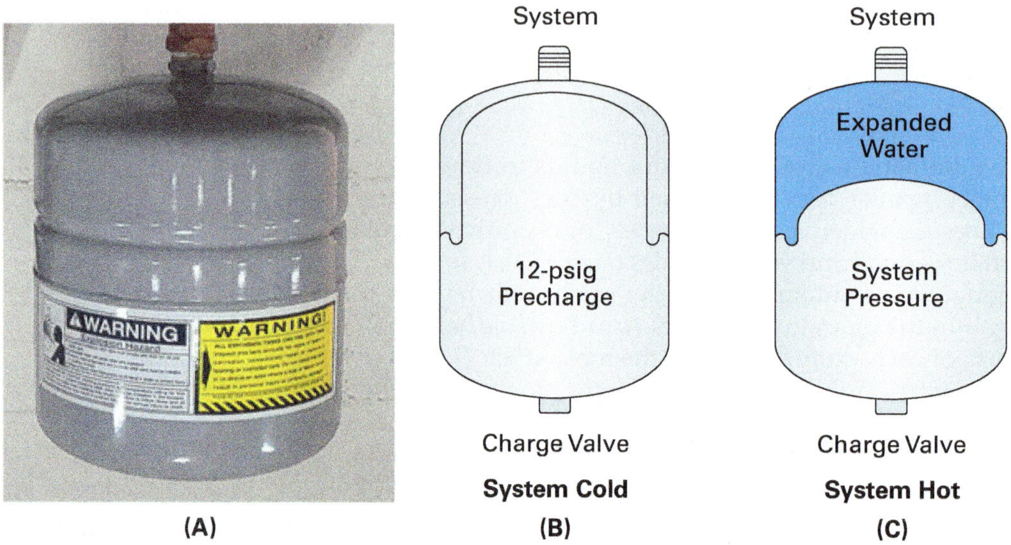

Figure 23 Expansion tanks.
Source: (a) Wendy Becovitz-McKinzie

WARNING!

Hot water under pressure can escape with explosive force, and it can cause dangerous scalding. At 125°F (52°C), it takes about 1½ to 2 minutes for hot water to cause scalding. At 155°F (68°C), it takes about one second. Wear appropriate personal protective equipment. Review and follow all construction drawings closely.

2.3.0 Installing and Maintaining Recirculation Systems

Plumbers install and repair hot water recirculation systems. Many of the materials and the methods used to install recirculation systems are common to other types of plumbing installations. You will need to read the construction drawings before constructing the system. Your knowledge and experience will be helpful when you build a recirculation system, but you also need to be aware of some problems that are unique to recirculation systems. These problems are discussed below.

2.3.1 Installing a Hot Water Recirculation System

Depending on the design of the recirculation system, you may need to connect more than one riser to the hot water supply main. Ensure that the connection to each riser is correct. Otherwise, problems may develop with the hot water flow in some of the risers. Correct connections are also important in systems that have been altered by previous repairs. Each riser may require a slightly different connection because of the connection's location and the riser's length and slope. Commonly used fittings for this application include the following:

- Tee fittings (the most typical)
- 90-degree fittings
- Couplings

Knowing when to install each type of connection comes from practical experience. The International Plumbing Code (IPC) and the International Energy Conservation Code (IECC) specify limits on the length from the hot water source to the fixture location. Consult these codes and your local code for guidelines, and also seek advice from experienced plumbers.

When installing more than one water heater or storage tank in a high-rise building, install check valves on the return risers. These valves will prevent the backflow of cold water into the hot water supply system. Install check valves near the water heater to prevent hot water from re-entering the cold water system. Check your local codes for specifications. Tempering mixing valves can be located in several places in the system. One effective location is outside the heat exchanger on the discharge line.

Install centrifugal pumps on the hot water return line. Locate them as close as possible to the water heater (refer to *Figure 17*). Install pumps with a bypass or a union. In some high-rise buildings, it may be necessary to install several pumps on different floors. If so, follow the plans and specifications closely.

2.3.2 Troubleshooting and Maintaining a Recirculation System

Make sure water pressure booster systems are designed and installed correctly. A malfunctioning booster pump usually means an entire water supply system must be shut down until the pump can be repaired. In a home this can be a nuisance, but in a hospital it can cost lives. Quick and efficient troubleshooting and repair or replacement is essential.

Recirculation systems are made up of familiar components. You have learned how to troubleshoot problems with water tanks, water heaters, pipes, and valves. This knowledge will enable you to troubleshoot a recirculation system. In some cases, a manufacturer's service technician may be required to do this work.

Consider the use of balancing valves to prevent pressure differentials among the stacks in a recirculation system with more than one stack. Balancing valves compensate for different lengths and heights of pipe in various parts of the system. Open and close the individual valves until the pressure is uniform throughout the system.

Scale buildup is a common occurrence in a recirculation system. Scale buildup is caused by deposits that can create problems, such as clogged pipes. Clogged pipes will need to be cleaned or replaced.

Going Green

Solar Hot Water Systems

City, state, and federal government programs are encouraging homeowners and businesses to use solar hot water heating. Solar energy is a renewable resource and creates few greenhouse gas by-products. Since 1992, the federal government has issued tax credits to businesses that use solar power. Many states offer similar incentives for homeowners and businesses.

Some systems convert sunlight into electricity. Others collect solar energy and use it to heat fresh water supplies and swimming pools directly. These direct systems, called solar thermal systems, are the least expensive form of solar energy for applications connected to power grids. Solar power is gaining popularity as people become more environmentally conscious. Plumbers can expect to install an increasing number of solar hot water systems in the near future.

2.0.0 Section Review

1. When laying out a forced circulation system, install the pump on the _____.
   a. supply line before the water heater
   b. supply line after the water heater
   c. return line before the water heater
   d. return line after the water heater

2. Tempering mixing valves must be installed in systems that use _____.
   a. tankless heaters
   b. indirect heaters
   c. demand heaters
   d. storage heaters

3. Balancing valves are designed to compensate for different _____.
   a. pipe lengths and heights
   b. pipe diameters and fixtures
   c. water demands among fixtures
   d. operating pressures in a single system

Module 02403 Review Questions

1. A booster system that uses air pressure to force water from a tank to various parts of a building is called a(n) _____.
   a. constant-capacity system
   b. hydropneumatic tank system
   c. variable-capacity system
   d. elevated-tank system

2. A shortcoming of the constant-speed system is _____.
   a. excessive electricity use
   b. variations in water pressure
   c. excessively noisy operation
   d. temperature fluctuations

3. A pressurized air-bladder storage tank is sometimes used with a(n) _____.
   a. hydropneumatic tank system
   b. elevated-tank system
   c. variable-capacity system
   d. constant-capacity system

4. The most widely used type of water pump used in plumbing systems is the _____.
   a. positive-displacement pump
   b. peristaltic pump
   c. hydrostatic pump
   d. centrifugal pump

5. A centrifugal pump can be classified by the _____.
   a. number of vanes on the impeller
   b. impeller rotational speed
   c. number of intakes the impeller has
   d. outlet size

6. Excessive wear of single-suction pump seals and impellers is avoided by installing a _____.
   a. flow restrictor
   b. venturi
   c. volute
   d. backflow preventer

7. Booster systems are frequently designed to provide _____.
   a. low flow at high pressure
   b. low flow at low pressure
   c. high flow at high pressure
   d. high flow at low pressure

8. Pumps in a water pressure booster system supplied by a well should be provided with a _____.
   a. thermal sensor
   b. pressure switch
   c. flow sensor
   d. water hammer arrester

9. The most important factor in sizing water pressure booster systems is _____.
   a. average flow
   b. minimum probable flow
   c. maximum probable flow
   d. adjusted mean flow

10. Items classified as intermittent flow typically have a usage time shorter than _____.
    a. 2 minutes
    b. 5 minutes
    c. 7 minutes
    d. 10 minutes

11. To locate leaks in a water pressure booster system tank, use a _____.
    a. soap-and-water solution
    b. fluorescent indicator solution
    c. pressure tester
    d. leak-detector kit

12. If a pump motor overheats, it will be turned off by a device called a(n) _____.
    a. heat switch
    b. safety cutout
    c. overload protector
    d. flow controller

13. Recirculation systems are less effective when the distance from the water heater to the farthest hot water tap is less than _____.
    a. 50'
    b. 100'
    c. 150'
    d. 200'

14. The need for multiple heaters and storage tanks in a large recirculation system is determined by the _____.
    a. plumbing engineer
    b. architect
    c. building inspector
    d. plumbing contractor

15. To allow air to be purged from piping, a manual or automatic air-relief valve is installed _____.
    a. at the water-heater outlet
    b. at the lowest point in the system
    c. at the highest point in the system
    d. ahead of the circulating pump

Answers to odd-numbered review questions are found in the Module Review Answer Key at the back of this book.

Answers to Section Review Questions

Answer	Section	Objective
Section One		
1. d	1.1.3	
2. a	1.2.3	
3. d	1.3.3	
4. b	1.4.2	
Section Two		
1. c	2.1.1	
2. b	2.2.4	
3. a	2.3.2	

MODULE 02404

Indirect and Special Waste

Objectives

Successful completion of this module prepares you to do the following:

1. Identify indirect waste systems.
   a. Identify the installation requirements for indirect waste piping.
2. Identify special waste systems.
   a. Identify the various types of special waste.
   b. Explain how to install interceptors.

Performance Tasks

Under supervision, you should be able to do the following:

1. Install an indirect waste system.
2. Size and install an interceptor using information provided by the instructor.

Overview

Indirect waste systems prevent sewer gas and waste from contaminating fixtures and appliances. Indirect waste systems enable fixtures and appliances to drain through pipes that are not directly connected to the building waste system. Wastes that require treatment before being discharged into the drain, waste, and vent (DWV) system are called special wastes. They are handled using special techniques and drain installations. Interceptors are used to slow, separate, and contain special wastes before they can enter the sewer system. Each type of interceptor has its own special requirements, and codes may vary widely on installation specifications.

CODE NOTE

Codes vary among jurisdictions. Because of the variations in code, consult the applicable code whenever regulations are in question. Referring to an incorrect set of codes can cause as much trouble as failing to reference codes altogether. Obtain, review, and familiarize yourself with your local adopted code.

Industry-Recognized Credentials

If you are training through an NCCER-accredited sponsor, you may be eligible for credentials from NCCER. The ID number for this module is 02404. Note that this module may have been used in other NCCER curricula and may apply to other level completions. Contact NCCER at 1.888.622.3720 or go to **www.nccer.org** for more information.

You can also show off your industry-recognized credentials online with NCCER's digital credentials. Transform your knowledge, skills, and achievements into credentials that you can share across social media platforms, send to your network, and add to your resume. For more information, visit **www.nccer.org**.

Digital Resources for Plumbing

Scan this code using the camera on your phone or mobile device to view the digital resources related to this craft.

SCAN ME

1.0.0 Indirect Waste Systems

Performance Task	Objective
1. Install an indirect waste system.	Identify indirect waste systems. a. Identify the installation requirements for indirect waste piping.

Indirect waste system: A drainpipe attached to a fixture or appliance that is separated from the regular drainage system by a backflow preventer.

Indirect waste: Wastewater that flows through an indirect waste pipe.

Air gap: The unobstructed vertical distance through the free atmosphere between the outlet of the waste pipe and the flood level rim of the receptacle into which the waste pipe is discharging.

Air break: A piping arrangement (in a drainage system) in which a drain from a fixture, appliance, or device discharges indirectly into another fixture, receptacle, or interceptor at a point below the flood level rim and above the trap seal.

All fixtures and appliances drain away waste products, thus becoming unsanitary. Fixtures are designed to be cleaned. Certain public-use fixtures need extra protection from sewage backup. If a backup occurred in a restaurant ice bin, for example, it could cause illness and even death. One way to protect fixtures and appliances is to have them drain through pipes that are not directly connected to the building waste system. Such pipes are called **indirect waste systems**, and the water that flows through them, not surprisingly, is called **indirect waste**. An indirect waste pipe does not connect directly with the drainage system, instead discharging into the drainage system through a physical air separation.

Indirect waste systems are used to keep sewer waste from contaminating fixtures if drains back up. An **air gap**, which is commonly used to protect against indirect waste, is simply a space between the indirect waste disposal pipe and the building waste system (*Figure 1*). The air gap between the indirect waste pipe and the flood level rim of the waste receptor must be at least twice the effective opening of the indirect waste pipe, with a minimum of 1".

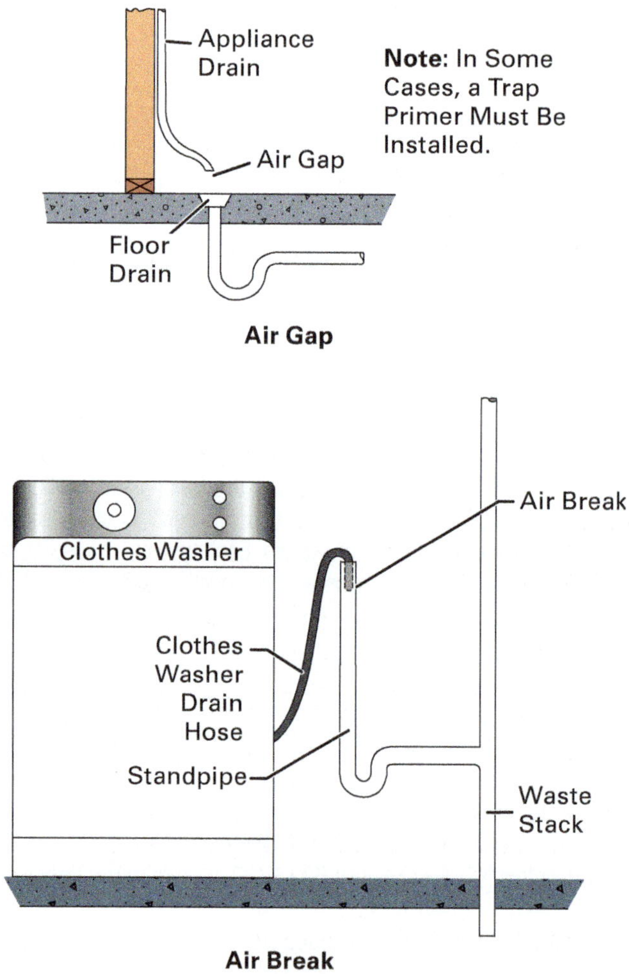

Figure 1 Air gap and air break.

The **air break** is another type of indirect waste device. In an air break, a smaller-diameter pipe drains into a larger-diameter pipe. The air break prevents the water in the larger pipe from returning to the smaller pipe. Note that the waste line must be below the flood rim of the **receptor**, but above the weir of the trap or the seal. Use air breaks where there is no possibility of siphonage, such as in a laundry tray. Washing machines are often installed with air breaks. Codes vary on the requirement for air gaps and air breaks. Always consult your local code before installing either one.

Receptor: A floor, sink, standpipe, hub drain, or floor drain that receives the discharge of one or more indirect waste pipes.

Indirect wastes are produced by appliances and fixtures in the following:

- Commercial food-preparation facilities
- Food-handling equipment, humidifiers, and dishwashers
- Food storage areas
- Medical and dental offices
- Air conditioning condensate and other potable and nonpotable clear-water waste installations

Indirect waste systems have many features in common. They also have their own special requirements. The following section discusses the requirements for the installation of piping for indirect waste systems. Always refer to the local applicable code for the requirements that apply to your area.

1.1.0 Installing Indirect Waste Piping

The air gap between an appliance waste pipe and the drain must be at least twice the diameter of the appliance waste pipe. For example, a waste pipe of 2" in diameter must have an air gap of at least 4". Some indirect waste pipes empty into a sink basin. Measure the air gap from the flood level of the sink to the terminal of the waste pipe.

To prevent splashing, install a receptor on the inlet to the waste system. Receptors can be funnels, floor drains, trapped and vented slop sinks, or other similar fixtures. Ensure that the receptor has a removable strainer or basket that will keep solids out of the waste pipe. Waste receptors must not be installed in unoccupied portions of the building, in concealed spaces, in plenums or crawl spaces, in attics, in interstitial spaces above ceilings, or in interstitial spaces below floors. Make sure waste receptacles are readily accessible for cleaning. An **open-hub waste receptor** (often referred to simply as a "hub drain") is a pipe or pipe hub extending at least 1" above the floor. This receptor does not require a strainer.

Another way to prevent splashing is to install a **modified waste line**. Using a sanitary tee, attach a vertical pipe to the terminal of the waste line (*Figure 2*). The pipe should be open at both ends. A modified waste line allows the indirect waste to mix with air as it exits the pipe. This helps the wastewater flow smoothly into the drain.

Open-hub waste receptor: A waste drainpipe or pipe hub that extends at least 1" above the floor line.

Modified waste line: A vertical pipe at the terminal of a waste line that allows indirect waste to mix with air.

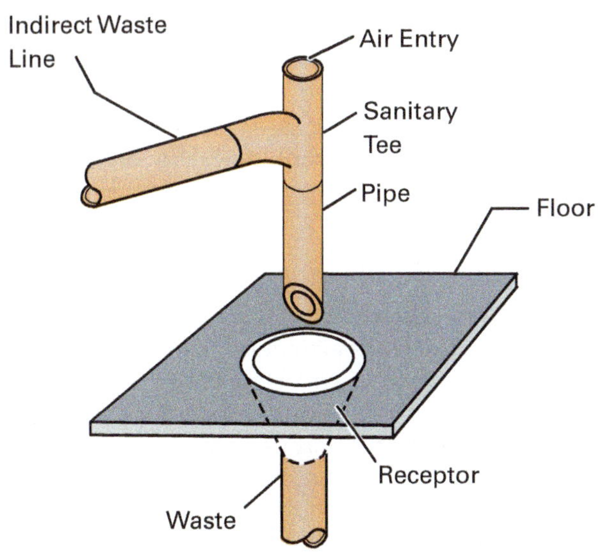

Figure 2 Modified indirect waste line.

Many indirect waste systems are used only intermittently. They may drain only small amounts of waste at a time. As a result, trap seals can dry up if not periodically refreshed. If that happens, sewer gases can escape into the surrounding air, causing odors and presenting a potential health hazard. Using an automatic trap primer can help prevent the loss of the trap seal. Automatic trap primers can be supplied by domestic water, wastewater, reclaimed (graywater), or polymer trap inserts. Automatic trap primers may use either gravity or electricity to operate. Use a potable-water **automatic trap primer**, wastewater-supplied trap primer device, or reclaimed (or graywater-supplied) trap seal primer valve to maintain a constant seal in a low-use trap (*Figure 3*). Automatic trap primers may be designed to close automatically, or to operate electronically or under the pressure of gravity.

Automatic trap primer: A device that feeds water into a low-use trap, allowing the trap to maintain a constant seal.

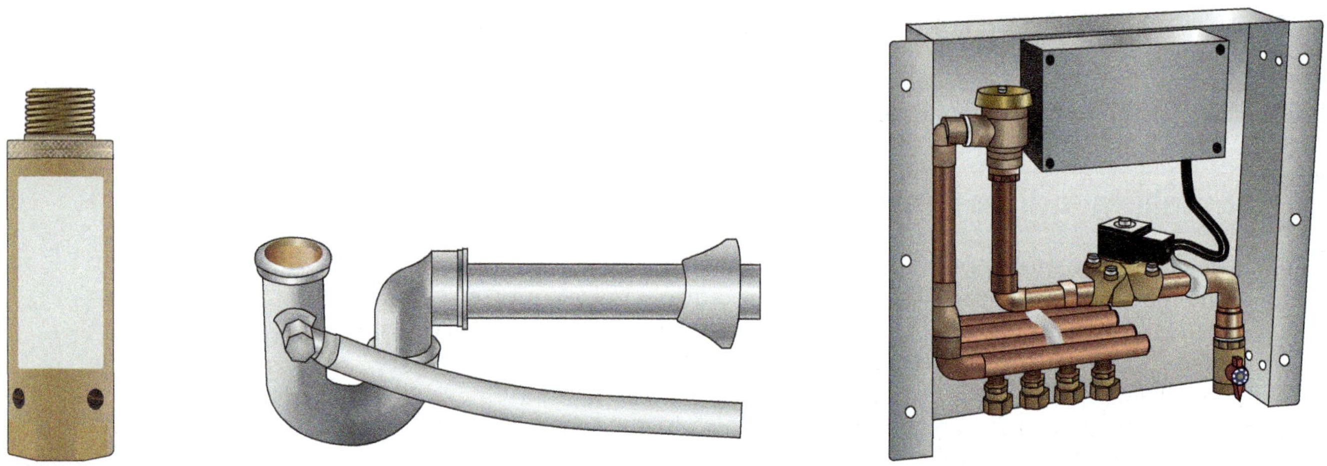

(A) Automatic Dosing **(B) Gravity** **(C) Electronic**

Figure 3 Automatic trap primers.

A potable-water-supplied trap seal primer valve, wastewater-supplied trap primer device, or reclaimed or graywater-supplied trap seal primer valve must supply water to the trap, per applicable ASSE codes. The discharge pipe from the trap seal primer valve or device must connect to the trap above the trap seal on the inlet side of the trap, again, per code.

Automatic trap primers may be designed to dose automatically due to pressure differences, or to operate electronically or by gravity. The type of automatic trap primer will be specified in the project specifications. To prevent the buildup of odors from sewer gases, install a trap primer. Ensure that indirect waste installations can be easily flushed. A hose can be used to flush pipes not equipped with a hot water connection.

Do not connect steam lines to a plumbing system. Steam or other waste that is hotter than 140°F should be discharged into a cooling sump or basin. After the wastewater has cooled, it is discharged into the sewer system.

1.1.1 Food-Preparation, Handling, and Storage Installations

Codes require indirect waste lines with air gaps for walk-in refrigerators and freezers, icemakers, and ice compartments. Wastewater from multiple ice compartments and icemakers can be collected at one point, which can be a floor sink or a floor drain (*Figure 4*). This reduces the amount of piping required to drain multiple fixtures. Many codes require indirect waste lines that are longer than 30" to have traps. The International Plumbing Code® (IPC®) specifies that for floor drains within areas subject to freezing, the floor drain should not be trapped; it should instead discharge into a waste receptor located outside the area subject to freezing.

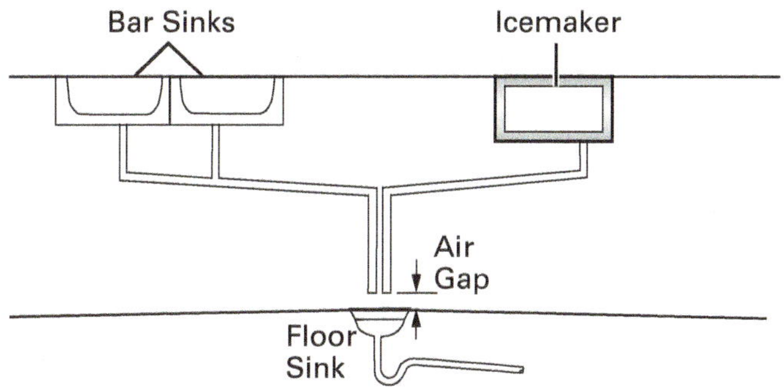

Figure 4 Indirect waste piping for bar sinks and icemakers.

Commercial dishwashers and dish-washing sinks typically cannot be attached directly to the building drain. Use an indirect waste system in these scenarios. Because large quantities of grease may be discharged from dish-washing fixtures, install a grease interceptor on the indirect waste line before the air gap. Refer to your local plumbing code for the requirements in your area.

Many commercial kitchen appliances, such as potato peelers, create solid wastes. Install floor sinks with a strainer and grate. These arrangements were once referred to as **sediment buckets** to trap solid wastes (*Figure 5*). A sediment bucket is a removable basin inside a drain that holds solid waste but allows water to drain.

Sediment bucket: A removable basin in a drain that prevents solid wastes from entering a sanitary system.

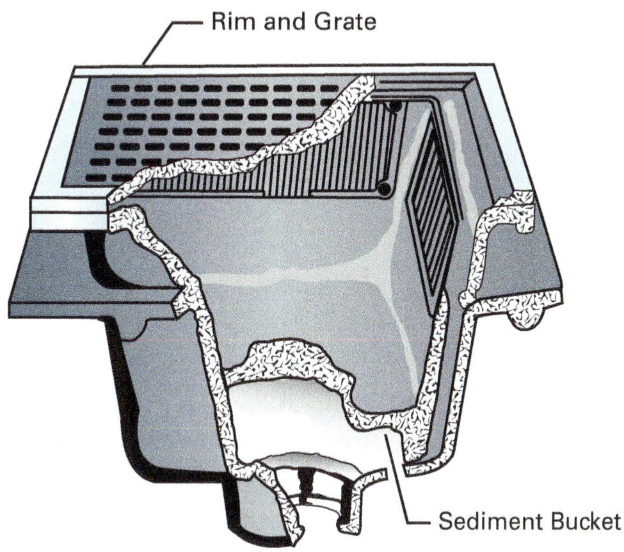

Figure 5 Floor sink with sediment bucket.

Floor and area drains are often set before the concrete floor is poured. Ensure that the elevation of the drain is correct. Locate the drain at the lowest point in the floor or area. Refer to the floor plan, plumbing drawings, or specifications for the drain's elevation, the horizontal distances from walls, and the floor finishes.

Floor sinks may be used in installations that require a high degree of sanitation, such as hospitals, laboratories, cafeterias, and restaurants. Floor sinks have a larger holding area to accept waste before it goes into the drain-waste-vent system and are typically cast-iron coated with porcelain enamel to allow for ease of cleaning. Floor drains are flush with the floor and do not have a basin to accept waste before it enters the drainage system.

The basic difference between a floor drain and a floor sink, therefore, is that the floor sink is coated with porcelain, or another easily cleaned finish. Floor sinks reduce the amount of contaminant that can collect in the grate and drain body. Consult the local applicable code or the project specifications to determine if the installation of a reduced pressure zone (RPZ) principle backflow preventer assembly is required on the same line as the floor sink.

1.1.2 Health Care Installations

Health care facilities need sanitary conditions to protect patients' health. Effective waste disposal helps limit the spread of disease. Indirect waste systems are used in a wide variety of health care facilities, including the following:

- Surgical and dental theaters
- Clinics and infirmaries
- Pharmaceutical and research laboratories
- Youth and elder care institutions

Consult your local code before installing health care waste systems.

Install traps on indirect waste lines for bedpan steamers. Batteries of fixtures, such as sterilizers, can drain into a single receptor. Attach clinical sinks directly to the building waste system. Bedpan washers are connected directly to the waste-and-vent system.

WARNING!

Wastewater from medical and dental facilities is hazardous. It can contain bacteria and viruses that can cause illness and death. Ensure that the design of the DWV system adheres strictly to code before installing it.

1.1.3 Clear-Water Waste Installations

Many other fixtures and appliances create indirect waste. Some installations that produce clear-water wastes follow:

- Swimming pools
- Relief valves
- Air conditioning condensate drains
- Ice machines
- Refrigerators
- Water storage tanks
- Water heaters
- Drinking fountains

Water-heater relief lines (*Figure 6*) create indirect waste only when there is an issue with the water heater. Three-compartment sinks used for food services are required to have an indirect waste receptor (*Figure 7*). Measure the

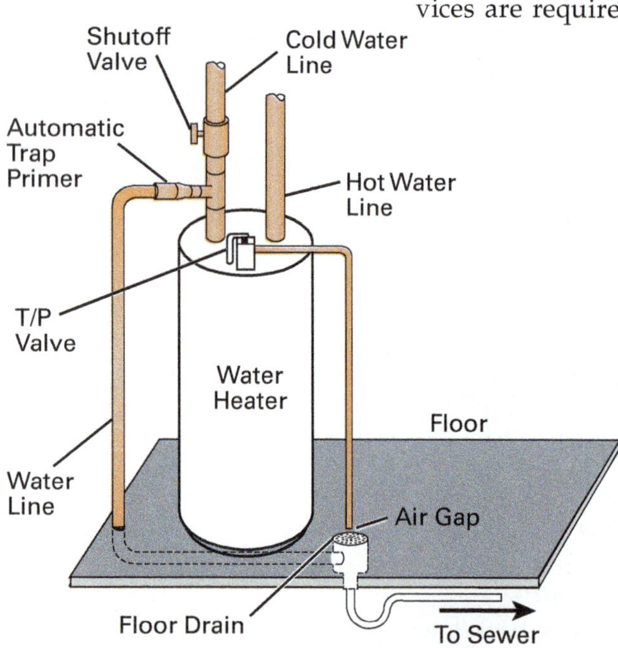

Figure 6 Water-heater unit installed in a basement.

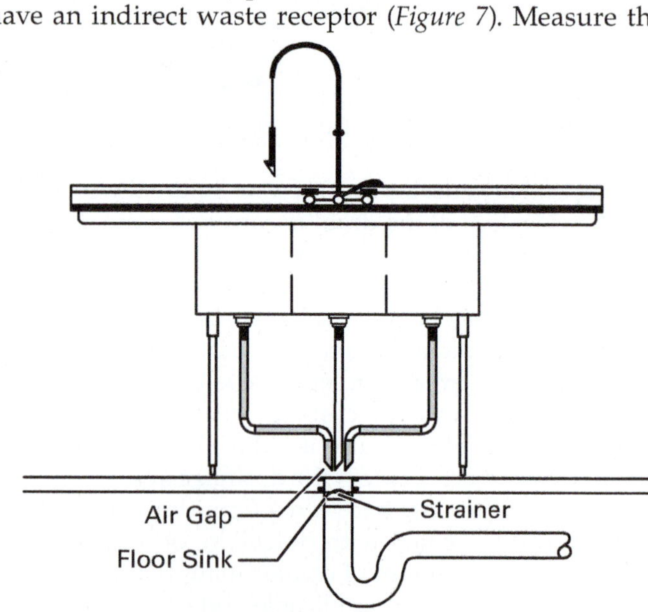

Figure 7 Three-compartment sink used for food services.

air gap between the flood line of the receptor and the waste-pipe terminal. Refer to your local code when installing these and other indirect wastewater systems.

1.0.0 Section Review

1. Many codes require indirect waste lines to have traps if their length exceeds _____.
   a. 48"
   b. 36"
   c. 30"
   d. 12"

2.0.0 Special Waste Systems

Objective

Identify special waste systems.

a. Identify the various types of special waste.

b. Explain how to install interceptors.

Performance Task

2. Size and install an interceptor using information provided by the instructor.

Indirect waste is not the only kind of waste that requires special drain installations. **Special wastes** must be treated before they can be discharged safely into the system, due to one or more of the following conditions:

Special waste: Waste that must be removed, diluted, or neutralized before it can be discharged into a sanitary system.

- Their temperature is too high for the septic or sewer system, such as water above 140°F.
- They contain solid and partially solid matter that will settle out.
- They contain acids that must be neutralized.
- They contain harmful or potentially harmful substances, such as grease, oil, and gasoline.

Some special wastes are treated by evaporation, while others must be separated from the wastewater. Still others must be diluted or chemically neutralized first.

2.1.0 Identifying Types of Waste

Wastes include grease, oil, and gasoline; solids, such as sediment, glass, and hair; corrosive wastes (which are considered special waste); and high-temperature wastes. Grease or animal fat is generated in large quantities where food is prepared and served commercially. Grease will clog sewer and septic pipes if it is allowed to enter the system. Oil and gasoline are volatile petroleum products. They are used to operate, clean, repair, and store motor vehicles. These fluids could ignite and damage sewer, septic, or stormwater drainage systems. Sediment can clog drains and pipe. Corrosive chemicals like acids can damage pipes and drains, generate toxic fumes, and interfere with the waste treatment process. High-temperature waste can cause pipes to expand and contract. This weakens pipe joints and causes leaks.

Treatments vary for each type of waste. Improper treatment could damage the sewer or septic system. It could also contaminate fresh-water supplies and cause sickness or even death. Review the proper treatment method before treating waste. Your local code will also provide specific guidelines.

Health care facilities use special waste systems. As with indirect waste systems in health care facilities, install traps on special waste lines. Batteries of fixtures can be drained into a single receptor. An interceptor is often required in special waste piping, depending on the types and concentrations of the waste materials in the system. In some installations, the waste may be treated prior to its entry into the waste system. Corrosive liquids or harmful chemicals must be diluted, neutralized, or treated using a code-approved neutralizing device. Always refer to the local applicable code before installing special waste systems in health care facilities.

2.2.0 Installing Interceptors

Interceptor: A device designed and installed to separate and retain for removal, by automatic or manual means, deleterious, hazardous, or undesirable matter from normal wastes, while permitting normal sewage or wastes to discharge into the drainage system by gravity.

Baffle: Partition inside the body of an interceptor. Baffles are designed to reduce turbulence caused by incoming wastewater and to block waste from escaping through the outlet.

Interceptors separate special wastes before they can enter a sewer, septic, or stormwater system. Small interceptors are about the size of an ordinary P-trap. Some custom-built underground interceptors are large enough for a person to stand in. No matter what size they are, most interceptors operate the same basic way.

The internal volume of an interceptor is larger than that of the drainage pipes. This causes wastewater to slow down as it enters, allowing the waste materials time to settle, float, or evaporate out of the wastewater. **Baffles** are partitions in the interceptor that reduce turbulence caused by flowing wastewater. Baffles also keep wastes from escaping through the outlet. Free of the contaminants, the wastewater discharges into the sewer system by gravity.

Interceptors work a lot like fixture traps. Both are designed to seal one part of a plumbing installation from waste products in another part. However, interceptors violate one of the basic principles of traps: Interceptors are not self-cleaning and must be manually cleaned.

Site plans specify where large interceptors should be installed. An interceptor can be above or below floor level. The installation's design and the available space determine the location. Provide enough space to allow cleaning equipment, such as pumps and augers, to fit. Ensure that covers and sediment buckets can be removed easily.

Locate the interceptor close to the fixture, because this will help prevent the inlet piping from clogging. Install exterior interceptors below the frost line to keep them from freezing. Support large interceptors with a solid foundation, because the weight of aboveground interceptors can cause structural problems in some buildings. Even interceptors installed below grade need support. Otherwise, pipe joints may fail and damage the tank. Packed sand over undisturbed earth is a good foundation.

Think About It

The Trap "Trap"

Interceptors are sometimes called traps, such as a grease trap. You have already learned that the term *trap* also applies to installations that maintain water seals in fixtures. Both types of traps are vital components in DWV systems because they protect people from exposure to noxious and harmful substances. However, they operate very differently. Interceptors and traps have their own unique requirements. Do not confuse the two! The results could be dangerous. Take the time to review your local code requirements for both interceptor and trap installations.

If the waste includes toxic gases, steam, high-pressure wastes, or chemicals, vent the interceptor through a roof vent. Ensure that the inlet and outlet piping is installed properly, and seal all piping joints tightly to prevent liquid or gas leaks. For concrete and masonry interceptors, carefully grout the joints between the piping and the tank.

2.2.1 Grease Interceptors

For small amounts of grease, install a small hydromechanical grease interceptor inside the building (*Figure 8*). These grease interceptors are typically made of steel and have a bolt-on lid. Ensure that the interceptor is easy to reach for

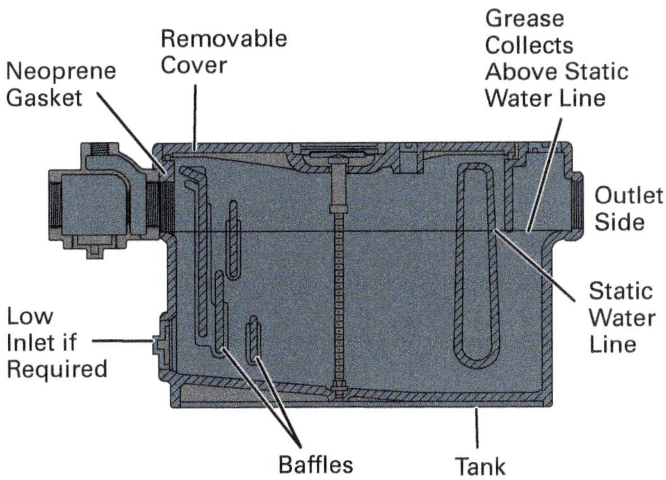

Figure 8 Typical interior grease interceptor.

Did You Know?

Maintaining Grease Interceptors

Grease interceptors require regular service. Provide the building managers or occupants with a maintenance schedule. Explain that regular servicing protects the health of the public and helps protect the owners from fines, penalties, and lawsuits.

cleaning, and locate it as close to the sink or appliance as possible. Otherwise, foul odors may occur. Odors are the biggest problem associated with indoor grease interceptors. Proper venting will help reduce odors. Vent grease interceptors using the guidelines for venting a trap. Cleaning can be a problem when lid removal is required.

One cleaning option is to use a **draw-off hose** (*Figure 9*). Draw-off hoses are used to siphon the grease out of an interior grease interceptor; they can be either gravity-fed or connected to a pump. Ensure that a draw-off valve is installed at the outlet of the interceptor. Closing the valve allows the water level in the interceptor to rise, which forces the grease to flow into the draw-off hose. Consult your local code for the proper procedure.

Draw-off hose: A hose used to siphon grease and oil from interceptors.

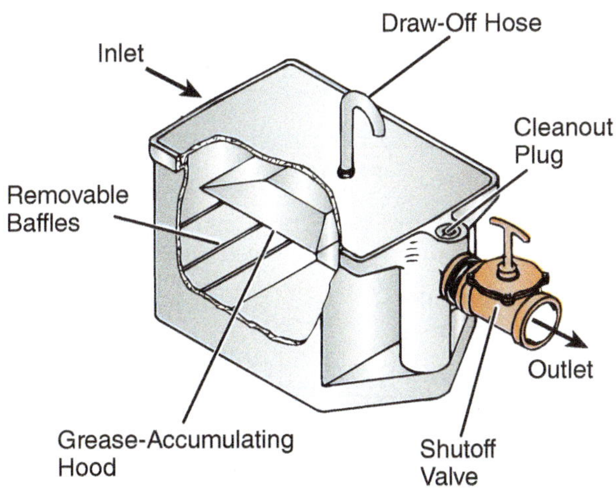

Figure 9 Draw-off hose installed on a grease interceptor.

Excessive flow into an interceptor can cause turbulence and reduces the amount of time grease has to coagulate and float to the top for draw-off. These conditions will permit grease to escape through the outlet. Install a **flow control fitting** on the inlet pipe to slow the flow of wastewater into an interceptor (*Figure 10*). The flow control's inlet is carefully sized, limiting the amount of

Flow control fitting: A device that regulates the flow of wastewater into grease and oil interceptors.

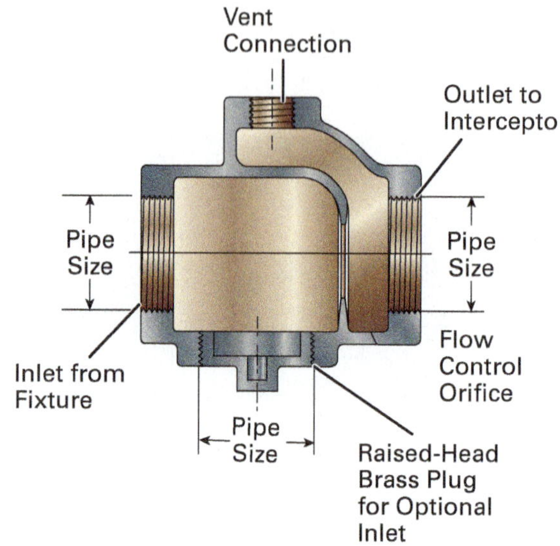

Figure 10 Flow control fitting.

wastewater that can flow into the interceptor. Flow control devices with removable or adjustable parts are not permitted in grease interceptors.

Large interceptors are located outside and underground. Precast concrete tanks, like septic tanks, can be used as large grease interceptors. They can be site-built from concrete, masonry, and pipe fittings (*Figure 11*). They may also be pre-constructed off-site. The dimensions can be varied to meet size requirements. Ensure that the inlet is positioned correctly in relation to the outlet and the static waterline. Increasing the height of the outlet pipe allows more solids to collect. This reduces the number of times the interceptor will need to be serviced. Access to the interceptor is provided by a manhole or through an opening for a draw-off hose. Two- and three-chamber grease interceptors are more efficient than single-chamber designs (*Figure 12*).

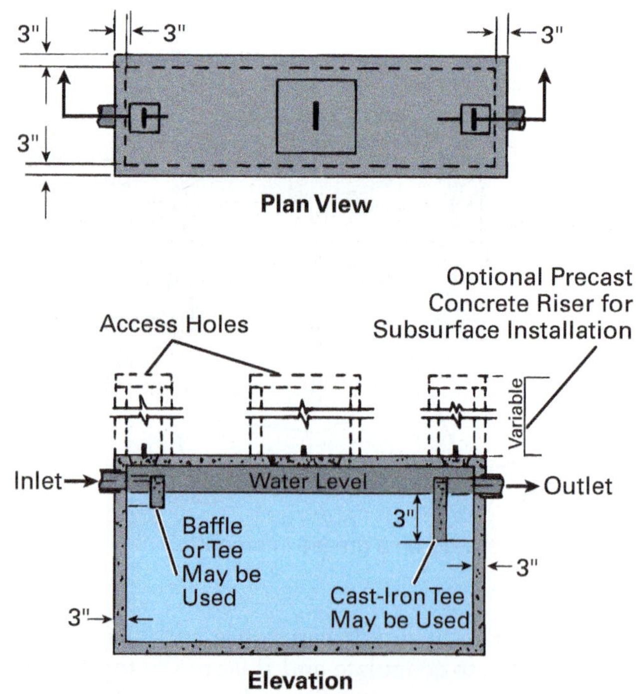

Figure 11 Typical large site-built grease interceptor.

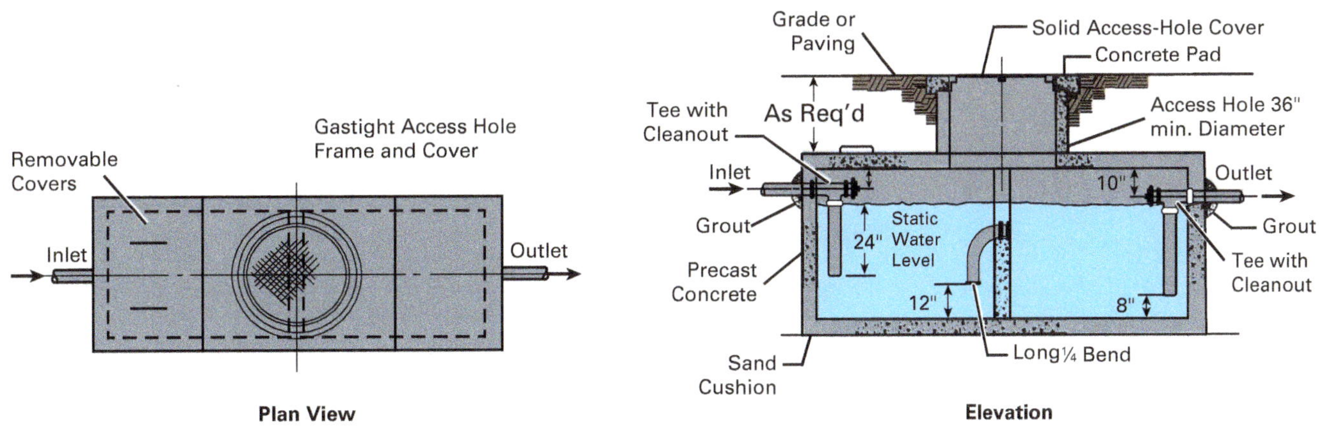

Figure 12 Two-chamber grease interceptor.

WARNING!

Climb-in interceptors may contain corrosive, flammable, or odorous substances. Personal injury could result, including poisoning or an infection, such as hepatitis. Obtain thorough confined-space training before servicing climb-in interceptors. Use appropriate personal protective equipment, and wear hand protection while working.

Codes provide formulas to allow you to calculate the right size for grease traps. The formulas in *Table 1* are based on the Environmental Protection Agency (EPA)-2 model; always refer to your local code. First, determine the total maximum drainage flow from each fixture. This number is estimated in gallons per minute (gpm). Next, factor in the estimated load factors. Multiply the total drainage flow by the load factor. Multiply that number by 60 (minutes) to get the maximum flow in one hour. Finally, multiply the maximum flow by 2 (representing the two-hour retention time specified in the code). This number gives you the total required volume of the grease trap in gallons.

Table 1 Grease-Interceptor Sizing Formula for a Restaurant

A. Determine maximum drainage flow from fixtures:		
Type of Fixture	**Flow Rate**	**Amount**
Restaurant kitchen sink	15 gpm	
Single-compartment sink	20 gpm	
Double-compartment sink	25 gpm	
2 single-compartment sinks	25 gpm	
2 double-compartment sinks	35 gpm	
Triple sink, $1^1/_2$" drain	35 gpm	
Triple sink, 2" drain	35 gpm	
30-gallon dishwasher	15 gpm	
50-gallon dishwasher	25 gpm	
50- to 100-gallon dishwasher	40 gpm	
B. Total		
C. Estimate Loading Factors:		
Restaurant type	Fast food/paper delivery	= .50
	Low volume	= .50
	Medium volume	= .75
	High volume	= 1.00
D. Step B × Step C:		**(Subtotal)**
E. Subtotal × 60 min. × max. flow for 1 hour		
F. Step E × 2 hours' retention time × vol. of trap (gal.)		

Grease interceptors may also be sized using the capacity method. This is done by calculating the flow rate of each sink that drains into the interceptor. Use the following steps to calculate grease-interceptor capacity using flow-rate calculations:

Step 1 Determine the capacity of each sink in cubic inches:

$$\text{length} \times \text{width} \times \text{height} = \text{capacity in cubic inches}$$

Step 2 Convert the capacity from cubic inches into gallons per minute (gpm) using the following formula:

$$\text{capacity in cubic inches}/231 = \text{capacity in gpm}$$

Step 3 Account for displacement, which is the sink's actual usable volume, using the following formula:

$$\text{capacity in gpm} \times 0.75 = \text{actual capacity in gpm}$$

Step 4 Add the actual capacities of the sinks together. This is the total capacity that the grease interceptor should be able to handle in one minute.

Note that if drain-down time is not a critical consideration, the project specifications may call for an interceptor that is sized for as much as half the calculated flow rate.

To clean small grease interceptors, use the following procedure:

Step 1 Remove the lid.

Step 2 Scoop the grease out of the interceptor, and place it into a watertight container (wear proper protective equipment for the hands and face).

Step 3 Thoroughly clean the sides and inside lid of the interceptor.

Step 4 Make sure the container is tightly sealed before properly disposing of it according to local code requirements.

Step 5 Re-install and seal the lid.

Grease disposal varies from city to city. Some areas forbid placing grease in landfills. Other codes allow some grease to be placed in garbage cans in sealed containers. Professional contractors perform service and repair on exterior grease interceptors. They use large pumper trucks to suck the grease out of the interceptor. Some recycling companies collect grease and convert it into fuel and industrial products. Rendering companies also collect grease and use it in the manufacture of consumer products, such as soap and perfume. As a plumber, you may be called on to recommend contractors, recyclers, and rendering companies. Only recommend those that follow environmental standards.

Ensure that food-waste grinders and garbage disposals do not discharge into a grease trap. Most codes place restrictions on garbage-disposal waste entering sewer systems. Check your local code for the requirements.

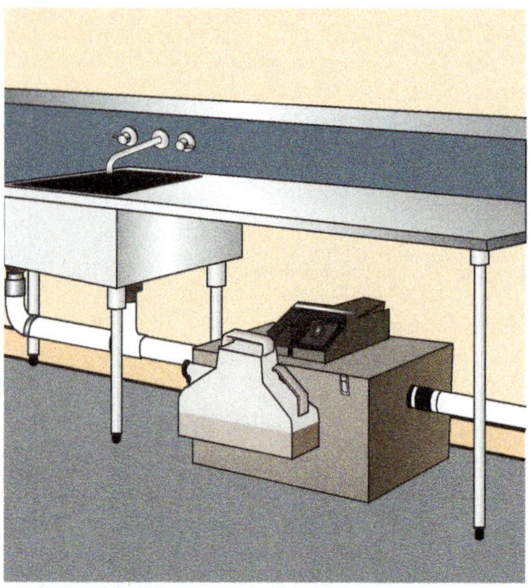

Figure 13 Mechanical grease interceptor.

Mechanical grease interceptor: A plumbing appurtenance, fitted with a timer-activated self-cleaning capability, installed in the sanitary drainage system to separate free-floating fats, oils, and grease from wastewater.

2.2.2 Mechanical Grease Interceptors

Mechanical grease interceptors, also called automatic grease removal devices (GRDs), are used in systems where grease, oils, and fats exceed the amount that small grease interceptors can handle (*Figure 13*). As with a regular grease interceptor, a mechanical grease interceptor works by passing wastewater through a strainer that separates fats, oils, and grease from the wastewater. The wastewater is then allowed to drain through the interceptor's outlet, while the greasy wastes are stored until they are removed for recycling.

Mechanical grease interceptors are also designed to be self-cleaning. At regular intervals, an internal heater warms the interceptor to 115°F–130°F, which causes any remaining grease to turn into liquids that then float to the top of the interceptor's retention tank. A skimming wheel collects the floating liquid waste, and a wiper blade slides the waste off the wheel into a special outlet that is connected to a separate waste container (*Figure 14*). Mechanical grease interceptors are designed to operate automatically. They only require attention when removing or replacing the waste container, or for maintenance.

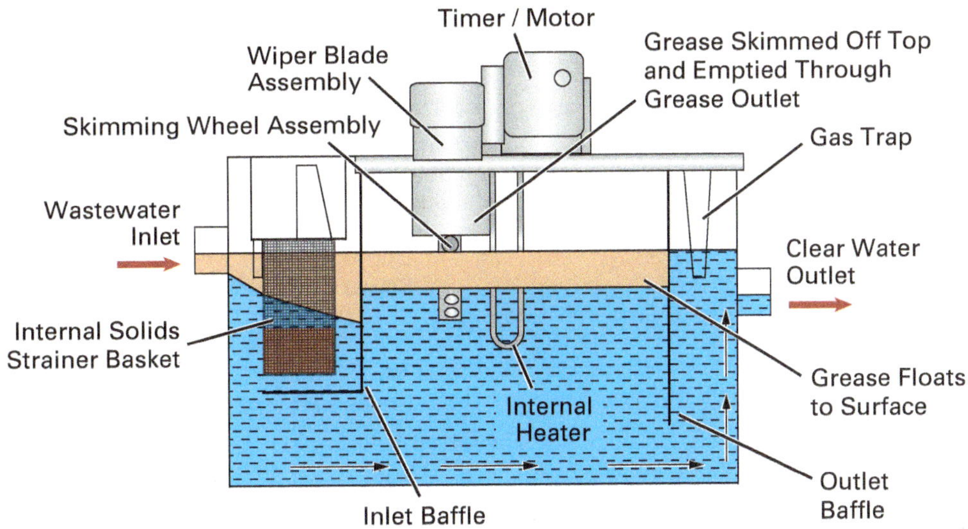

Figure 14 Cut-away illustration of a mechanical grease interceptor showing the self-cleaning process in operation.

American National Standards Institute (ANSI) Standard 112.14, *Grease Interceptors*, covers the sizing, design, installation, and testing requirements for mechanical grease interceptors. Ensure that the interceptor has been correctly sized for the calculated wastewater flows for all connected fixtures. Refer to the manufacturer's instructions and the local applicable code for sizing requirements. Prefabricated interceptors are based on the manufacturer's design and specifications. The inlet and outlet may vary based on manufacturer. Install the interceptor so that it can be accessed easily for inspection and maintenance.

Install mechanical grease interceptors downstream from the fixtures to which they are connected, as close to the fixtures as possible. Use flexible-sleeve pipe couplings to connect the interceptor inlet and outlet to the DWV system piping. Ensure that the unit has been filled with water before starting the first self-cleaning cycle. Otherwise, the heating element can be damaged. To prevent the risk of back siphonage, install an outlet vent or vacuum breaker as close as possible to the interceptor outlet. The vent or vacuum breaker should be at least one-half the diameter of the system outlet connection. Where permitted by code, air admittance vents may also be used to vent the interceptor.

When the installation has been completed, set the interceptor's self-cleaning timer for an appropriate interval, as specified in the manufacturer's instructions. Typically, a single cleaning cycle should not be allowed to exceed 2 hours. The internal strainer and external grease container should both be emptied at least daily. The skimming wheel and wiper blade should be inspected at least weekly to ensure it is free of buildup. The manufacturer's instructions will specify other regular and periodic maintenance requirements that operators should observe.

CAUTION

Do not install mechanical grease interceptors on waste lines connected to food-waste disposers, potato peelers, or solid-waste grinders. The solid waste can clog the interceptor and cause the motor to burn out. Also, the self-cleaning cycle could cause the solid waste to catch fire.

2.2.3 Oil Interceptors

The design of an oil interceptor is similar to that of a grease interceptor. However, oil and gasoline are harder to separate from wastewater than grease is. Codes vary widely on how and when to install oil interceptors. Consult the local code authority before construction.

Use draw-off hoses to remove oil from an interceptor. Otherwise, oil will have to be skimmed out by hand—a messy operation. There are two ways to use draw-off hoses. **Gravity draw-off systems** (*Figure 15*) drain oil into a storage tank. The oil drains out when it reaches a certain height within the interceptor. The height is determined by an adjustable draw-off sleeve inside the interceptor. For gravity systems, set the terminal of the sleeve to $^1/_8$" above the normal water operating line. In **manual draw-off systems** (refer to *Figure 15*), oil drains out of the interceptor by operating a draw-off valve. The terminal of the adjustable draw-off sleeve must be set to $^1/_4$" below the static waterline.

Gravity draw-off system: A system that automatically drains oil from an interceptor into a storage tank.

Manual draw-off system: A system that drains oil out of an interceptor through a valve-operated draw-off hose.

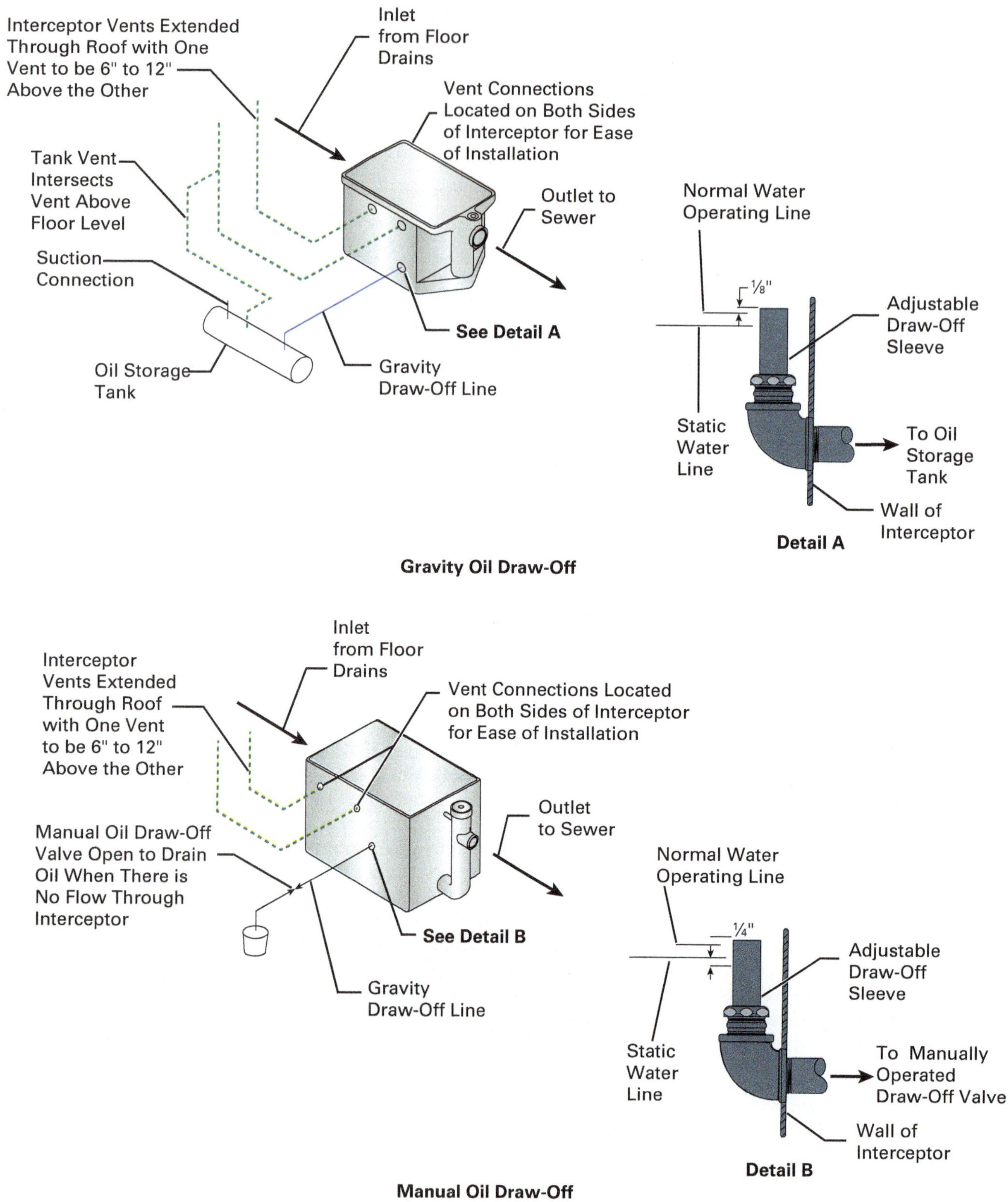

Figure 15 Gravity and manual draw-off oil interceptors.

Be sure to close the flow control valve before operating the draw-off valve. Both draw-off systems use sediment buckets or similar devices to catch dirt and grit before they enter the drain lines.

Install vents on oil interceptors. Vents allow flammable and noxious gases to escape harmlessly and prevent back pressure from building up in the interceptor. Oil interceptors require two vents; terminate one vent 6" to 12" above the other. Oil storage tanks on gravity draw-off systems must be vented.

Many businesses produce special wastes that must be handled according to the local applicable health and plumbing codes. For example, dentists' offices require the installation of sediment buckets on waste lines connected to suction devices. Waste produced by the processing of X-ray plates must be disposed of separately from regular wastewater. Some codes treat the wastes produced by funeral homes during the embalming process as special wastes. Always refer to the local applicable health and plumbing codes when installing DWV systems in facilities that produce unusual waste.

2.2.4 Sediment Interceptors

Sediment interceptors prevent sand, dirt, and grit from entering the waste system (*Figure 16*). Small interceptors are installed in place of the P-trap. The top and bottom of sediment interceptors are designed to be removed. This allows for regular cleaning no matter how they are installed. Large sediment interceptors can be installed below, on, or above the floor (*Figure 17*). They can be either precast or site-built (*Figure 18*). Larger inlets may require the installation of a vent.

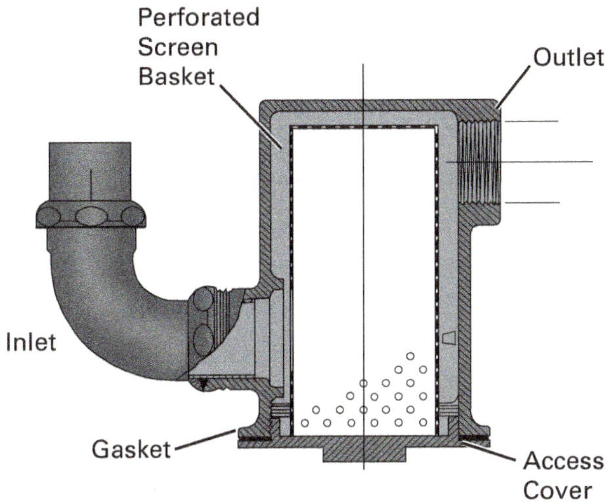

Figure 16 Small sediment interceptor.

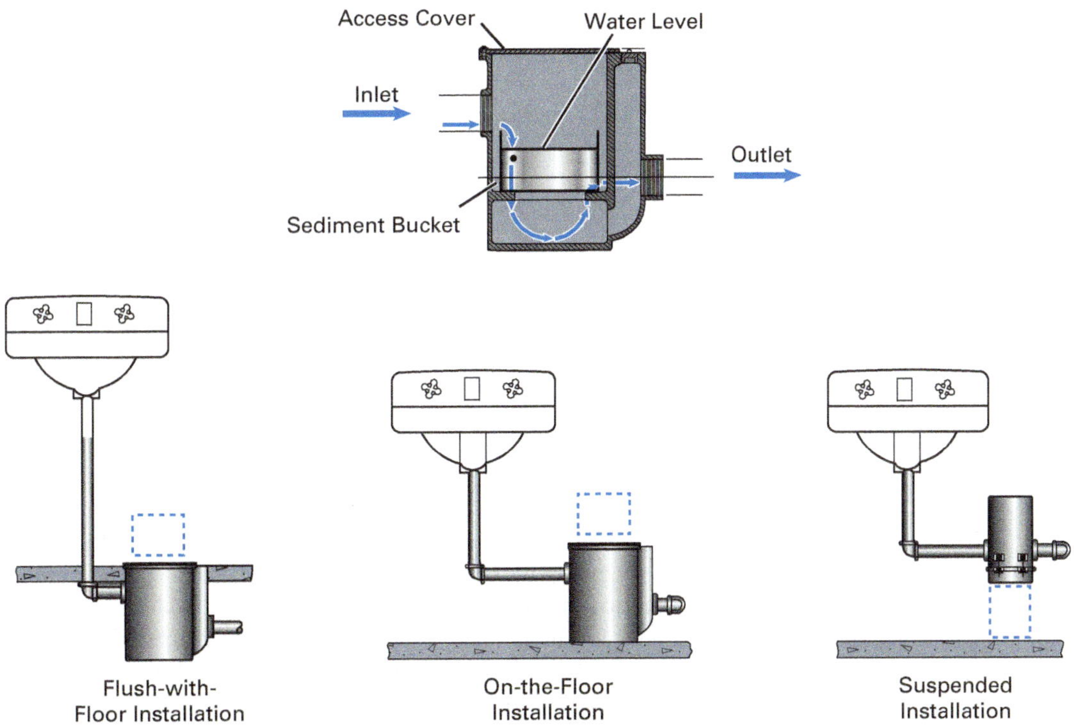

Figure 17 Large sediment interceptors installed on wall-hung hand sinks.

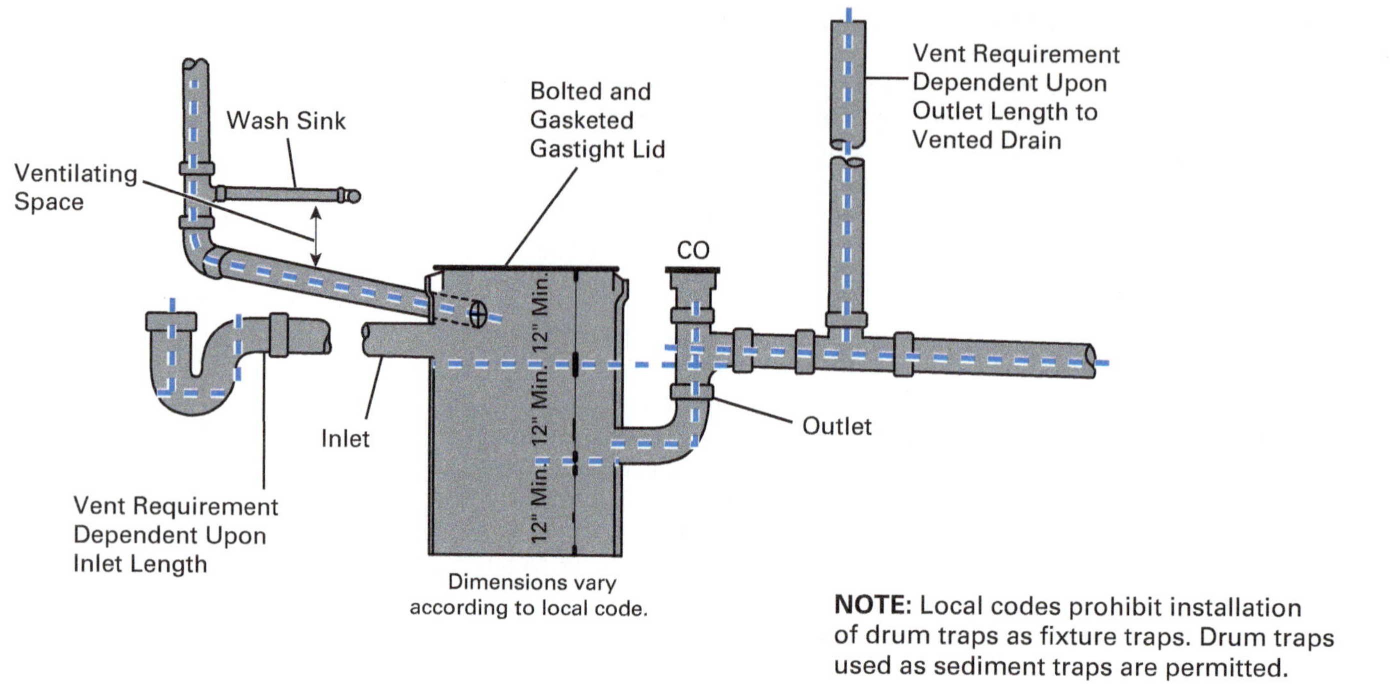

Figure 18 Example of a site-built sediment interceptor with typical information provided in design drawings.

Specially designed sediment interceptors can also be installed in a variety of other facilities, including the following:

- Car washes
- Parking garages
- Swimming pools
- Commercial laundries (*Figure 19*)
- Jewelry shops
- Computer-chip manufacturing plants

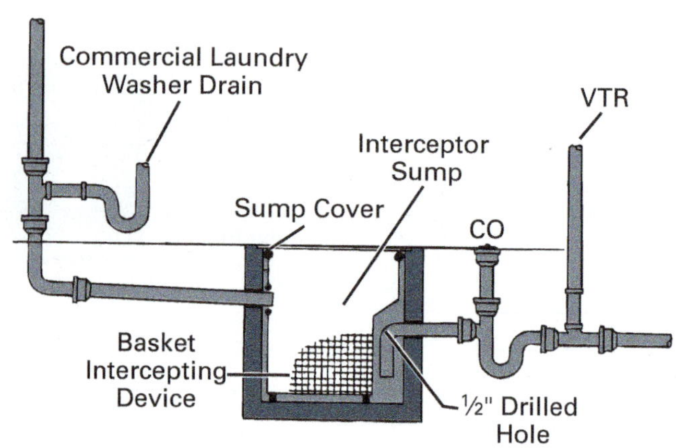

Figure 19 Typical laundry interceptor.

Each type of interceptor has its own special requirements. Before installing a special interceptor, be sure to review all applicable local codes, and seek qualified instruction if you are uncertain about installation.

On small sediment interceptors, the lower opening should be the inlet. Otherwise, the interceptor will not hold enough liquid to form a trap seal. In that case, sewer gas can escape, and special wastes can enter the sewer or septic system. Provide enough space for a drain pan to fit below the interceptor. This will prevent liquid waste from spilling when the interceptor is cleaned.

Did You Know?
Best Practices for Water Disposal

Special waste drainage systems must be designed and built according to strict standards to ensure that the environment is protected from accidental contamination and pollution. For example, a contractor installing a stormwater interceptor must be able to show how stormwater is cleaned, the types of special waste removed, and where the wastewater is drained. In effect, stormwater is treated as a hazardous material. Local codes require these systems to be documented, particularly those that dispose of sediment.

2.2.5 Catch Basins

As you have seen, sediment buckets installed in floor drains intercept sand and grit deposited on the floor. **Catch basins** perform a similar function, but on a larger scale. Catch basins are reservoirs that allow sediment to settle before wastewater discharges into the drain system. Install catch basins in parking lots, area drains, and yard drains. Catch basins can be customized to match installation needs. Review local codes for applicable requirements.

Catch basins are also used to cool high-temperature waste before it enters the waste system. Do not install catch basins to intercept sanitary sewage. Catch basins do not require traps if they drain into a storm sewer or drainage field. Ensure that cover plates on the catch basin are easy to access so that built-up sediment can be removed.

Catch basin: A reservoir installed before the sanitary piping. It allows sediment to settle out of wastewater.

CAUTION

Do not use catch basins or sediment interceptors to trap sanitary waste. Install traps on all interceptors that discharge into sanitary sewer systems. This will allow the interceptor to drain properly.

Think About It
Neutralization Sumps

Chemical waste systems are a type of special waste. Corrosive liquids, spent acids, and toxic chemicals pose serious health risks. They can't discharge directly into the drainage system. They must first be diluted, chemically treated, or neutralized. Neutralization sumps, or tanks, are used to treat these types of waste. Only then can these wastes discharge into the sewer. Consult your local code for specific requirements.

Some waste must pass through a neutralizing agent before it can be discharged. Neutralizing agents react chemically with special wastes. The reaction turns the waste into substances that can be discharged safely. One of the most commonly used agents is limestone.

Crushed limestone is used primarily in acid waste systems. Battery-charging stations, nearby emergency showers, and other highly acidic chemicals use special acid pipe as well as a limestone neutralizing pit to treat waste before it can enter the drain-waste-vent system. There are codes governing the disposal of corrosive and other harmful chemical wastewater; these codes also govern temperature of the wastewater before it may enter the drain-waste-vent system.

2.0.0 Section Review

1. The type of waste that can cause pipes to expand and contract is _____.
   a. corrosive waste
   b. high-temperature waste
   c. volatile waste
   d. solid waste

2. When installing an outlet vent or vacuum breaker on a mechanical grease interceptor, it should be sized at least _____.
   a. twice the diameter of the system inlet connection
   b. one-half the diameter of the system inlet connection
   c. twice the diameter of the system outlet connection
   d. one-half the diameter of the system outlet connection

Module 02404 Review Questions

1. Systems used to keep sewer wastes from contaminating fixtures if drains back up are known as _____.
   a. safety drain systems
   b. indirect waste systems
   c. sanitary drain systems
   d. isolated waste systems

2. In an air break, _____.
   a. a minimum 1" gap is required
   b. a larger pipe drains into a smaller pipe
   c. a partial vacuum is formed
   d. a smaller pipe drains into a larger pipe

3. An open-hub waste receptor must extend above the floor a distance of at least _____.
   a. 1"
   b. 3"
   c. 6"
   d. 12"

4. To prevent the escape of sewer gases from a low-use trap, install a(n) _____.
   a. adjustable-level valve
   b. continuous drip system
   c. automatic trap primer
   d. anti-venting sealer

5. A grease interceptor must be installed in the indirect waste line serving a _____.
   a. commercial dishwasher
   b. vegetable-preparation sink
   c. commercial clothes washer
   d. medical laboratory

6. A receptor is _____.
   a. any pipe or container into which an indirect waste line empties
   b. another name for a grease trap
   c. a type of strainer
   d. a floor, sink, standpipe, hub drain, or floor drain that receives the discharge of one or more indirect waste pipes

7. Treatment methods for indirect and special wastes _____.
   a. vary for each type of waste
   b. are typically identical from substance to substance
   c. are a gray area in the code
   d. are negligible, because they do no damage if applied incorrectly

8. A device or structure that slows wastewater flow so that waste materials can be separated by settling, floating, or evaporating is referred to as a(n) _____.
   a. isolator
   b. separator
   c. interceptor
   d. receptor

9. A draw-off hose is used to clean a _____.
   a. catch basin
   b. grease interceptor
   c. water filter
   d. sediment interceptor

10. The automatic cleaning cycle for a mechanical grease interceptor should be limited to a maximum of _____.
    a. 30 minutes
    b. 1 hour
    c. 90 minutes
    d. 2 hours

Answers to odd-numbered review questions are found in the Module Review Answer Key at the back of this book.

Answers to Section Review Questions

Answer	Section	Objective
Section One		
1. c	1.1.1	1a
Section Two		
1. b	2.1.0	2a
2. d	2.2.2	2b

MODULE 02405

Hydronic and Solar Heating Systems

Source: Courtesy of Apricus Solar Co., Ltd.

Objectives

Successful completion of this module prepares you to do the following:

1. Describe the principles of hydronic and solar heating systems.
   a. Describe water heaters, boilers, and collectors.
   b. Describe hot water circulation.
   c. Describe radiators and radiant loops.
2. Describe the basic types of hydronic and solar heating systems and their components.
   a. Describe the basic types of hydronic systems.
   b. Identify the basic types of solar heating systems.
3. Describe the procedures for roughing-in, installing, and testing hydronic and solar heating system piping.
   a. Describe how to rough-in the piping.
   b. Describe how to install boilers and heating units.
   c. Explain how to install controls.
   d. Explain how to test and balance the system.
   e. Describe corrosion prevention techniques.

Performance Task

Under supervision, you should be able to do the following:

1. Lay out a hydronic or solar heating system.

Overview

Hydronic systems use heated water or steam to heat rooms or an entire building through either radiant heat or forced heat. Most hydronic systems are powered by electricity or gas; some systems, called solar heating systems, are powered by the sun's heat. Hydronic and solar heating systems typically consist of a water heater or boiler (or, in solar heating systems, a collector), a circulation system, radiators, and controls. When installing hydronic or solar heating systems, plumbers first rough-in the piping; then install the heater, boiler or collector, and controls; and, finally, test, balance, and start the system.

CODE NOTE

Codes vary among jurisdictions. Because of the variations in code, consult the applicable code whenever regulations are in question. Referring to an incorrect set of codes can cause as much trouble as failing to reference codes altogether. Obtain, review, and familiarize yourself with your local adopted code.

Digital Resources for Plumbing

NCCER

SCAN ME

Scan this code using the camera on your phone or mobile device to view the digital resources related to this craft.

NCCER Industry Recognized Credentials

If you're training through an NCCER-accredited sponsor, you may be eligible for credentials from NCCER's Registry. The ID number for this module is 02405. Note that this module may have been used in other NCCER curricula and may apply to other level completions. Contact NCCER's Registry at 1.888.622.3720 or go to **www.nccer.org** for more information.

You can also show off your industry-recognized credentials online with NCCER's digital credentials. Transform your knowledge, skills, and achievements into credentials that you can share across social media platforms, send to your network, and add to your resume. For more information, visit **www.nccer.org**.

1.0.0 Principles of Hydronic and Solar Heating Systems

Performance Tasks

There are no Performance Tasks in this section.

Objective

Describe the principles of hydronic and solar heating systems.

a. Describe water heaters, boilers, and collectors.
b. Describe hot water circulation.
c. Describe radiators and radiant loops.

Hydronic system: A plumbing system that uses water or steam to heat a building. It is also called a radiant system.

Solar thermal heating system: A hydronic system that uses the sun to heat water. Also referred to simply as a solar heating system.

Radiant system: Another name for a hydronic system.

Forced heat system: A type of hydronic system that uses fans and ducts to circulate air heated by passing it over coils containing hot water.

Plumbing systems that use water or steam to heat a room or an entire building are called **hydronic systems**. While most hydronic systems use electricity or gas to heat the water in the system, **solar thermal heating systems** use the sun to heat water instead. Hydronic and solar heating systems can be divided into two broad categories: **radiant systems** and **forced heat systems**. Radiant hydronic systems transfer the heat from hot water to objects in a room, a principle called radiant heat. In forced heat systems, also called water-sourced heat pumps, blowers circulate air over coils containing hot water to transfer the heat from the coils to the air. The warmed air is then circulated throughout the building using air handlers and ducts, much like a conventional heating, ventilating, and air conditioning (HVAC) system. Forced heat systems are hybrid systems that are used mostly in large buildings.

Hydronic and solar heating systems offer many advantages over forced heat systems, because they do the following:

- Eliminate drafts, cold spots, and dry air
- Do not spread dust, mold, or airborne diseases
- Heat all the objects in a room, resulting in greater heat retention
- Operate cost-effectively

Hydronic and solar heating systems work somewhat like recirculation systems. The major difference is that hydronic and solar fixtures provide heat instead of water. You know that hot water recirculation systems must be insulated to prevent heat loss, but in hydronic and solar heating systems, heat loss at the correct locations is precisely the goal. The heat from the water in the system heats rooms in a building.

Specific heat: A measure of how much heat a liquid can hold per pound, measured in British thermal units (Btus).

Hydronic and solar heating systems use the heat given off by cooling water or steam to heat rooms. Heating occurs because water has a higher **specific heat** than almost any other liquid. Specific heat is a measure of how much heat a liquid can hold per pound. Specific heat is measured in British thermal units (Btus).

Did You Know?

Early Hot-Water and Steam Hydronic Systems

The earliest hydronic systems used steam to provide radiant heat. The famous Roman baths relied on steam heat, for example. By the mid-19th century, steam systems were in common use in homes and commercial buildings. Because many early steam systems operated under pressure, their boilers often exploded. Hot water systems were introduced into the United States from Canada in the late 1880s. Hot water was safer to use than steam, because hot water systems operated at normal air pressure. Hot water systems were also easier to maintain and operate. By the end of World War II, hot water replaced steam as the heating medium of choice.

You have learned that 1 Btu is the amount of energy required to raise 1 pound of water 1°F. When 1 pound of hot water cools from 190°F to 150°F, it gives off 40 Btus of heat. Steam is even more efficient than water. When 1 pound of steam condenses back into water, it gives off 970 Btus.

The components of a typical hydronic system are illustrated in *Figure 1.*

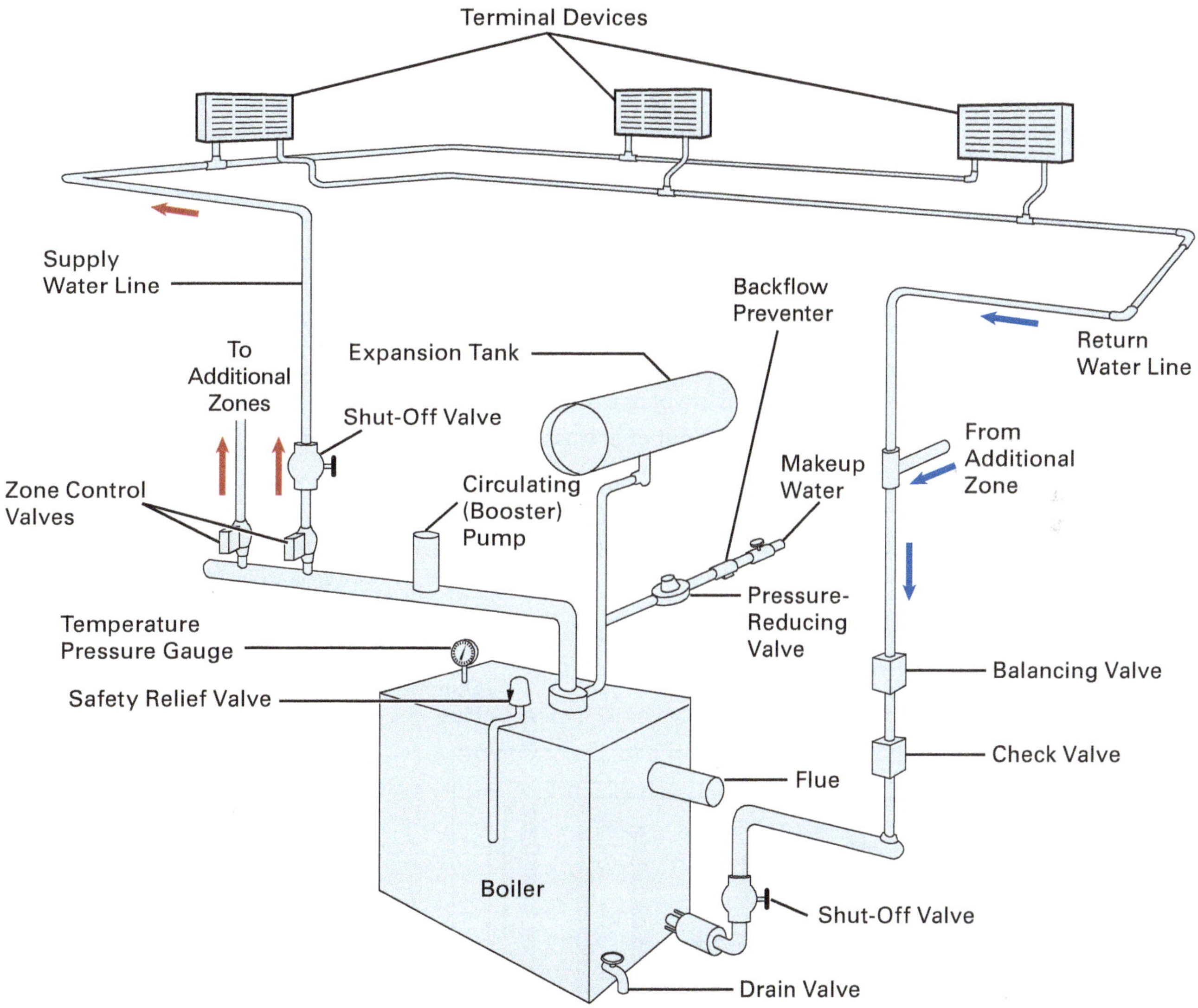

Figure 1 Basic components of a hydronic heating system.

A basic hydronic heating system must do the following:

- Create hot water or steam
- Distribute hot water or steam
- Transfer the heat from the hot water or steam into the room

Did You Know?

The Franklin Stove

In 1741, Benjamin Franklin invented a metal-lined stove that produced more heat and less smoke than other types of stoves in use at the time. The stove uses the principle of radiant heat to warm a room. Stoves based on Franklin's design are still widely used. The Franklin stove is considered the first radiant heat system.

1.1.0 Understanding How Water Heaters, Boilers, and Collectors Work

Most hydronic systems use potable water from the water supply system. The system is topped off as needed when water is lost through evaporation. Water in hydronic systems typically circulates at a temperature between 180°F and 210°F. Steam systems usually operate between 212°F and 220°F. When steam cools, it turns back into water; this water is called condensate. Condensate returns to the heating system and is reheated into steam. Steam systems can usually be found in houses that were built before World War II.

Some small hydronic systems use water heaters to supply the hot water, but most systems use a **boiler** to provide the hot water or steam. Boilers have a larger capacity than water heaters. They also generate more heat than water heaters. Often, the terms *water heater* and *boiler* are used to mean the same thing, but they have very different requirements and applications.

Boiler: In a hydronic system, a device used to heat or boil water. Boilers differ from typical water heaters in that they have a greater capacity, operate at higher temperatures and pressures, and have a greater heat output.

According to many state and local plumbing and mechanical codes, a boiler has the following characteristics:

- A capacity of at least 120 gallons
- Water temperature at or above 210°F
- Pressure of at least 150 pounds per square inch gauge (psig)
- Heat input of at least 200,000 Btus per hour

Another way to heat water is to use solar energy. In a solar heating system, water is pumped through a **collector**, where it is heated by sunlight. A typical collector is a narrow rectangular box with a metal plate, a water pipe, and a transparent cover (*Figure 2*). The collector is set up so that sunlight enters

Collector: A device used in a solar heating system to receive heat from sunlight and transfer it to water or another liquid.

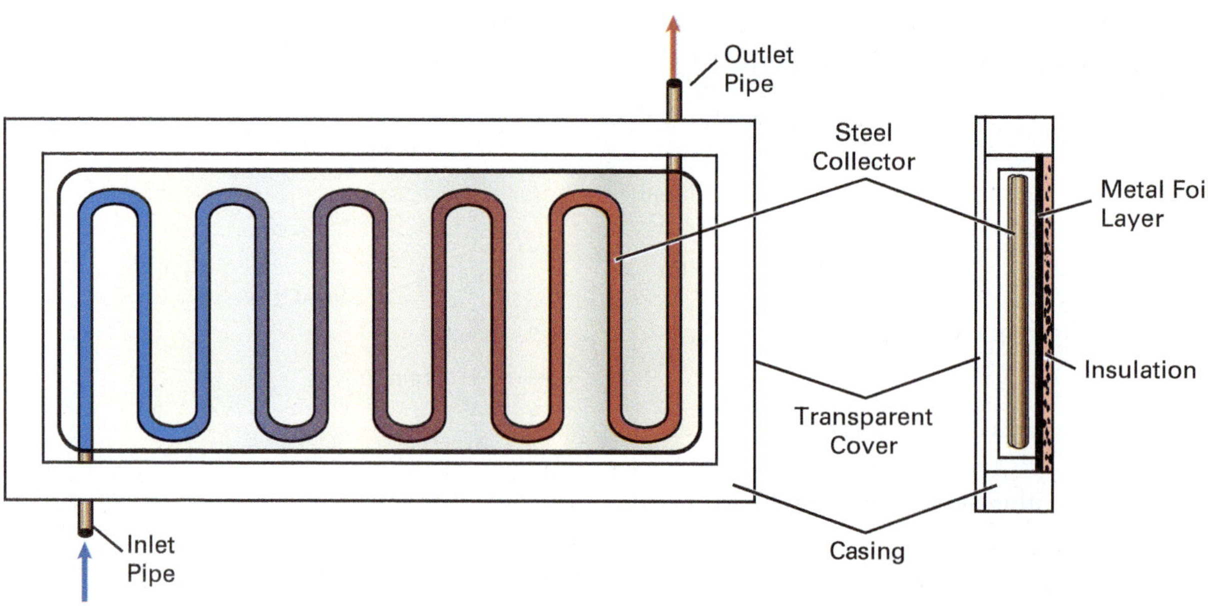

Figure 2 Typical solar collector.

through the transparent cover. Sunlight heats the metal plate inside the collector. The heat is transferred to the water pipe, which runs through the collector in a series of S-curves. The transparent cover acts as a reflector to keep the heat from radiating back out. The hot water is then pumped throughout the system. Some systems use a liquid other than water to absorb the sun's heat.

Hydronic and solar heating systems require the installation of an expansion tank. The tank absorbs excess pressure caused by hot water expansion. Expansion tanks prevent hot water from being wasted through the temperature and pressure (T/P) relief valve. The size of the expansion tank will vary based on the capacity of the system. Some hydronic systems may use a bladder-type expansion tank.

Going Green

Geothermal Energy

Heat that is created deep within Earth is called geothermal energy. Natural hot water sources, such as hot springs, are heated by geothermal energy. Humans, too, can tap geothermal energy to heat water, objects, and the air. It is an environmentally friendly source of energy, and it is potentially inexhaustible as well. Today, geothermal energy accounts for less than 0.1 percent of our energy use around the world; as the demand for alternative energy sources grows and technologies for extracting and using geothermal energy improve, that percentage will increase steadily.

1.2.0 Understanding How Hot Water Circulation Works

In many systems, once the water is heated, a pump circulates the water throughout the system. Centrifugal pumps are the most commonly used type of circulating pump. Use a stainless steel or bronze pump to handle potable water. These metals do not rust; rust will contaminate the water and may clog the pipes. Consult local codes for additional requirements about circulating pumps.

Did You Know?

Guidelines for Water Heaters and Boilers

In many jurisdictions, water heaters and boilers are governed by separate sets of guidelines. The model building codes published by the International Code Council (ICC) reflect this separate treatment. The *International Plumbing Code®* (IPC®) covers water heaters but not boilers, while the *International Mechanical Code®* (IMC®) covers boilers in some detail. Many state and local codes are based on the ICC codes.

Some jurisdictions also require a special permit for the installation of boilers.

Some hydronic and solar heating systems do not use circulating pumps. These systems rely on **convection** to circulate the water. Convection is circulation caused by the different densities of hot and cold water. Hot water rises, and cold water sinks. The system is designed so that the heavier cold water flows back to the tank or collector. This forces the hot water in the tank or collector to circulate.

Convection: Circulation caused by the sinking of dense cold water and the rising of hot water.

1.3.0 Understanding How Radiators and Radiant Loops Work

Sometimes when you pick up a cup of hot coffee, the cup is too hot to hold. That transfer of heat from the coffee cup to your hand is called **radiation**. Radiation is the discharge of heat through space from one body to another. Radiation does

Radiation: The transfer of heat between bodies through space.

not heat the air between bodies very much. Hydronic systems transfer the heat from hot water to objects in a room in the same way that the coffee cup warms your hand. Because hydronic systems use radiation to distribute heat, they are often called radiant systems.

Radiator: A heating unit that emits heat from water via a series of metal fins.

Radiant loop: A heating unit that consists of a network of hot water pipes in the floor or ceiling.

Hydronic systems use **radiators** and **radiant loops** to transfer heat from the hot water or steam into a room. Radiators emit heat from a series of metal fins. Water circulating through the hydronic system heats the fins (*Figure 3*). Objects in the room, including people, then absorb this heat. The fins also set up natural convection air currents, which draw cool air from the floor of the room, heat it as it passes between the fins, and discharge the heated air into the room.

Radiators, the most common method of heat transfer in a hydronic system, can be built into the baseboards of a room. They can also be installed as freestanding units (*Figure 4*). Most radiators use copper tubing and aluminum fins. Steel is sometimes used for pipe and fins. Radiators are the most common method of heat transfer in a hydronic system. They are also relatively simple to install and service.

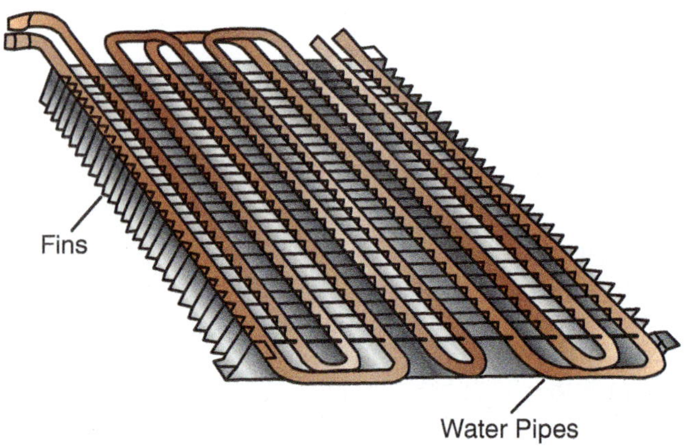

Figure 3 Components of a typical radiator.

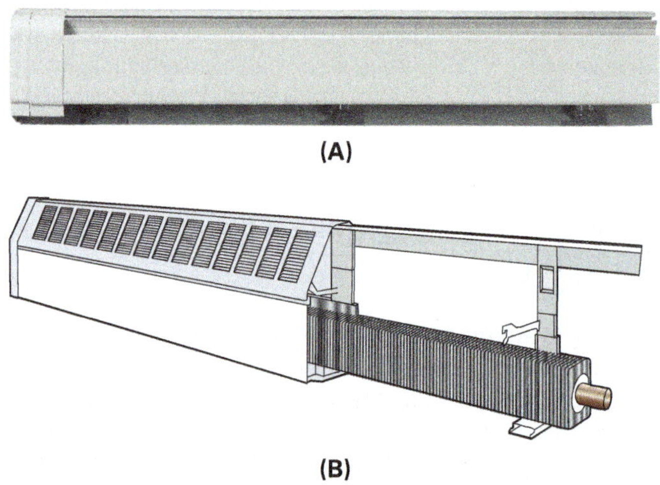

Figure 4 Examples of freestanding radiators.
Source: (A) Courtesy of Slant/Fin Baseboard.

A radiant loop is a network of hot water pipes embedded in the ceiling or floor (*Figure 5*). Radiant loops turn the floor or ceiling into a radiator. Pipe in a radiant loop can be made from a variety of materials. These include, but are not limited to, copper, CPVC (chlorinated polyvinyl chloride), polyethylene, and PEX (cross-linked polyethylene). Tubing for the loop can be laid in a concrete slab or hung underneath a wood floor. Because they are located in the floor or ceiling, radiant loops do not get in the way when the room is painted or cleaned.

Radiant Floor-Heating Installations

Thanks to improved piping materials—particularly PEX (cross-linked polyethylene), PEX-AL-PEX (a layer of aluminum between two layers of PEX), and O_2B Oxygen Barrier PEX (PEX tubing lined with an oxygen barrier)—and innovative pre-engineered solutions from manufacturers, radiant floor-heating installation is becoming easier and less costly. Some plumbing firms are even specializing in these installations. While radiant floor heating is still more expensive to install than forced heat, it offers superior comfort, energy efficiency, and long-term operating-cost savings. It also is quiet, clean, and easy to maintain, and it enables zoned temperature control—making it an increasingly frequent choice for custom homes and retrofits of existing homes, especially in the northern and western states. Radiant floor-heating systems are more complex to design and install compared to radiator systems, however. They also require more piping.

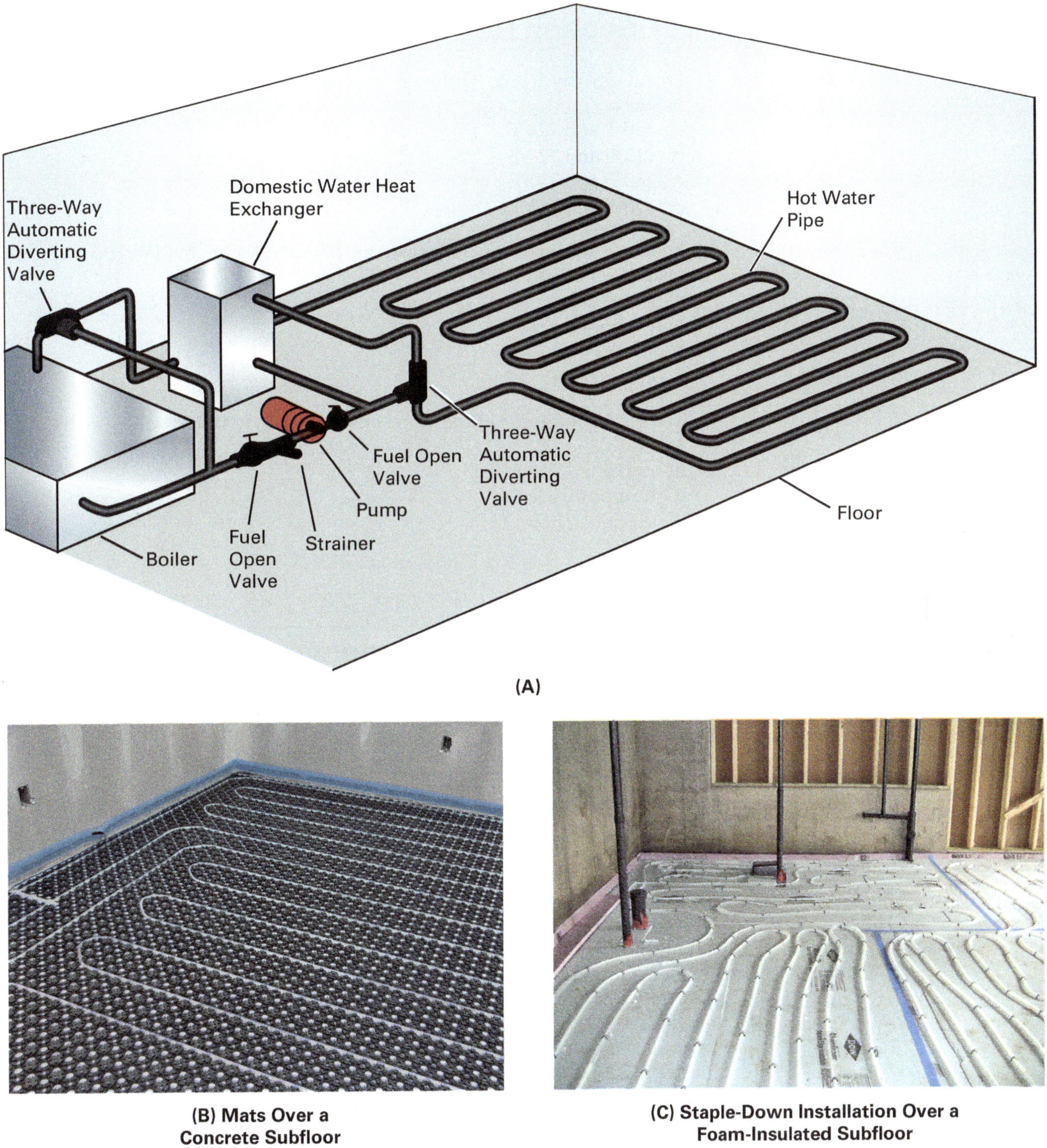

Figure 5 Radiant loops installed in floors.
Source: (B, C) Courtesy of Uponor

1.0.0 Section Review

1. Steam in hydronic systems typically circulates at a temperature between _____.
   a. 220°F and 228°F
   b. 212°F and 220°F
   c. 180°F and 210°F
   d. 150°F and 180°F

2. Hydronic and solar heating systems that do not use circulating pumps rely instead on the principle of _____.
   a. radiation
   b. evaporation
   c. convection
   d. conduction

3. The most common method of heat transfer in a hydronic system is the _____.
   a. fan
   b. radiator
   c. heat pump
   d. thermosiphon

2.0.0 Types and Components of Hydronic and Solar Heating Systems

Performance Task

There are no Performance Tasks in this section.

Objective

Describe the basic types of hydronic and solar heating systems and their components.

a. Describe the basic types of hydronic systems.
b. Identify the basic types of solar heating systems.

Hydronic and solar heating systems can be designed to meet the heating needs of almost any building. The same basic components can be assembled to heat a three-bedroom home or a multi-story office complex. Your skills as a plumber will allow you to design a system that suits both the type of building and the client's needs. In this section, you will learn about the different types of hydronic and solar heating systems. Each of these systems is designed for specific applications. Local codes specify the heating systems that can be installed in your area.

2.1.0 Types of Hydronic Systems

One-pipe system: A hydronic system in which water or steam is fed to the heating units and returned to the boiler through a single run of pipe.

Two-pipe system: A hydronic system that uses separate supply and return lines.

Some hydronic systems supply water or steam to the heating units and return the water to the boiler through a single run of pipe. They are called **one-pipe systems**. Other systems use separate supply and return lines. They are called **two-pipe systems**. Each of these systems can be built in two different ways, depending on the type of building and its heating needs. At first, these different types of systems may seem confusing. Take the time to review the different designs, and become familiar with them. Each has advantages for different applications. Discuss these systems with an experienced plumber to learn more about their strengths and weaknesses.

2.1.1 One-Pipe Systems

Heating units in a one-pipe system can be installed in either **series** or **parallel**. In a series arrangement, all the water in the loop goes through each heating unit (*Figure 6*). Series systems are simple to install and relatively inexpensive. However, series arrangements have several drawbacks. Because all the water flows through each heating unit, each unit must be sized to carry the total flow. The water temperature drops as the water flows through each heating unit. This means that heating units farther down the line cannot provide as much heat as units closer to the boiler. This type of hydronic system is practical only for the smallest applications.

Series: A one-pipe system in which all the water in the supply loop goes through each heating unit.

Parallel: A one-pipe system in which each heating unit draws only the amount of water it requires from the supply line.

In a parallel arrangement, not all the water in the loop flows through each heating unit. Each heating unit draws only what is required from the main supply (*Figure 7*). One-pipe systems with a parallel arrangement are also called **primary-secondary systems**. Parallel systems distribute heat more evenly throughout the building than series systems. Heating units can be sized to take less water.

Primary-secondary system: A one-pipe system with a parallel arrangement.

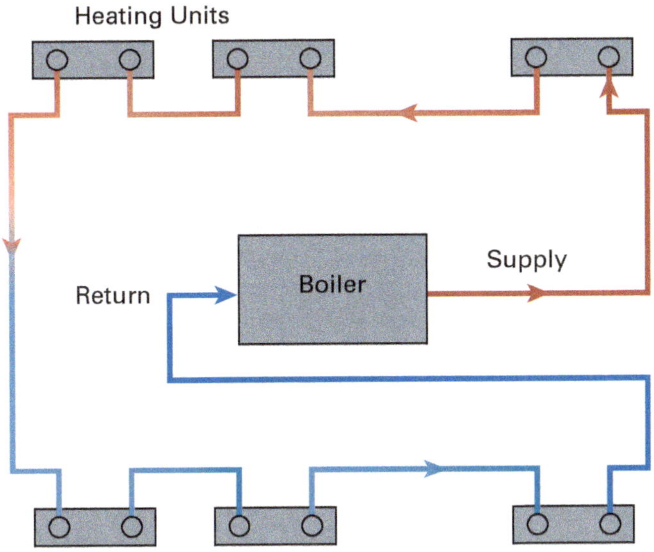

Figure 6 Series arrangement of heating units in a one-pipe hydronic system.

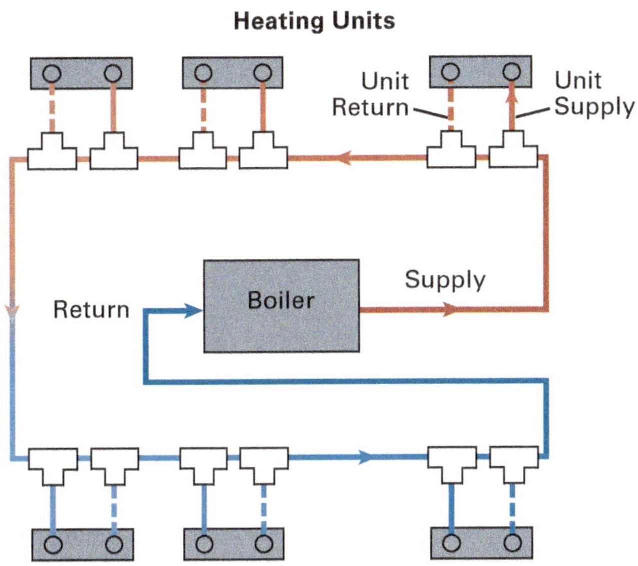

Figure 7 Parallel arrangement of heating units in a one-pipe hydronic system.

For one-pipe steam systems, supply pipes should be at least 2" in diameter. This allows steam to flow to the heating units while the condensate drains back to the boiler down the same pipe. This two-way flow often makes steam systems noisier than one-pipe water systems.

2.1.2 Two-Pipe Systems

Two-pipe systems have separate supply and return lines. This results in better temperature control than one-pipe systems. In a two-pipe system, the diameter of the supply pipe decreases after it passes each heating unit. Likewise, the return pipe's diameter increases after each unit. The heating units are sized only for their individual load. Because they require more pipe and fittings, two-pipe systems are more expensive to install.

There are two types of two-pipe systems. The difference has to do with the direction of the return flow. In a **direct return system**, the return pipe is installed parallel to the supply pipe. The return water or condensate flows back to the boiler in the opposite direction from the flow in the supply pipe (*Figure 8*). This means that the water from the first heating unit in the system is the first to return to the boiler. In a **reverse return system**, the return and supply pipes are also parallel. However, the return water flows in the same direction as the flow in the supply pipe (*Figure 9*). In a reverse return system, water from the last heating unit is the first to return to the boiler.

Direct return system: A two-pipe system in which the return water flows in the opposite direction from the flow in the supply pipe.

Reverse return system: A two-pipe system in which the return water flows in the same direction as the flow in the supply pipe.

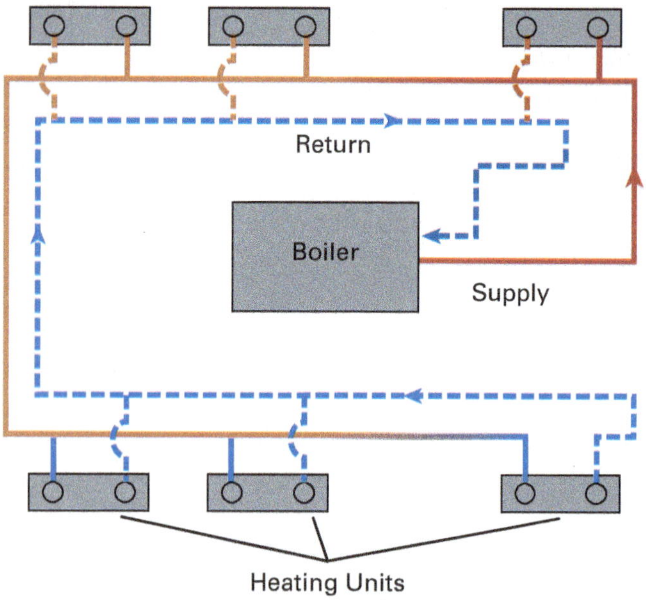

Figure 8 Direct return arrangement in a two-pipe system.

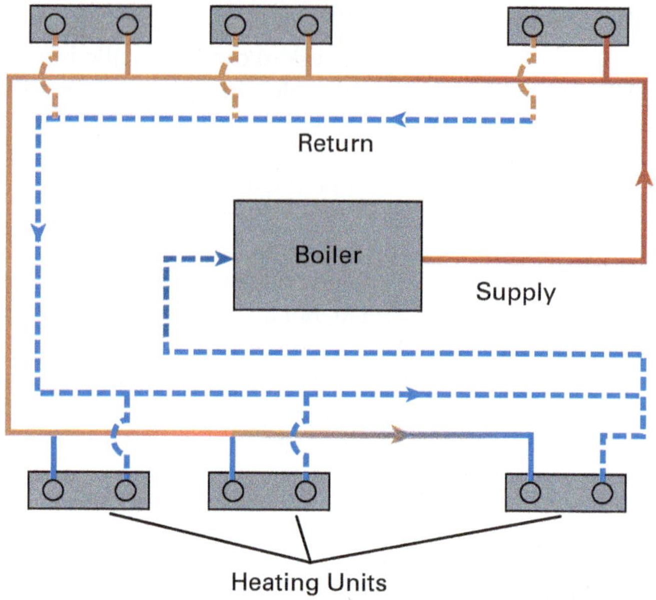

Figure 9 Reverse return arrangement in a two-pipe system.

2.2.0 Types of Solar Heating Systems

For buildings that receive direct sunlight year-round, solar power is a cost-effective heating method. In a solar heating system, the fuel—sunlight—is free. This can offset the added expense of installing special solar heating equipment. A boiler or water heater is often used to supplement solar heating so that a building can still be heated when there is little or no sunlight. Solar heating systems are available in many different designs. This module will focus on the systems you are most likely to encounter. Solar heating systems can be grouped into the following categories:

- **Passive systems**, which include **batch systems** and **thermosiphon systems**
- **Active systems**, which include **direct systems** and **indirect systems**

Passive systems rely on convection to move water through the system, whereas active systems use pumps and controls. Each of these systems is discussed in more detail in the following sections.

Passive system: A solar heating system that relies on convection to circulate water.

Batch system: A solar heating system in which water is both heated and stored in the collector. The system uses the principle of convection to circulate water.

Thermosiphon system: A passive system in which hot water is stored in a tank located higher than the solar collector.

Active system: A solar heating system that uses a pump and controls to circulate hot water through the system.

Direct system: A solar heating system that uses a pump and temperature controls to control the flow of hot water. It is also called an open loop system.

Indirect system: An active system that uses a liquid other than water to distribute heat from the collector. It is also called a closed loop system.

Know the Code™

Solar thermal equipment and appliances must conform to the requirements of the International Mechanical Code and ICC 900/SRCC 300. Solar thermal systems shall be listed and labeled in accordance with ICC 900/SRCC 300 and shall be installed in accordance with the manufacturer's instructions and ICC 900/SRCC 300.

2.2.1 Batch Systems

A batch system is the simplest way to use solar energy to provide heat. A batch system consists of a solar collector with one or more water tanks inside it. The collector heats the water in the tanks. When hot water is drawn, water flows from the collector tanks. Cold water then flows into the collector to make up for the hot water that was used. A conventional water heater supplements the collector supply as needed (*Figure 10*).

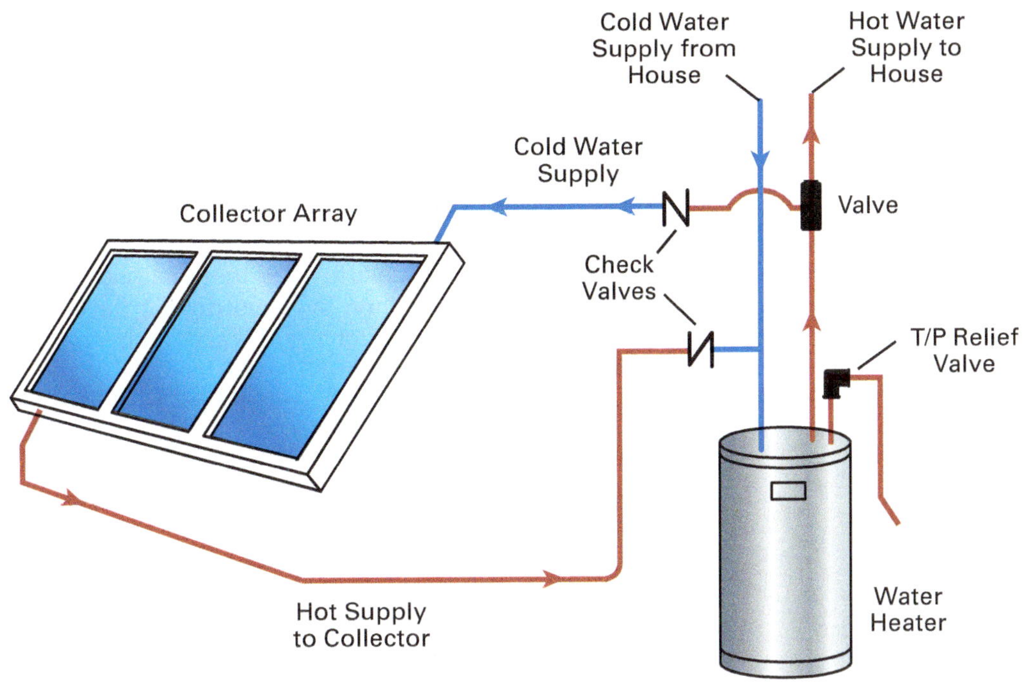

Figure 10 Typical batch system.

Batch systems are less effective than other solar heating systems for space heating. They can be built from supplies available at most hardware stores. Manufactured batch systems are also available. Because the water in the system does not move when there is no demand, batch systems are vulnerable to freezing. Ensure that pipes are insulated thoroughly, and install a cover on the collector to prevent heat loss after sunset.

2.2.2 Thermosiphon Systems

Thermosiphon systems also use convection to move hot water. Hot water is stored in a tank located higher than the solar collector (*Figure 11*). After water is heated in the collector, it flows into the top of the storage tank. As the water in the tank cools, it settles to the bottom of the tank. From there, the cool water flows back to the collector, where it is warmed again. This cycle is called **thermosiphoning**.

Thermosiphoning: In a thermosiphon system, the process whereby cold water flows back to the collector to be reheated.

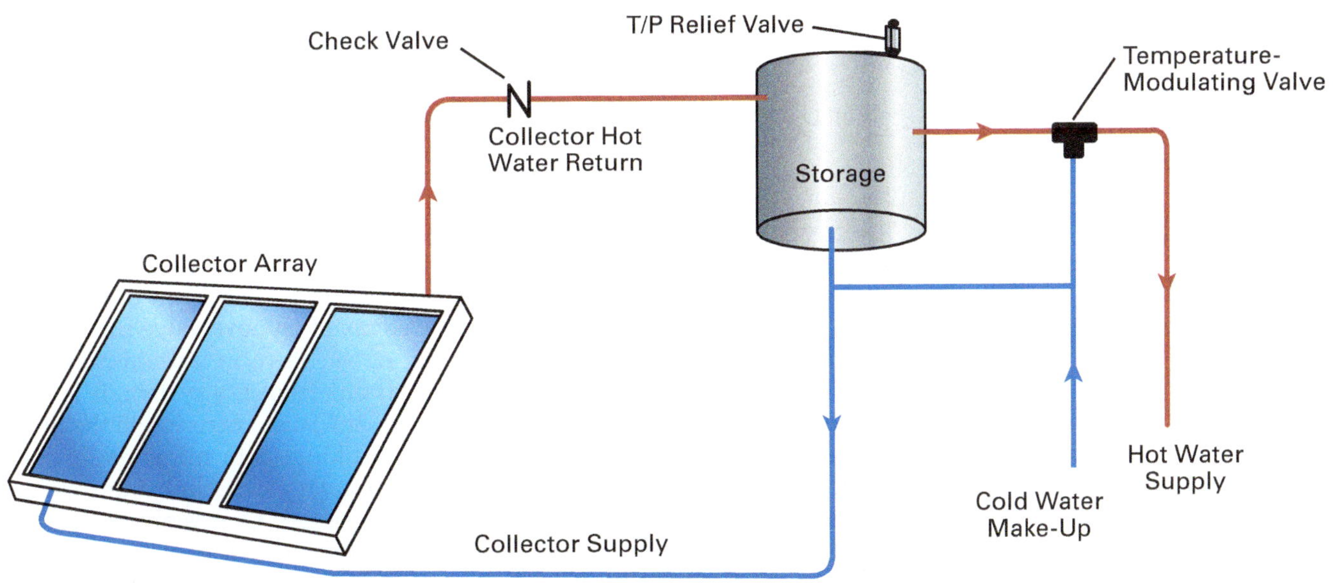

Figure 11 Typical thermosiphon system.

Reverse thermosiphoning: In a thermosiphon system, a condition in which water flows backward through the system.

Thermosiphon systems operate by the difference in density of the water in the storage tank and the water in the collector. Cool water in the storage tank and hot water in the collector means rapid flow. Hot water in the storage tank and cool water in the collector means little or no flow. This condition could even lead to **reverse thermosiphoning**. Reverse thermosiphoning means that water flows backward through the system. If this happens, the system will lose heat. Install check valves to prevent reverse thermosiphoning. Reverse thermosiphoning also can occur in other solar heating systems.

As with batch systems, ensure that thermosiphon systems are insulated against heat loss. Thermosiphon systems do not have temperature controls. Water temperature in the system will vary with changes in the weather. Thermosiphon systems do not provide 100 percent of the heating needs of a typical building. Backup heating is required for most installations.

2.2.3 Direct Systems

Open loop system: Another name for a direct system.

Like thermosiphon systems, direct systems have a collector and a storage tank. Direct systems use pumps to circulate the hot water (*Figure 12*). In a direct system, water temperature can be controlled. The pump turns on when the water in the collector reaches a preset temperature higher than the temperature of the water in the storage tank. Direct systems are simple to construct. They are also very efficient to operate. Direct systems are also called **open loop systems**.

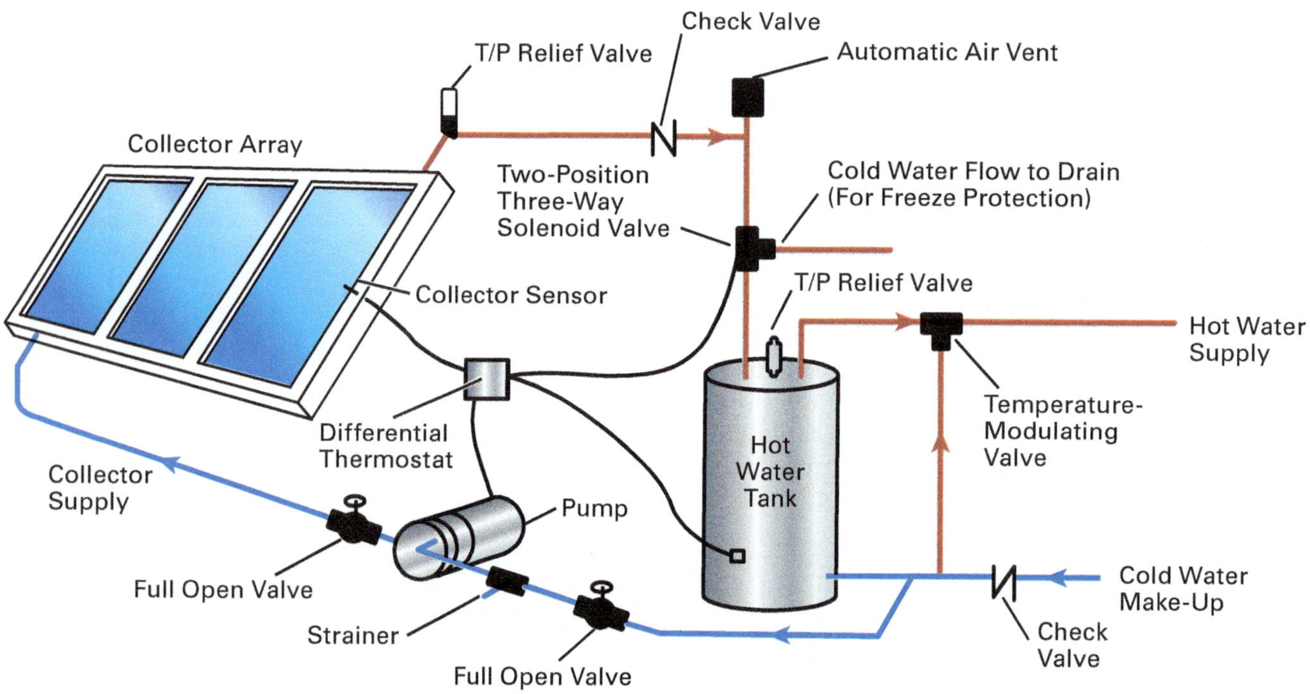

Figure 12 Typical direct system.

Drain-down system: A direct system in which water drains out of the collector when the pump shuts off.

In some direct systems, water drains out of the collector when the pump shuts off. The system will not freeze when it is not in use. Such systems are called **drain-down systems** (*Figure 13*). Drain-down systems often have valve problems if the valve is not used for a long time. Failure of the valve in a drain-down system could damage the collector. When water drains from the system, it is replaced with air. Repeated exposure to air and water can cause corrosion in the pipes.

Direct systems are vulnerable to scale buildup from minerals in domestic water. Many direct systems use aluminum collectors. Aluminum eventually corrodes when exposed to domestic water supplies. Corrosion can cause scale buildup in the pipes and other components exposed to the water. As a result of this buildup, the system cannot operate as efficiently as before. These problems limit the applications of direct systems.

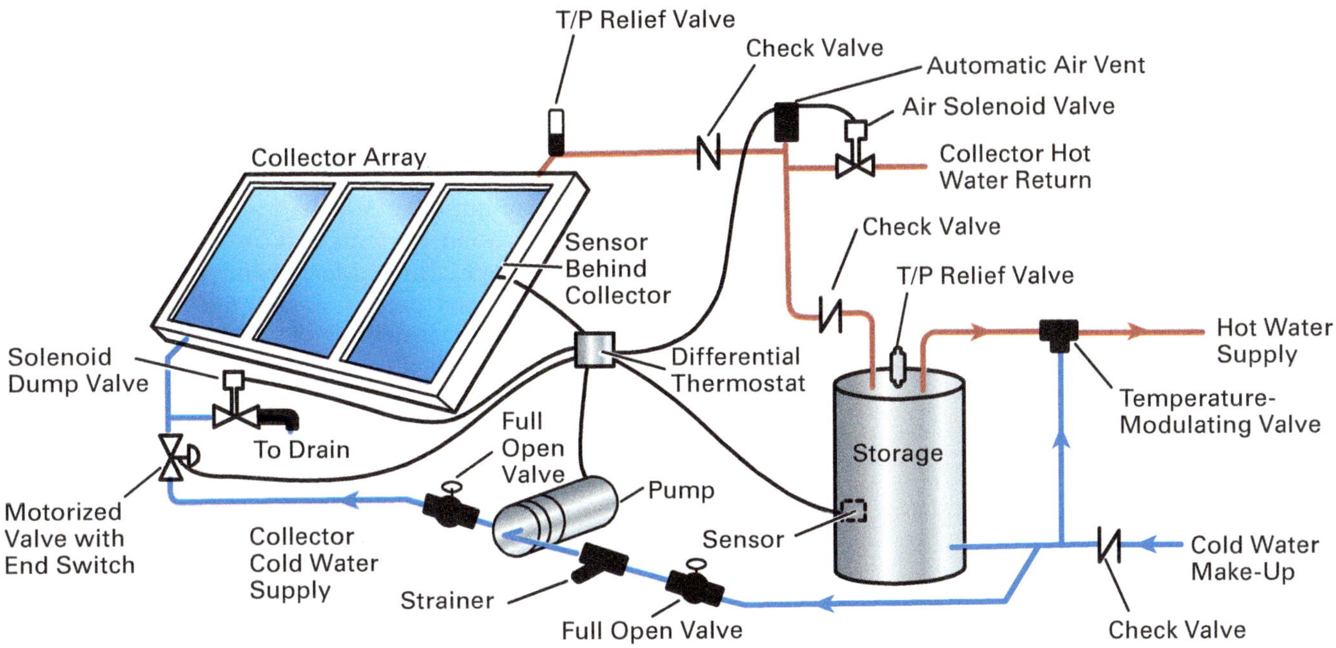

Figure 13 Typical drain-down system.

2.2.4 Indirect Systems

Batch, thermosiphon, and direct solar heating systems use water to circulate heat. Other systems can use fluids other than water, such as **ethylene glycol**. Ethylene glycol is an alcohol-based substance commonly used as antifreeze in car engines. Systems that use liquids other than water are called indirect systems. Indirect systems are often called **closed loop systems** because the heat-transfer liquid is kept separate from the water supply.

In an indirect system, the glycol or other liquid passes from the collector to a storage tank filled with water (*Figure 14*). As the heated liquid passes through the tank, it transfers its heat to the water with the aid of a **heat exchanger**, after which the heated water circulates through the heating system. A heat exchanger

Ethylene glycol: An alcohol-based liquid used in an indirect system to transfer heat from a collector to a heat exchanger.

Closed loop system: Another name for an indirect system.

Heat exchanger: A device used in an indirect system to transfer heat from one liquid to another in such a way that the two liquids do not come into contact with each other.

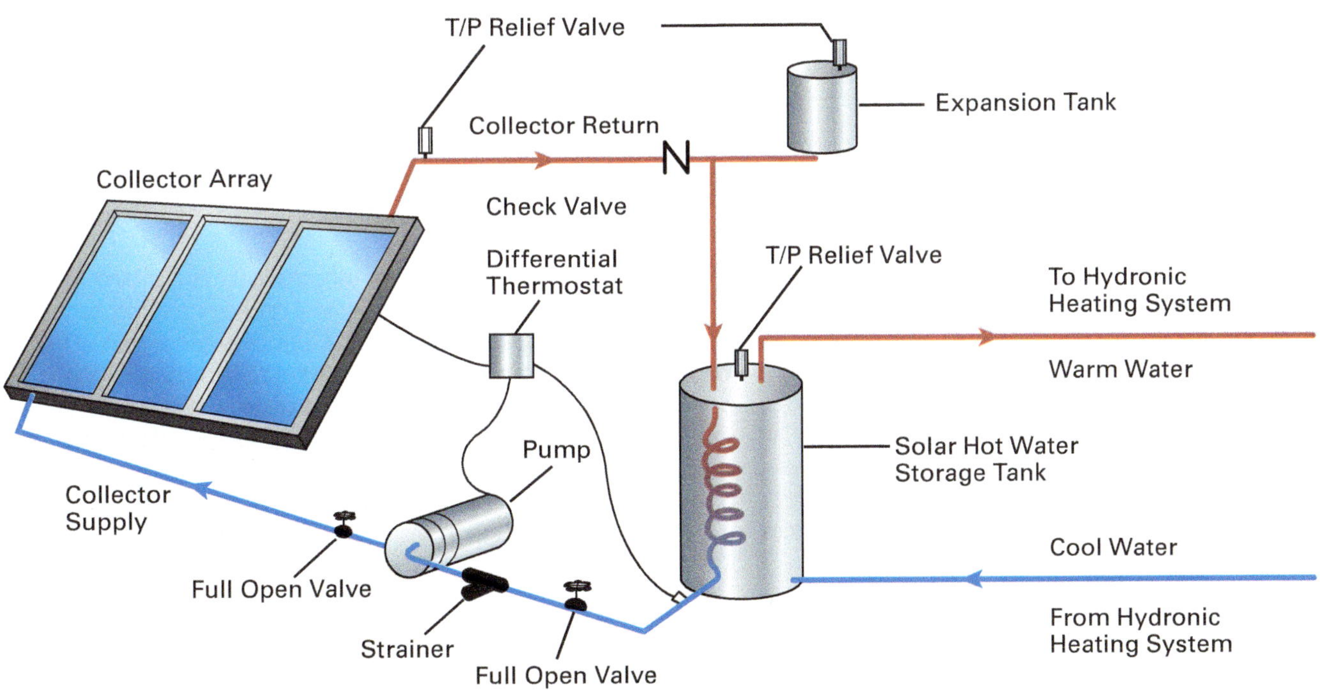

Figure 14 Typical indirect system.

is a device that allows two liquids to flow past each other in separate tubes. Heat is transferred from one liquid to the other through the tube walls without the liquids coming into contact. Most heat exchangers are designed with a double wall. The double wall ensures that the liquid in the collector loop does not mix with the water. Codes require this protection to prevent contamination of the potable water supply by toxic liquids such as ethylene glycol (*Figure 15*).

Drain-down systems that use liquids other than water in the collector loop are less likely to suffer from corrosion. Systems that use ethylene glycol will not freeze when exposed to cold temperatures. Most liquids have a lower specific heat than water. In other words, they carry less heat per pound of fluid than water. Using liquids with low specific heat can reduce the efficiency of the heating system.

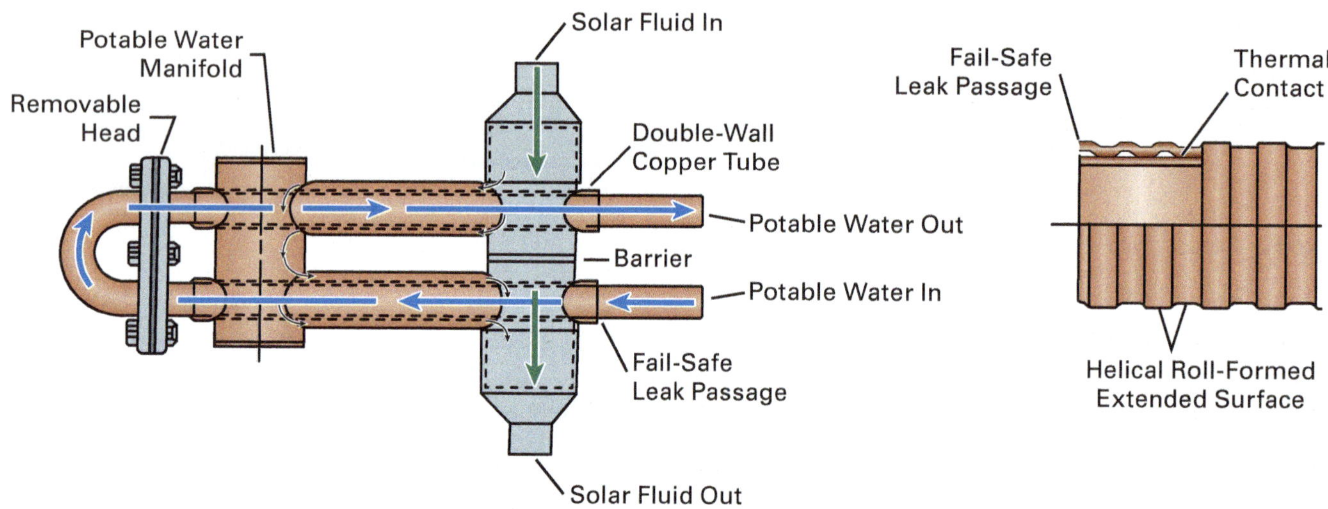

Figure 15 Double-walled heat exchanger.

WARNING!

Some indirect systems use single-wall heat exchangers, which are more efficient than double-wall versions. Many codes do not permit single-wall heat exchangers because they are prone to leaking toxic liquids into the system over time. Codes require that each line carrying nonpotable liquids be marked. Identify each outlet on a nonpotable water line.

Think About It

Making Solar Heating Cost-Effective

Solar heating systems often have high initial costs but lower operating costs than other types of heating systems. Operating costs can be reduced even further if other energy conservation steps are taken. Insulate piping, walls, and ceilings to reduce heat loss. Size the system to account for seasonal changes in demand and available sunlight. The client should not expect a solar heating system to provide 100 percent of the building's heating needs. Install a backup system for cloudy days.

Federal tax incentives for solar heating were eliminated in the mid-1980s. Many cities and states have followed suit. Some jurisdictions still provide rebates and tax incentives for homes that use solar heating systems. Ask an accountant about applicable local laws. Explain all the variables and options to a client before you install a solar heating system.

2.0.0 Section Review

1. Two-pipe systems offer better temperature control than one-pipe systems because they _____.
   a. require less insulation
   b. use two or more heat exchangers
   c. have separate supply and return lines
   d. allow for the installation of more radiators per line

2. After long periods of disuse, the devices in a drain-down system that will often experience problems are the _____.
   a. valves
   b. thermosiphons
   c. radiators
   d. heat exchangers

3.0.0 Roughing-In, Installing, and Testing Hydronic and Solar Heating System Piping

Objective

Describe the procedures for roughing-in, installing, and testing hydronic and solar heating system piping.

a. Describe how to rough-in the piping.
b. Describe how to install boilers and heating units.
c. Explain how to install controls.
d. Explain how to test and balance the system.
e. Describe corrosion prevention techniques.

Performance Task

1. Lay out a hydronic or solar heating system.

Like any other plumbing installation, hydronic and solar heating systems must be installed correctly to function. Systems are usually located before the drywall has been placed. Use standard professional techniques when installing hydronic and solar heating systems. Use approved piping materials and sizes. Locate the boiler, heating units, and piping to achieve the most efficient flow path. When hanging and supporting pipes, select the proper size of materials and use the correct tools.

The installation process can be broken down into the following four steps:

- Roughing-in the piping
- Installing the boiler
- Installing controls
- Starting, testing, and balancing the system

Note that these instructions are general guidelines only. Follow the manufacturers' instructions when installing boilers, pumps, and controls. Always refer to the local code when installing and testing hydronic and solar heating systems.

Did You Know?

Press-Fitting Technology

Solderless copper joints are an increasingly popular method for joining fittings and connections. Advantages include much faster prep and installation time, and reliable water- and gastight connections. Press fittings can be installed even with water present in the system. Press-fit connections are especially useful for maintenance and repair work, because they significantly reduce system downtime.

As you work on a heating system installation, you need to anticipate potential problems and provide ways to either prevent them or allow for quick repair. Locate shut-off valves behind an access panel, or in basements and crawl spaces. Do not run pipes along exterior walls or near air vents. Run pipes through the center of walls to avoid nail punctures. Some codes require the installation of backflow prevention devices (*Figure 16*) on hydronic heating system supply lines to protect against backflow from the heated part of the system. Refer to your local applicable code to determine whether backflow preventers are required. If so, install the appropriate backflow prevention devices according to the manufacturer's instructions.

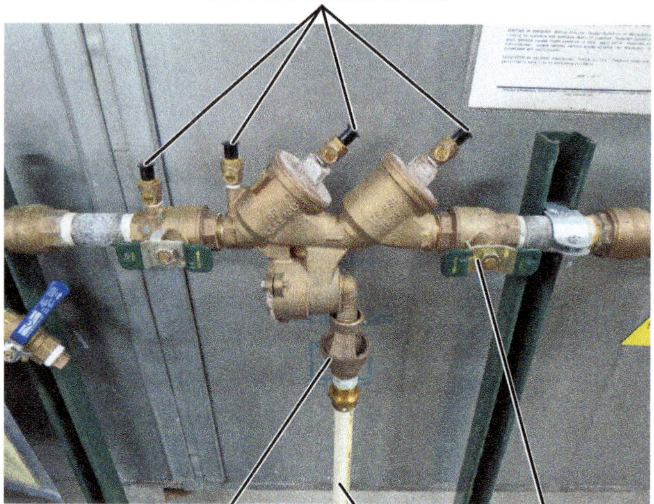

Figure 16 Backflow preventer installed on a supply line.

Think About It

Heating Unit Rough-In

You have already learned how to rough-in a fixture. Use the same technique to locate the heating units in a hydronic or solar heating system. Measurements are available in the manufacturers' catalogs. Catalogs are available from plumbing distributors or on the Internet. Note that critical dimensions may vary between styles of the same brand. Rough-in measurements tell you exactly where the piping should exit the walls or floors. You will use these dimensions to run the water supply piping and to attach the fixtures to the floor or wall.

3.1.0 Roughing-In the Piping

Begin by determining the type of heating unit to be used. If radiators are to be installed, lay out the location of each heating unit. Determine the distance between units and the distance of each unit from the finished wall. Rough-in the entry points for the inlet and outlet piping in the wall. For a radiant loop, determine the location of the supply and return lines in the floor or ceiling. Refer to the manufacturer's instructions for installation requirements. Consult your local code for materials and other guidelines.

For a hydronic system, begin by establishing the location of the boiler. Plan the pipe runs between the boiler and the heating units. Inspect the planned path of the pipe runs for obstacles, structural members, and other concerns. Determine the location of different heating zones within the building. Locate **zone control valves** at each of these zones. These are valves that allow the owner to set the temperature in various parts of the building. Next, determine

Zone control valve: In a hydronic system, a valve that allows the temperature in various parts of the building to be preset.

the most efficient flow path. Ensure that there are no overlaps. A typical hydronic system installation with boiler, water heater, control valves, and piping is shown in *Figure 17*.

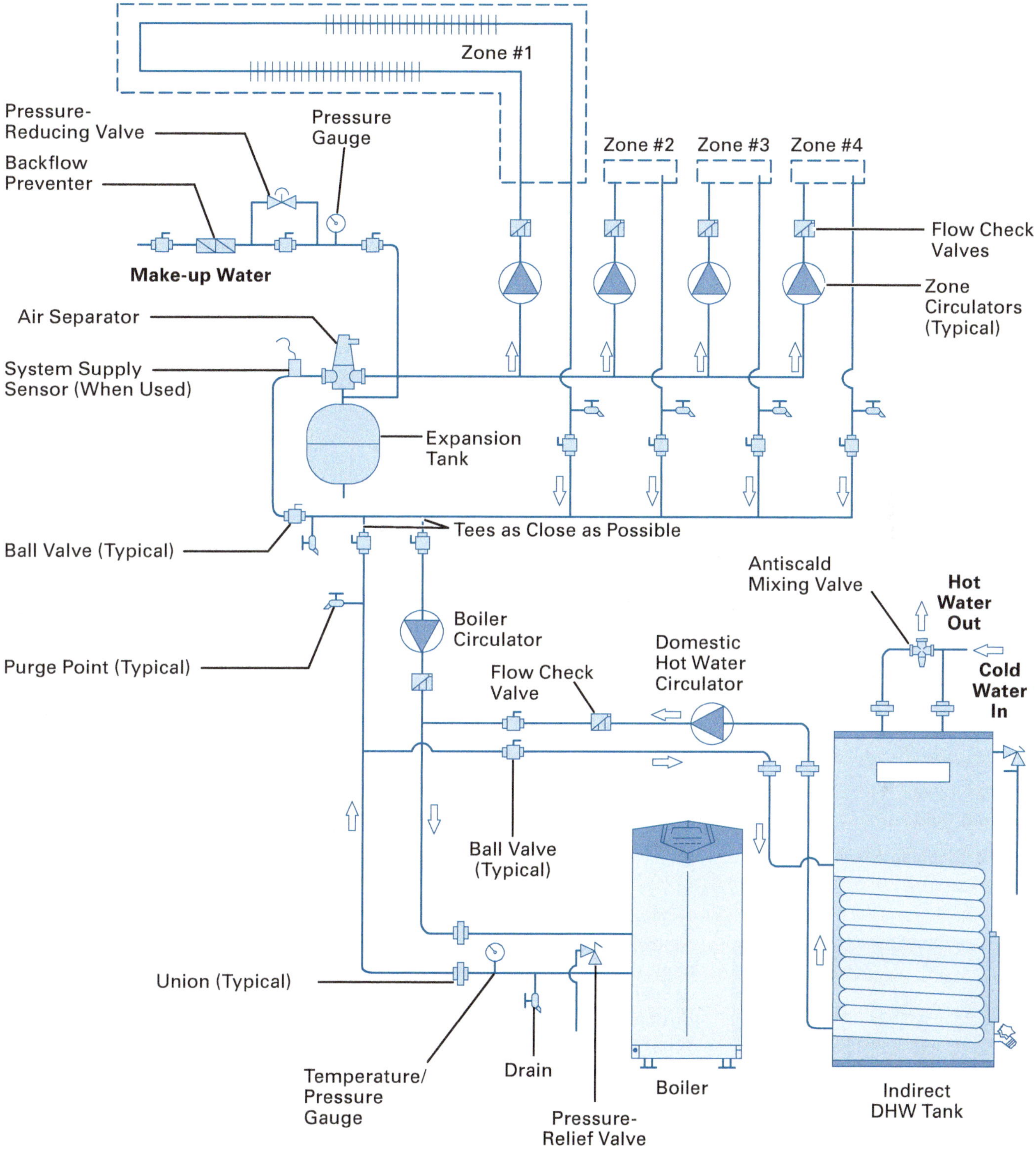

Figure 17 Boiler, water heater, valve, and piping installation for a typical hydronic system.

After you determine the layout of the supply and return piping, install the pipe runs. Use the skills and techniques you learned earlier in this curriculum to install the supply and return piping. Install the mains first, followed by the zone valves. Drill holes through the wall for the supply and return pipes. For pipe that is not insulated, drill the hole one size larger than the pipe's outside diameter. Place a plastic isolator on the pipe to reduce the noise of pipe expansion. Then run branches from the top of the mains to and from the radiation units.

Pressurized, or diaphragm, expansion tanks are also in common use in closed systems. They contain a flexible diaphragm, or bladder, that separates the system water from the air. *Figure 18* shows the action of the diaphragm as the water expands. It should be noted that tank pressure must be adjusted and checked with the tank isolated from the water system—tank and system pressure will change as thermal expansion takes place. Pressurized expansion tanks often come precharged with air, but the pressure can be adjusted through the charge valve, if needed, to fit system design conditions. Pressurized expansion tanks must be installed according to the manufacturer's instructions. This is because the piping between the system and tank differs from that used with a standard expansion tank. It is also important that any air in the system be purged to the atmosphere and not allowed to enter the tank(s).

Pressure-reducing valves are used in hydronic systems anywhere it is necessary to reduce a higher water or steam pressure to a lower one for input to a device. All systems have a feedwater pressure-reducing valve installed in the boiler water make-up line (*Figure 19*). This device automatically replenishes any water lost through leaks in the system. It also reduces the pressure of the cold water supplied from the water utility to a pressure suitable for use with the boiler. The valve maintains the water supplied to the boiler at a pressure less than that of the boiler relief valve. Feedwater pressure-reducing valves have a built-in strainer and a low-inlet-pressure check valve.

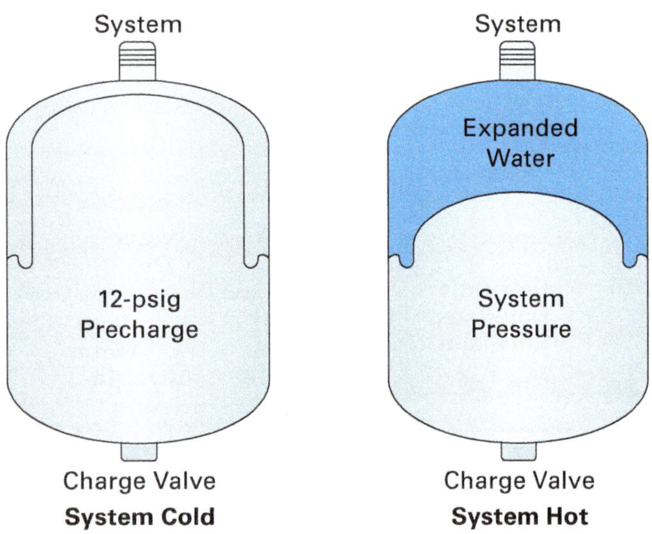

Figure 18 Diaphragm tank operation.

Figure 19 Feedwater pressure-reducing valve.
Source: Courtesy of Xylem

Figure 20 Pressure-relief valve.

The pressure-reducing valve typically used as a boiler feedwater line valve controls the output pressure and flow by keeping a balance between the pressure of an internal spring applied to one side of a diaphragm and the pressure of the delivered water applied to the other side of the diaphragm. It allows water to pass through the valve whenever the pressure at its outlet side drops below its pressure setting. The spring pressure is manually adjusted to set the desired output pressure level. Typically, it is factory-set at 12 pounds per square inch (psi), but it is adjustable between 10 and 25 psi. When adjusted, care must be taken to make sure the setting provides the needed pressure at all elevated points in the system. The model shown in *Figure 19* has a lever that can be pulled up, allowing the water to bypass the regulator and flow through quickly to fill an empty system.

A pressure-relief valve (*Figure 20*) is used to protect the boiler and the system from high pressures caused by either water thermal conditions or steam pressure conditions in the boiler. It does not operate unless an overpressure condition exists. The typical hot water boiler is constructed for a maximum working pressure of less than 30 psi, at which point the relief valve is designed to open fully. Note that this is significantly less than the average household domestic water pressure. As a result, water fed to the boiler is admitted through a pressure–reducing valve, set at an appropriate pressure to accommodate proper system operation while protecting the boiler from damage. The safety relief valve is set to open at the same pressure as the boiler's specified maximum operating pressure and discharge hot water via a waste pipe attached to its outlet port. Safety relief valves should always be installed as directed by the manufacturer and in conformance with the local applicable code.

Check valves (*Figure 21*) come in several types and are used to allow water or steam to flow through a pipe in one direction, but not in the other. The most common type of check valve has a swinging gate or flapper that swings open to allow flow in one direction but closes if the flow is reversed. This type of valve must be installed horizontally with the hinge at the top of the valve. Some check valves are spring-loaded to help in closing the flapper. Another type, called a lift check valve, works so that the flapper lifts off the seat to allow flow.

Check valves are used anywhere in a system where it is necessary to prevent the reverse flow of water or steam. For example, when installed in the circulating pump line of a hot water system, the check valve allows hot water to flow only when the circulating pump is running. When the pump is off, it prevents any backflow of water caused by gravity to be returned to the boiler. In hot-water and steam heating systems, check valves are used in the domestic water supply lines to the boiler. They are also commonly used in the condensate return lines of steam systems.

Figure 21 Check valve.
Source: Courtesy Watts Water Technologies, Inc

3.2.0 Installing the Boiler and Heating Units

For a hydronic system, install the boiler after you rough-in the heating units and install the piping. Locate the boiler according to the manufacturer's instructions and local code. Ensure that the boiler is plumb, level, and accessible. Allow room for access covers. For gas boilers, ensure that there is access for the flue and combustion air. For steam boilers, note that the steam supply line must be precisely located at or near the top of the boiler. The supply line feeds steam from the boiler's **steam dome** at the top of the boiler. The steam dome is the part of the boiler where steam collects after it is formed.

Steam dome: In a steam hydronic system, the place in the top of the boiler where steam collects.

Shim the boiler to a level position, and follow the manufacturer's instructions to hold the boiler in place. Then install the supply and return pipes using approved connectors. Connect the piping for the boiler feedwater, T/P relief valve, and expansion tank, if one is used. Ensure that the relief valve is piped to a drain. Install the pump according to the manufacturer's instructions. Install a purging valve on the boiler's return line.

After you install the boiler, install the heating units. Follow the manufacturers' instructions when installing heating units. Install air vents on each heating unit, and install the covers on the units. Connect each unit to the supply and return risers.

3.3.0 Installing Controls

Controls play an important role in hydronic and solar heating systems. They allow the plumber and the customer to adjust the system according to the temperature, weather, time of day, and many other factors. Be sure to install controls correctly. Controls should be properly rated for the system's temperatures and pressures. Ensure that valves and tees are facing the correct way. A backward valve or tee could interrupt flow and reduce the system's efficiency.

WARNING!

The maximum operating pressure of a pump is stamped on its nameplate. Do not exceed this pressure, or the system could fail. Failure not only could damage the system, but also could cause injury or death.

CAUTION

When installing a pump in a hydronic or solar heating system, install a strainer on the inlet side; this will protect the pump from damage caused by solids entering the pump. Install a check valve on the discharge side.

Refer to your local code for specific guidelines for installing controls. Some types of controls and materials may be prohibited by code. Controls for hydronic and solar heating systems are covered in the following sections. Many of them are already familiar to you. Others are used only in hydronic or solar heating systems.

3.3.1 Installing Hydronic System Controls

Airtrol valve: In a hydronic system, a valve that purges air from the system into the expansion tank.

Figure 22 Pressure regulator valve.
Source: Courtesy Watts Water Technologies, Inc

Diverter tee: In a one-pipe system, a tee that diverts the water flow by using an internal baffle to create a pressure drop.

Like many other plumbing installations, hydronic systems use various types of controls to ensure smooth operation. You are already familiar with many of these controls. Others are specially designed for use in hydronic systems.

A pressure gauge monitors the system's pressure. Use a T/P relief valve or safety valve to keep the system pressure below the boiler's maximum working pressure. Attach an **airtrol valve** to the expansion tank to purge air from the system into the expansion tank. Airtrol is short for air control.

Flow control valves are check valves that allow hot water to flow only when the pump is on. The valve prevents the system from draining when the pump is turned off. This helps maintain high temperatures throughout the system. Install a pressure regulator valve (*Figure 22*) in the cold-water line. Pressure regulator valves ensure that the water pressure in the system remains steady. The valve will protect the heating equipment from damage caused by a surge in pressure. Install the circulating pump on the boiler's hot water outlet. Size the pump to overcome pipe friction in the system. Review the pump's specifications to ensure that the pump can supply the correct amount of water at the proper temperature to each heating unit.

For one-pipe parallel systems, install a **diverter tee** (*Figure 23*) on each supply or return riser leading to the hot water supply pipe. Also refer to *Figure 7*, which shows heating units connected to a one-pipe system via diverter tees. Diverter tees, also called venturi tees, use an internal baffle to create a pressure drop. The pressure drop draws the water along the desired direction of flow. Diverter tees allow a single pipe to act as both a supply and a return main. Diverter tees are often referred to as monoflo tees, after the popular Monoflo® brand manufactured by Bell & Gossett (a Xylem brand).

Diverter tees can be installed on supply risers or return risers. Consult the heating-unit specifications for the proper location. Install a balancing valve (*Figure 24*) to set, balance, and control the flow rate. Balancing valves can be used to control flow for a single heating unit or an entire zone.

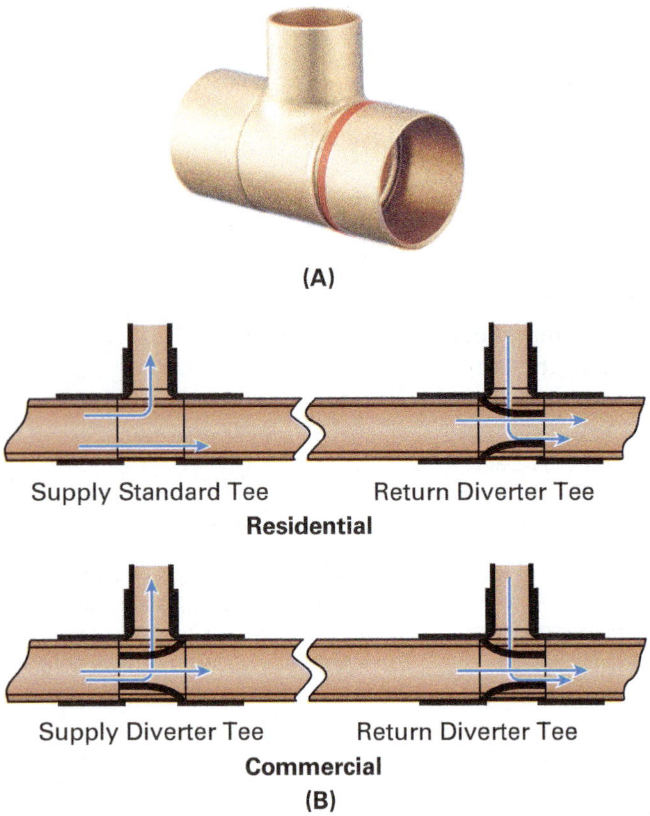

Figure 23 Diverter tees.
Source: Courtesy of Xylem Inc.

Figure 24 Balancing valve.

Proper Installation of Diverter Tees

Diverter tees allow a pipe to act as both a supply and a return line. If tees are not installed correctly, heat will not circulate efficiently throughout the system.

One of the most popular brands of diverter tees is the Monoflo® tee, invented in the early 1920s by the Bell & Gossett Company (now part of Xylem). Monoflo® tees have a red band on one side. Install the tee so that the side with the band faces the heating unit. This will ensure proper flow. If a one-pipe system is experiencing heating problems, ensure that the tees are installed correctly.

CAUTION

Despite a manufacturer's quality measures, controls may have defects. Be sure to examine controls for holes, cracks, or poor casting. Read and follow the manufacturer's installation instructions.

Some codes require the installation of reduced pressure zone principle (RPZ) backflow preventers on hydronic systems (*Figure 25*). RPZs protect the potable water supply in case of back pressure or back siphonage. Install the backflow preventer on the cold-water supply line before the pressure regulator valve. Install a low-water safety cutoff switch on the boiler's burner. The switch will turn off the burner if the water level falls below the boiler's heat exchangers. Ensure that all valves are installed correctly. Use accessories that are permitted by local code. Ensure that all controls are rated for the system.

Figure 25 A reduced pressure zone principle backflow preventer assembly.
Source: Courtesy of Zurn

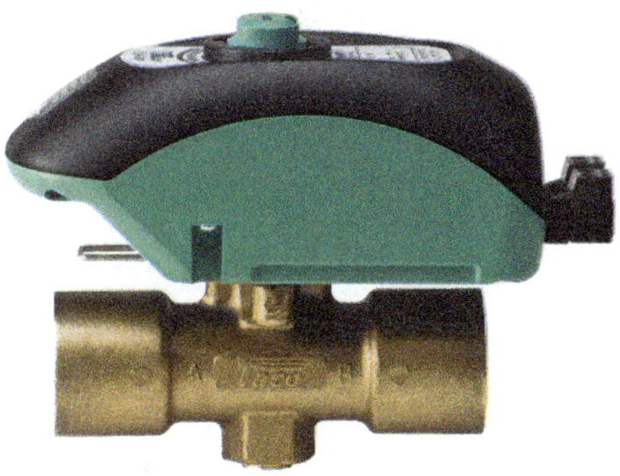

Figure 26 Zone control valve.
Source: Courtesy of Taco, Inc.

Zone control valves (*Figure 26*) manage the flow of water in each zone of a zoned system. They are usually two-position, thermostatically controlled valves. They can be either two-way or three-way valves. Zone control valves can be operated by heat or with an electric motor. In a heat-operated valve, a resistance wire around the valve heats the valve's bimetallic element when the zone thermostat calls for heat. This causes the bimetallic element to expand and slowly open the valve. When the zone thermostat is satisfied, the bimetallic element cools, allowing the valve to slowly close.

Electric-motor operated valves also operate to open and close the valve slowly, under control of the zone thermostat. They may contain a switch to operate the circulating pump. Slow opening and closing of zone control valves is necessary to reduce expansion noise and prevent water hammer. The time required for a typical zone control valve to open and close ranges from about 15 to 30 seconds. Some zone control valves have a feature that enables the valve to be opened manually in the event of a power failure. This feature is also useful when troubleshooting heating problems in a zoned system. Some zone control valves have a flow indicator dial that aids in system balancing.

Figure 27 Aquastat.

The aquastat (*Figure 27*) is a control that is used almost exclusively on boilers. This thermally operated switch is used to control hot-water boilers. An aquastat works basically the same way as a thermostat with the exception that it is designed to control water temperatures instead of air temperatures. Aquastats may have the sensing element directly or indirectly immersed in the water. Most use a well that is first placed into the boiler, and then the aquastat probe is inserted into the well. This gives the sensing element excellent exposure to the water temperature, while allowing it to be replaced without draining the water.

CAUTION

Use a tee to connect the house supply line to the pipe that runs between the boiler and the expansion tank. Install a globe valve, pressure-reducing valve, and check valve on the supply line before the tee. Follow the manufacturer's instructions or the local applicable code, whichever is stricter. Install a globe valve on the expansion tank.

3.3.2 Installing Solar Heating System Controls

Controls play an important part in solar heating systems. They maintain the system's temperatures and pressures. One of the most common controls is a **differential thermostat**. This device activates the pump when the water in the collector reaches a preset temperature above that of the water in the tank. As the temperature in the tank rises, so does the temperature at which the pump turns on. The differential thermostat improves the system's efficiency. For example, imagine that the differential thermostat in a solar hot water system is set to activate at a 15°F differential. When the temperature of the water in the collector is 15° hotter than the water in the tank, the pump will turn on.

Differential thermostat: In a solar heating system, a control that activates a pump when water in the collector reaches a specified temperature above that of the water in the storage tank.

Limit switches work much the same way as differential thermostats. They activate only when the water in the collector reaches a preset temperature. Using the example mentioned above, imagine that instead of turning on the pump when the water in the collector is 15° hotter than the water in the tank, the limit switch turns on the pump when the water reaches 140°F and shuts off when it reaches 125°F.

Limit switch: In a solar heating system, a switch that activates when the water in the collector reaches a preset temperature.

Limit switches control temperature and pressure more efficiently than differential thermostats. However, they can cause problems if they are not set properly. The upper and lower temperatures of a limit switch must be manually set. Limit switches cannot adjust to daily temperature fluctuations. Afternoon sunlight might warm the system above the switch's upper setting. The switch will then try to lower the temperature to the lower limit. As a result, the pump will be forced to run for a long time. The action will waste all the heat collected in the system. A new system must be carefully adjusted to keep pump cycling to a minimum and stored heat at a maximum.

NOTE

Refer to the local applicable code to identify whether plumbers or electricians are responsible for installing low-voltage wiring in solar heating systems.

Install a **thermistor** to control pump cycling. A thermistor adjusts a pump's speed as the water temperature rises and falls. The term *thermistor* is short for thermal resistor. A thermistor allows a pump to increase flow in proportion to a drop in temperature. As the temperature rises, the pump operates more slowly. As a result, the water temperature in the system remains steady throughout the day.

Thermistor: In a solar heating system, a control that adjusts a pump's speed according to changes in the water temperature.

Also, install check valves to prevent reverse thermosiphoning of hot water. Ensure that they are installed along the direction of flow. Install a T/P relief valve in the hot water loop, especially on a closed system, and install a T/P valve near the expansion tank.

3.4.0 Testing and Balancing the System

After you install the piping, boiler, heating units, and controls, test the system. Begin by charging the system with water through the filler inlet. Connect a garden hose to the system drain outlet. Run the hose to a laundry sink or other container. Then purge the entire system through the relief valve. Open the boiler's purge valve to drain the boiler. Do not operate the pump or boiler during the purge. Allow the system to purge all water and air before you shut off the filler inlet.

When the system has been completely purged through the relief valve, open the pressure regulator valve and the boiler's return line. Allow the boiler to fill. Then open the boiler's supply line. Allow a small amount of air to flow through the expansion tank. Then turn on the pump and fire the boiler. Ensure that all valves are working correctly. Set the thermostat and temperature controls.

WARNING!

Hot water under pressure can escape with explosive force. It can also cause dangerous scalding. At 125°F, it takes about 1 1/2 to 2 minutes for hot water to cause scalding. At 155°F, it takes only about 1 second. Wear appropriate personal protective equipment. Review and follow all construction drawings.

Allow air to bleed out from the highest points in the system. This step may require several cycles of purging and venting. When the system operates quietly, the air has been completely bled from the system. Finally, install all covers. Balance all zones using the zone control valves.

3.5.0 Preventing Corrosion in Hydronic and Solar Heating Systems

Once a heating system has been installed, tested, and balanced, it will operate reliably for years. Take the time to ensure that the components of the heating system are protected against corrosion.

The most common type of deterioration in a hydronic or solar heating system is galvanic corrosion. Galvanic corrosion can occur when two different types of metal are joined. Contact between the two metals can create a small electrical current. This current causes metal ions to flow from one pipe to the other. If the pipe is filled with mineral-rich water, the process will be accelerated. *Figure 28* shows the galvanic scale.

There are several ways to reduce or eliminate the threat of galvanic corrosion in a heating system:

- Use pipes that have chemically similar metals. Similar metals will reduce the strength of the electrical current between the two pipes.
- Join pipes using dielectric fittings.
- Install a sacrificial anode made of magnesium (*Figure 29*). Magnesium has a high positive charge. It will attract the electrical current in the system's pipes. Sacrificial anodes must be replaced periodically.
- Apply an **anodic inhibitor** to the pipes. Anodic inhibitors are chemicals that prevent corrosion. Follow the manufacturer's directions when you apply an anodic inhibitor.

Anodic inhibitor: A chemical applied to a pipe to prevent galvanic corrosion.

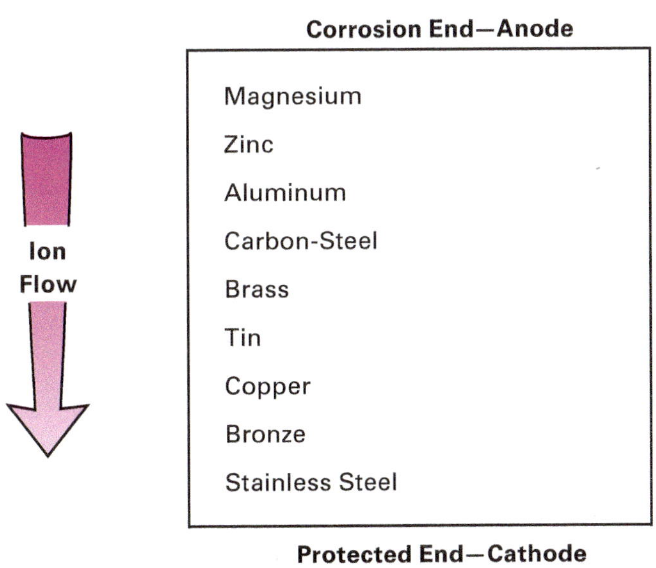

Figure 28 The galvanic scale.

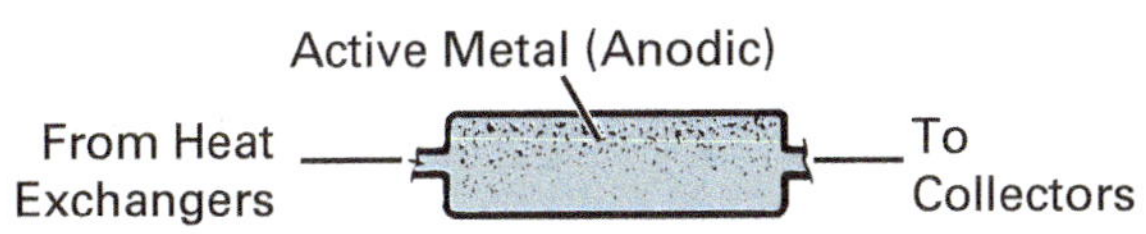

Figure 29 Magnesium sacrificial anode.

CAUTION

Antifreeze liquids are extremely toxic in any form. When used in closed systems, they can cause corrosion if they chemically decompose. A collector will reach 400°F if a pump breaks down. Ethylene glycol will break down if it is heated to that temperature. The resulting chemicals—glycolic acid and oxalic acid—are highly corrosive. Periodic maintenance will ensure that the system is operating correctly and that there is no corrosion damage.

Removing Air from a Hydronic System

When installing pipes, pitch them in the direction of flow. This will allow the air to bleed out of the system faster. Inspect the pitch of existing systems to ensure that pipes have not sagged over time.

Pump placement can also affect the way air escapes from the system. Install the pump on the supply line upstream of the expansion tank. This allows the pump to add a small amount of pressure to the line. This pressure will help air dissolve into the water. If air remains in the system after startup, raise the static fill pressure. The increased pressure will cause the air to dissolve into the water. Once the system begins to operate, lower the pressure to normal. Normal pressure will ensure that the relief valve does not open.

3.0.0 Section Review

1. Pressurized expansion tanks often come precharged with _____.
   a. helium
   b. water
   c. nitrogen
   d. air

2. When installing a boiler in a hydronic system, the return line is fitted with _____.
   a. a purging valve
   b. a T/P relief valve
   c. high-temperature adhesive
   d. $^3/_4$" noncorroding straps

3. A device that controls temperature and pressure more efficiently than a differential thermostat is the _____.
   a. thermistor
   b. thermosiphon
   c. aquastat
   d. limit switch

4. The first step in testing a hydronic or solar heating system is to _____.
   a. charge the system with water through the filler inlet
   b. purge the entire system through the relief valve
   c. connect a garden hose to the system drain outlet
   d. open the boiler's purge valve to drain the boiler

5. The most common type of deterioration in a hydronic or solar heating system is _____.
   a. oxidation
   b. galvanic corrosion
   c. precipitation of salts
   d. backflow

Module 02405 Review Questions

1. Hydronic and solar thermal heating systems can by classified as radiant systems or _____.
   a. gravity systems
   b. forced heat systems
   c. heat-sink systems
   d. blower-assisted systems

2. When 1 pound of steam condenses back to water, it gives off _____.
   a. 40 Btus
   b. 520 Btus
   c. 970 Btus
   d. 1,240 Btus

3. Steam heating systems are usually found in houses built _____.
   a. before World War I
   b. after 1950
   c. before World War II
   d. after 1985

4. To circulate potable water in a hydronic heating system, a pump should be made of stainless steel or _____.
   a. aluminum
   b. brass
   c. cast iron
   d. bronze

5. In a hydronic system, heat is emitted from a series of metal _____.
   a. louvers
   b. fins
   c. baffles
   d. dispersers

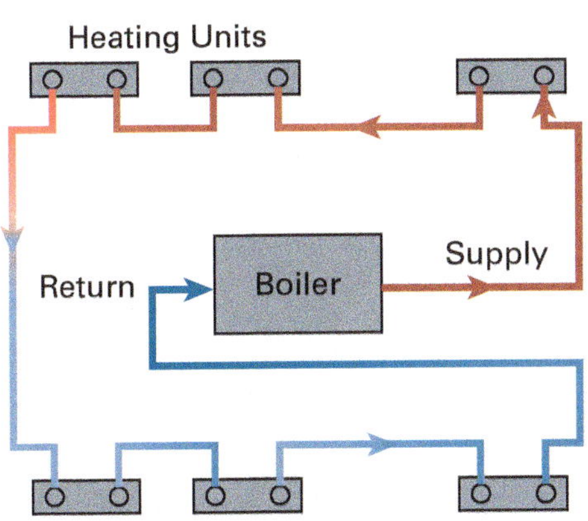

Figure RQ01

6. The one-pipe hydronic system illustrated in *Figure RQ01* has a _____.
   a. series arrangement
   b. primary-secondary arrangement
   c. parallel arrangement
   d. closed loop arrangement

7. A passive solar installation moves water through the system by means of _____.
   a. gravity
   b. conduction
   c. convection
   d. radiation

8. The simplest type of solar heating system is the _____.
   a. batch system
   b. direct system
   c. thermosiphon system
   d. indirect system

9. Water in a batch solar heating system does not move when there is no demand, making it vulnerable to _____.
   a. algae and mold growth
   b. evaporation loss
   c. stagnation
   d. freezing

10. Open loop systems are also called _____.
    a. direct systems
    b. batch systems
    c. indirect systems
    d. thermosiphon systems

11. Most heat exchangers are designed with a double wall to _____.
    a. prevent reverse thermosiphoning
    b. improve efficiency of heat transfer
    c. prevent contamination of potable water by toxic liquids
    d. increase flow rates through the system

12. When installing a hydronic system, begin by _____.
    a. identifying locations for the zone control valves
    b. establishing the location of the boiler
    c. planning the pipe runs between system components
    d. determining the most efficient flow path

13. In a pressurized expansion tank, system water is kept separate from the air by a(n) _____.
    a. bladder
    b. impermeable membrane
    c. bulkhead
    d. inflatable barrier

14. Check valves that allow hot water to flow only when the pump is operating are called _____.
    a. directional control valves
    b. diverter valves
    c. flow control valves
    d. safety valves

15. A chemical that can be applied to solar or hydronic system pipes to prevent corrosion is called a(n) _____.
    a. cathodic protector
    b. neutralizer coating
    c. antigalvanic solution
    d. anodic inhibitor

Answers to odd-numbered review questions are found in the Module Review Answer Key at the back of this book.

Answers to Section Review Questions

Answer	Section	Objective
Section One		
1. b	1.1.0	1a
2. c	1.2.0	1b
3. b	1.3.0	1c
Section Two		
1. c	2.1.2	2a
2. a	2.2.3	2b
Section Three		
1. d	3.1.0	3a
2. a	3.2.0	3b
3. d	3.3.2	3c
4. a	3.4.0	3d
5. b	3.5.0	3e

MODULE 02406

Codes

Objectives

Successful completion of this module prepares you to do the following:

1. Describe the model and local plumbing codes and their purposes.
   a. Describe the ICC model plumbing code.
   b. Describe the IAPMO model plumbing code.
2. Explain how plumbing codes are developed and revised.
   a. Explain model code standards.
   b. Explain how model codes are revised and adopted.
   c. Explain typical code changes.

Performance Task

Under supervision, you should be able to do the following:

1. Use the local applicable plumbing code to find and cite references for the questions in *Appendix 02406A*.

Overview

Codes and standards are adopted to protect public health and safety and to ensure that contractors perform to the minimum requirements prescribed by codes and standards. Many jurisdictions develop their codes from templates called model codes. Professional organizations work closely with government agencies, administrators, and industry professionals to develop and revise model codes. Codes must change over time to include new materials and technologies, and to reflect changing standards and practices. Model code-development cycles occur every 3 years, at which time proposed revisions to the codes are submitted and voted on during the development process. Jurisdictions revise model codes to meet local requirements for items such as frost depth, snow loads, wind speeds, and seismic zones. They may incorporate other requirements relevant to local industry practices. Professional plumbers must keep up with changes to the local code and model codes.

CODE NOTE

Codes vary among jurisdictions. Because of the variations in code, consult the applicable code whenever regulations are in question. Referring to an incorrect set of codes can cause as much trouble as failing to reference codes altogether. Obtain, review, and familiarize yourself with your local adopted code.

NCCER Industry-Recognized Credentials

If you are training through an NCCER-accredited sponsor, you may be eligible for credentials from NCCER. The ID number for this module is 02406. Note that this module may have been used in other NCCER curricula and may apply to other level completions. Contact NCCER at 1.888.622.3720 or go to **www.nccer.org** for more information.

You can also show off your industry-recognized credentials online with NCCER's digital credentials. Transform your knowledge, skills, and achievements into digital credentials that you can share across social media platforms, send to your network, and add to your resume. For more information, visit **www.nccer.org**.

Digital Resources for Plumbing

SCAN ME

Scan this code using the camera on your phone or mobile device to view the digital resources related to this craft.

1.0.0 Model and Local Plumbing Codes

Performance Task

1. Use the local applicable plumbing code to find and cite references for the questions in *Appendix 02406A*.

Objective

Describe the model and local plumbing codes and their purposes.

a. Describe the ICC model plumbing code.
b. Describe the IAPMO model plumbing code.

Code: A legal document enacted to protect the public and property, establishing the minimum standards for materials, practices, and installations.

Model code: A set of comprehensive, general guidelines that establishes and defines acceptable plumbing practices and materials, and lists prohibited installations.

All construction work is governed by **codes**. A code is a legal document adopted in a jurisdiction that establishes the minimum acceptable standards, rules, and regulations for all materials, practices, and installations used in buildings and building systems. In this module, you will learn about the different types of codes that affect plumbing installations, materials, and processes. You will also learn how codes are modified.

There are thousands of local codes in existence in the United States, and the potential total number of city, county, and state jurisdictions that could adopt codes is in the tens of thousands. Obviously, not all of these codes have been created entirely from scratch. Many of them were developed from templates called **model codes**.

Model codes are comprehensive sets of general guidelines that establish and define acceptable plumbing practices and materials. They also list prohibited installations and clarify obvious plumbing hazards. Model codes do not have the force of law, but they serve as the basis for more detailed, legally binding local codes developed by individual jurisdictions, which can adopt and amend the model code in whole or in part. Because they offer general guidance, model codes tend to focus more on processes than on specifics.

Model codes are created by professional organizations with the expertise and resources to undertake such a large-scale task. These organizations work closely with government agencies, code administrators, and building-industry professionals to develop, publish, and regularly revise the model codes. As with plumbing codes, model code organizations also publish codes to regulate the minimum safety provisions for topics including, but not limited to, the following:

- Residential and commercial building construction
- Fire prevention
- Mechanical and electrical installations
- Energy conservation
- Property maintenance
- Sewage disposal
- Swimming pools and spas
- Fuel gas
- Wildland-urban interface
- Green construction
- Solar energy
- Building codes and zoning

Most model code organizations are open to membership by professionals from the construction trades, architects and designers, regulation enforcement specialists, and industry representatives. These organizations have taken to the Internet to communicate with members and other interested parties. Websites feature the latest news releases, provide detailed information about pending and approved code changes, and offer membership information. Many organizations also offer the option to purchase model codes and other publications online.

In general, professional code-development organizations publish new editions of their codes every three years. However, states and local jurisdictions may choose to adopt new model codes only for some construction standards while keeping established codes for others. Some states even develop their own codes without reference to a model code, and some use more than one model code. Because all of these issues pose challenges for plumbing professionals trying to keep up with new developments, a summary review of the two major model plumbing-code organizations in the United States will help clarify this complicated situation.

Did You Know?

History of Codes

The predecessors of today's plumbing standards and practices can be traced to guidelines developed in ancient times to regulate safe building construction and public health in large urban settlements. Rudimentary regulations were incorporated into one of the earliest sets of codified laws. The Code of Hammurabi, written around 2100 BC for Hammurabi, King of Babylonia, declared that if a house collapsed, the builder should give his own house to the owner of the fallen house. If people died as a result of poor workmanship, the builder would lose his life, too. Modern building codes are not quite so drastic!

The plumbing standards and practices observed in the United States today are based on those originally developed in Europe beginning in the Renaissance in the 16th century. Beginning in 1519, for example, laws passed in various regions of France required that all houses should have indoor toilets or cesspools. In the mid-19th century, the British Parliament passed the 1848 Public Health Act, the first law of its kind in scale and scope. One of the first sanitation laws passed after Louis Pasteur discovered that infectious diseases were spread by microscopic organisms called germs was the New York Metropolitan Health Law, enacted in 1866, which served as a model for the health laws of other large cities.

Early in the 20th century, insurance companies in the United States called for the adoption of building codes to help reduce the excessive loss of lives due to fires in overcrowded inner cities. Over the next 40 years, code enforcement officials banded together in several regional professional organizations to develop and implement such codes, establishing the system that exists today.

1.1.0 Describing the ICC Model Plumbing Code

The **International Plumbing Code® (IPC®)** (*Figure 1*) is the model code published by the International Code Council, Inc. (ICC), located in Washington, DC. The IPC is a proven, comprehensive model plumbing code that works seamlessly with ICC's family of building codes. It sets minimum regulations for plumbing systems and components to protect life, health, and safety of building occupants and the public. The IPC is available for adoption by jurisdictions ranging from states to towns. It is currently adopted at the state or local level in 37 states in the United States, the District of Columbia, Guam, Jamaica, the Cayman Islands, and Puerto Rico.

International Plumbing Code® (IPC): The model plumbing code of the International Code Council (ICC). It was first issued in 1995.

The purpose of the IPC is to "establish minimum requirements to provide a reasonable level of safety, health, property protection, and general welfare by regulating and controlling the design, construction, installation, quality of materials, location, operation, and maintenance or use of plumbing equipment and systems." The IPC focuses primarily on laying out the specifications for various plumbing installations, but it also provides information that tells plumbers how to design and install systems that meet those specifications. For example, the code states simply that water supply lines and fittings must be installed in such a way as to prevent backflow. It goes on to lay out the procedures to be followed to ensure this, including the requirements for selecting and locating air gaps and other types of backflow preventers, protecting potable water outlets, and ensuring safe connections to the water supply and waste systems.

Figure 1 Cover of the 2024 International Plumbing Code®.

The IPC is built on the proven legacy of the BOCA National Plumbing Code, SBCCI Standard Plumbing Code, and ICBO Plumbing Code. The IPC sets minimum regulations for plumbing systems using both prescriptive and performance-related provisions, covering topics including, but not limited to, the following:

- Backflow prevention
- Fixtures and fittings
- Water supply and distribution piping
- Water heaters
- Sanitary drainage and venting
- Traps, grease interceptors, and separators
- Storm drainage
- Nonpotable water systems (rainwater, graywater, and reclaimed water)

The IPC is known for its dedication to safety, ease of use, embracing of new technology, and correlation with the ICC's family of codes. It is devoted to open and honest code development, and is revised on a three-year cycle through its highly respected code-development process.

1.1.1 The International Code Council

The ICC (*www.iccsafe.org*) was founded in 1994 by three model code-development organizations: the Building Officials and Code Administrators International, Inc. (BOCA), the International Conference of Building Officials (ICBO), and the Southern Building Code Conference International, Inc. (SBCCI). In 2003, the ICBO, BOCA, and SBCCI consolidated their services, products, and operations under the single entity of ICC.

The ICC issued its first IPC in 1995. The ICC also publishes model codes, nicknamed I-codes, for building construction, fire prevention, green construction, residential buildings, mechanical installations, fuel gas, energy conservation, property maintenance, private sewage disposal, swimming pools and spas, wildland-urban interfaces, and zoning.

The ICC's goal is to develop a consistent and uniform set of international codes. The ICC sees a variety of benefits arising from global uniform international codes and standards. Such beneficial results include, but are not limited to, the following:

- Construction professionals and code officials can work with a single set of requirements throughout the world.
- Manufacturers no longer have to design products for multiple standards.
- Professionals and manufacturers can better compete in worldwide markets, designing products to the same standard requirements.
- Education and certification programs are standardized.
- Construction and enforcement standards are consistent and in harmony with each other.

Did You Know?

A Brief History of BOCA and SBCCI

Two of the organizations that founded the International Code Council were the Building Officials and Code Administrators International, Inc. (BOCA) and the Southern Building Code Congress International, Inc. (SBCCI). Founded in 1915, BOCA established the National Plumbing Code® (NPC), as well as national codes for building construction, mechanical installations, fire prevention, property maintenance, private sewage disposal, and energy conservation. The NPC was adopted primarily by cities, counties, and states in the eastern and midwestern United States.

SBCCI was established in 1940. The first edition of its Standard Plumbing Code™ (SPC) appeared in 1955. The SPC was adopted by cities, counties, and states in the southern and southeastern United States. SBCCI also issued model codes for mechanical installations, excavation and grading, gas, fire prevention, housing, swimming pools, existing-building maintenance, building construction, and unsafe-building abatement.

With the advent of the ICC's International Plumbing Code®, both the NPC and the SPC have ceased publication.

Think About It

Local Codes Online

Many states and municipalities make their codes available online. These codes may be in the form of downloadable documents, smartphone apps, searchable databases, or simply the text of the code displayed on a web page. When referring to online versions of a local code, make sure that you are using the most recent and up to date version. Older versions of codes can remain on the web for many years after they have been superseded.

1.1.2 Chapters in the ICC Model Plumbing Code

The International Code Council's 2024 model plumbing code, the International Plumbing Code® (IPC), is divided into 15 chapters and six appendices (*Figure 2*). This section provides brief summaries of each chapter and appendix in the 2024 IPC. Note that the chapters and appendices may vary from edition to edition.

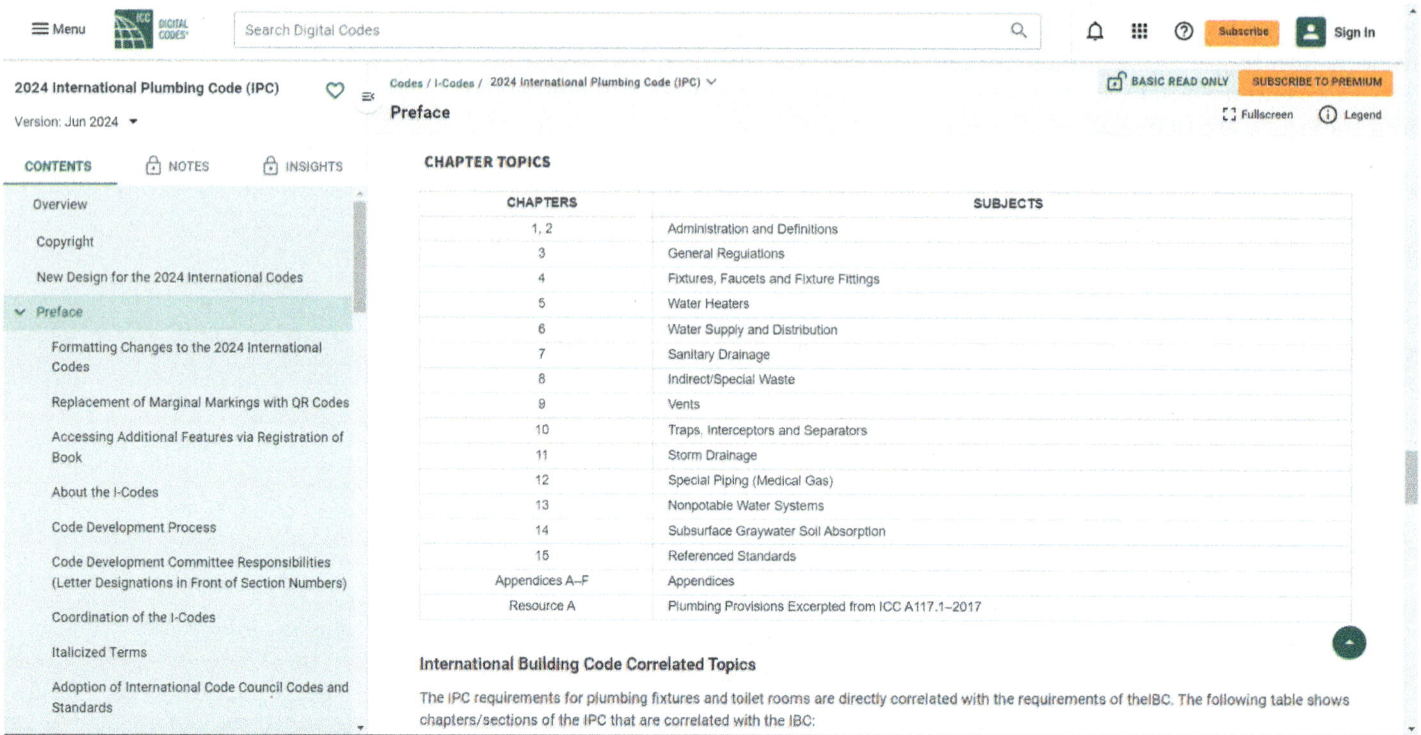

CHAPTERS	SUBJECTS
1, 2	Administration and Definitions
3	General Regulations
4	Fixtures, Faucets and Fixture Fittings
5	Water Heaters
6	Water Supply and Distribution
7	Sanitary Drainage
8	Indirect/Special Waste
9	Vents
10	Traps, Interceptors and Separators
11	Storm Drainage
12	Special Piping (Medical Gas)
13	Nonpotable Water Systems
14	Subsurface Graywater Soil Absorption
15	Referenced Standards
Appendices A–F	Appendices
Resource A	Plumbing Provisions Excerpted from ICC A117.1–2017

Figure 2 The International Plumbing Code® online.

Source: 2024 International Plumbing Code. Copyright © 2024. International Code Council, Inc. All rights reserved. Reprinted with permission.

Chapter 1, *Scope and Administration*, contains provisions for the application, enforcement, and administration of subsequent requirements of the code. In addition to establishing the scope of the code, Chapter 1 identifies which buildings and structures come under its purview. Chapter 1 is largely concerned with maintaining "due process of law" in enforcing the requirements contained in the body of this code. Only through careful observation of the administrative provisions can the code official reasonably expect to demonstrate that "equal protection under the law" has been provided. Chapter 2 is the repository of the definitions of terms used in the body of the code. Codes are technical documents; every word, term, and punctuation mark can impact the meaning of the code text and the intended results. The code often uses terms that have a unique meaning to the code, and the code meaning can differ substantially from the ordinarily understood meaning of the term as used outside of the code.

The terms defined in Chapter 2, *Definitions*, are of prime importance in establishing the meaning and intent of the code text using the terms. The user of the code should be familiar with and consult this chapter because the definitions are essential to the correct interpretation of the code. Additionally, the user may not be aware that a term is defined in the code, making familiarity with the code that much more important.

Where understanding a term's definition is especially important in understanding particular code provisions, the term will be shown in italics in the code. This includes only those terms that have a meaning unique to the code. In other words, the generally understood meaning of a term or phrase might not be sufficient or consistent with the meaning prescribed in the code; it is therefore essential that the code-defined meaning be known. Guidance regarding tense, gender, and plurality of defined terms, as well as guidance regarding terms not defined in the code, is also provided in this chapter.

While Chapter 3 is titled *General Regulations*, it is often referred to as "Miscellaneous." This is the only chapter in the code whose requirements do not interrelate. If a requirement cannot be located in another chapter, it should be located in this chapter.

Chapter 3 contains safety guidelines for the installation of plumbing and nonplumbing requirements for all types of fixtures. It also contains requirements for the identification of pipe, pipe fittings, traps, fixtures, and materials and devices used in plumbing systems. The safety requirements provide protection for the building's structural members, as well as prevent undue stress and strain on pipes. The building's structural stability is protected by the regulations for cutting and notching of structural members. Additional protection of the building occupants includes requirements to maintain the plumbing in a safe and sanitary condition, as well as privacy for those occupants.

Chapter 4 is titled *Fixtures, Faucets and Fixture Fittings*. This chapter regulates the minimum number of plumbing fixtures that must be provided for every type of building. This chapter also regulates the quality of fixtures and faucets by requiring those items to comply with nationally recognized standards. Because fixtures must be properly installed so they are usable by the occupants of the building, this chapter contains the requirements for the installation of fixtures. Since the requirements for the number of plumbing fixtures affect the design of the building, Chapter 29 of the International Building Code (IBC) includes, verbatim, many of the requirements listed in Chapter 4.

Chapter 5 is titled *Water Heaters*. It regulates the design, approval, and installation of water heaters and related safety devices. The intent is to minimize the hazards associated with the installation and operation of these devices. Although the code does not regulate the size of a water heater, it does regulate all other aspects of water-heater installation, such as temperature and pressure of relief valves, safety drip pans, installation, and connections. Where a water heater also supplies water for space heating, this chapter regulates the maximum water temperature supplied to the water distribution system.

Chapter 6 is titled *Water Supply and Distribution*. It regulates the supply of potable water from both public and individual sources to every fixture and outlet, so the water remains potable and uncontaminated. Chapter 6 also regulates the design of the water distribution system, which allows fixtures to function properly and helps prevent backflow conditions. The unique requirements of the water supply for health care facilities are addressed separately. It is critical that the potable water supply system remain free of actual or potential sanitary hazards by providing protection from backflow.

Chapter 7 is titled *Sanitary Drainage*. This chapter regulates the materials, design, and installation of sanitary drainage pipe systems, and connections made to the system. The intent is to design and install sanitary drainage systems that will function reliably and that are neither undersized nor oversized. These systems must be constructed from materials, fittings, and connections as prescribed in the code. Chapter 7 addresses the proper use of fittings for directing the flow into and within the sanitary drain-piping system. Materials and provisions necessary for servicing the drainage system are also included in this chapter.

Chapter 8 is titled *Indirect/Special Waste*. This chapter regulates drainage installations that require an indirect connection to the sanitary drainage system. Fixtures and plumbing appliances, such as those associated with food preparation or handling, health care facilities, and potable liquids, must be protected from contamination that can result from connection to the drainage system. An indirect connection prevents sewage from backing up into a fixture or appliance, providing protection against potential health hazards. The chapter also regulates special wastes containing hazardous chemicals. Special waste must be treated to prevent any damage to the sanitary drainage piping and to protect the sewage treatment processes.

Chapter 9 is titled *Vents.* This chapter covers the requirements for vents and venting. Knowing why venting is required makes it easier to understand the intent of the chapter. Venting protects every trap against the loss of its seal. Provisions set forth in this chapter are geared toward limiting the pressure differentials in the drainage system to a maximum of 1" of water column (249 Pa) above or below atmospheric pressure (i.e., positive or negative pressures).

Chapter 10 is titled *Traps, Interceptors and Separators.* It covers, as the name implies, the design requirements and installation limitations for traps. Prohibited types of traps are specifically identified.

Chapter 11 is titled *Storm Drainage.* It regulates the removal of storm water typically associated with rainfall. The proper installation of a storm drainage system reduces the possibility of structural collapse of a flat roof, prevents the leakage of water through the roof, prevents damage to the footings and foundation of the building, and prevents flooding of the lower levels of the structure.

Chapter 12 is titled *Special Piping and Storage Systems.* It contains the requirements for the design, installation, storage, handling, and use of nonflammable medical gas systems, including inhalation anesthetic and vacuum-piping systems, bulk oxygen storage systems, and the oxygen-fuel gas systems used for welding and cutting operations. The intent of these requirements is to minimize the potential fire and explosion hazards associated with the gases used in these systems.

Chapter 13 is titled *Nonpotable Water Systems.* This chapter regulates the design and installation of nonpotable water systems, including rainwater harvesting systems. The reduction of potable water use in buildings has led building designers in some jurisdictions to use nonpotable water for flushing of water closets and urinals. This chapter provides the overall requirements for such systems.

Chapter 14 is titled *Subsurface Graywater Soil Absorption.* It regulates the design and installation of subsurface graywater soil absorption systems for the disposal of on-site nonpotable water, such as graywater.

Chapter 15 is titled *Referenced Standards.* It contains a comprehensive list of all standards referenced in the code. The standards are part of the code to the extent of the reference to the standard. Compliance with the referenced standard is necessary for compliance with this code. By providing specifically adopted standards, the construction and installation requirements necessary for compliance with the code can be determined readily. The basis for code compliance is, therefore, established and available on an equal basis to the code official, contractor, designer, and owner.

Chapter 15 is organized in a manner that makes it easy to locate specific standards, listing all the referenced standards alphabetically, by acronym of the promulgating agency of the standard. Each agency's standards are then listed in either alphabetical or numeric order based on the standard identification. The list contains the title of the standard, the edition (date) of the standard referenced, any addenda included as part of the ICC adoption, and the section or sections of the code that reference the standard.

The six appendices cover plumbing permit-fee schedules, rainfall rates for major cities in the United States, structural safety, temperature information for plumbing systems in areas where freezing temperatures might exist, sizing of water-piping systems, and regulations for the setup and operation of a Board of Appeals.

The 2024 IPC® also incorporates provisions from the ICC A117.1 Standard for Accessible and Usable Buildings and Facilities. (ICC A117.1 is an approved American National Standard, or ANS, developed in accordance with the American National Standards Institute's requirements for ANS standard development.) Sections 404.2 and 410.3 reference ICC A117.1 for accessible plumbing facilities and fixtures. Key to these references are provisions that state, "Accessible plumbing fixtures shall be installed in accordance with ICC A117.1" (Section 404.2) and "where drinking fountains are primarily for children's use, the drinking fountains for people using wheelchairs shall be permitted to comply with the children's provisions in ICC A117.1 and drinking fountains for standing children shall be permitted to provide the spout at 30" (762 mm) minimum above the floor" (Exception 2 to Section 410.3).

For user convenience, and to ensure an accessible plumbing design, plumbing design-related excerpts from ICC A117.7–2017 have been included in the code.

1.1.3 Using the ICC Model Plumbing Code

The ICC's IPC model plumbing code is available in print in both loose-leaf and softcover formats, as well as a password-protected PDF document. It is also available online and through device apps.

Frost Depths

Model codes can be used by jurisdictions anywhere in the country. Because of this fact, they do not provide detailed guidance for specific concerns that vary from region to region due to weather, climate, and humidity.

One issue for which model codes do not provide specific guidance is frost depths. Wide variations in frost depths occur throughout the country. Model codes simply state that plumbing should be protected from freezing. It is up to local officials to define these depths in the local codes.

1.2.0 Describing the IAPMO Model Plumbing Code

The **Uniform Plumbing Code® (UPC)** (*Figure 3*) is the model code published by the International Association of Plumbing and Mechanical Officials (IAPMO) of Ontario, CA. According to IAPMO, the purpose of the UPC is to establish "minimum requirements and standards for the protection of the public health, safety and welfare." It is intended to apply to "the erection, installation, alteration, repair, relocation, replacement, addition to, use, or maintenance of plumbing systems within [the] jurisdiction." It covers piping for water supply and DWV systems, fixtures and traps, building drains and sewers, water heaters, and other systems, such as medical gas and vacuum systems and liquid- and fuel-gas systems. According to IAPMO, the goal of the UPC is "to provide consumers with safe and sanitary plumbing systems while, at the same time, allowing latitude for innovation and new technologies."

Uniform Plumbing Code® (UPC): The model code of the International Association of Plumbing and Mechanical Officials (IAPMO). It has been published since 1945.

1.2.1 The International Association of Plumbing and Mechanical Officials

The International Association of Plumbing and Mechanical Officials (*www.iapmo.org*) was established in 1926. The organization has published the Uniform Plumbing Code® (UPC) since 1945. The UPC provides minimum requirements and standards to protect public health, safety, and welfare. Its provisions are applicable to the erection, installation, alteration, repair, relocation, replacement, use, and maintenance of plumbing systems.

1.2.2 Chapters in the IAPMO Model Plumbing Code

The UPC is divided into 17 chapters and 19 appendices (*Figure 4*). This section provides brief summaries of each chapter and appendix in the UPC. Note that the chapters and appendices may vary from edition to edition.

Chapter 1 is titled *Administration*. It covers the scope and purpose of the model code, and establishes the procedures for resolving conflicts between codes and submitting plans to the authority having jurisdiction (AHJ). The requirements for repairs, alterations, and maintenance of existing and new construction are also covered. Chapter 1 also addresses organizational and enforcement issues, such as penalties for violations, issues related to permits, and requirements for inspections.

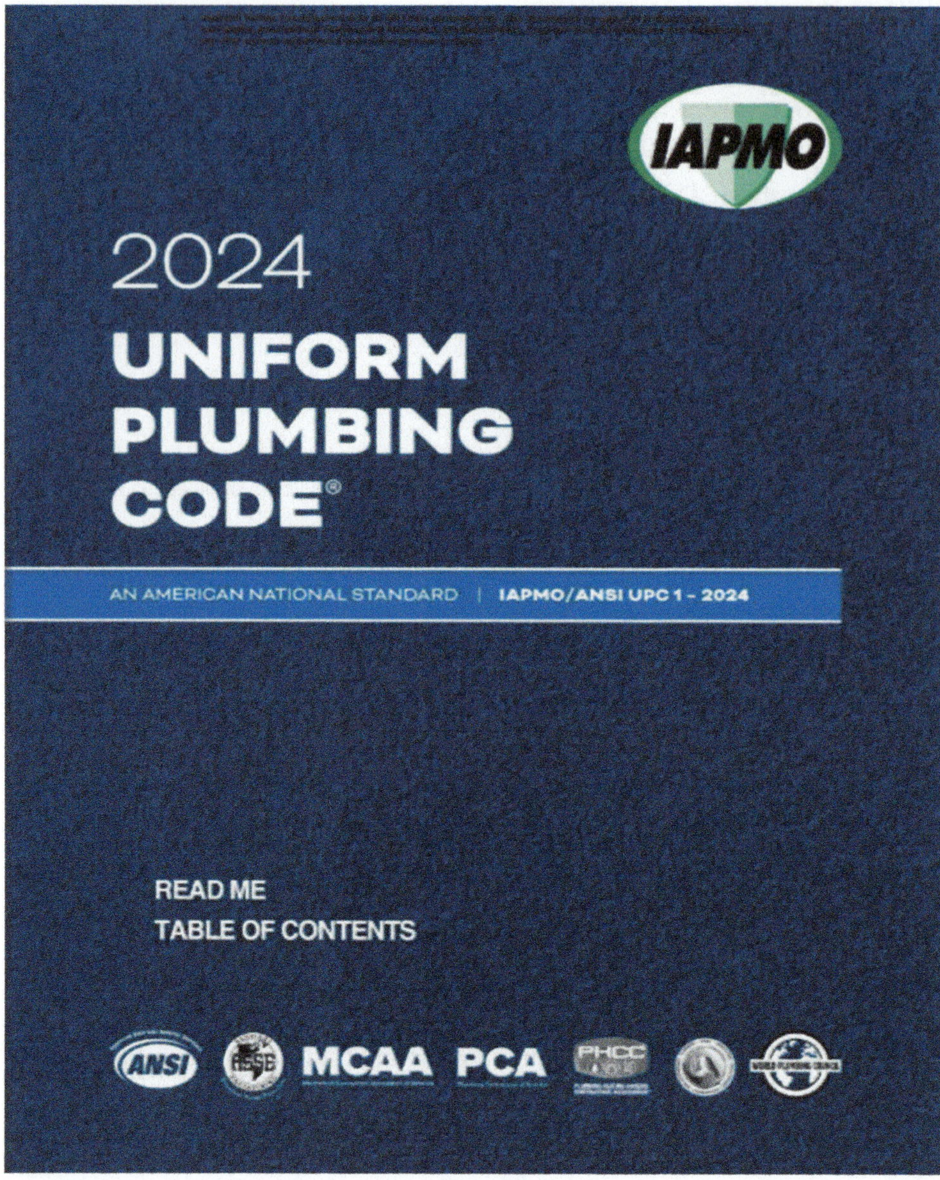

Figure 3 Cover of the Uniform Plumbing Code®.
Source: Courtesy of International Association of Plumbing and Mechanical Officials (IAPMO).

Chapter 2 is titled *Definitions.* As the title suggests, this section provides definitions for technical terms that are used throughout the code.

Chapter 3 is titled *General Regulations.* This chapter covers issues such as the minimum standards and alternates for materials, waste disposal, required connections, and the proper procedures to follow in the event of damage to the drainage system or the public sewer. Proper handling and discharge of industrial and detrimental wastes, proper and improper location, engineering practices, fittings, requirements for independent systems, standards for hanging and supporting pipe, and other specialized requirements are also covered in this chapter.

Chapter 4 is titled *Plumbing Fixtures and Fixture Fittings.* It covers general requirements for materials, including fixtures, joints, and the proper design and installation of water supply and DWV system fixtures and components. It also covers requirements for installing backflow prevention valves and floor drains, and covers the minimum requirements for water temperature, number of fixtures, and other plumbing facilities.

Chapter 5 is titled *Water Heaters.* This chapter, as its title suggests, addresses all aspects of sizing, locating, and installing water heaters. Subjects covered include safety devices, insulation, venting and chimneys, and accessibility.

Format of the Uniform Plumbing Code

The format of the *Uniform Plumbing Code* (UPC) arranges each chapter in accordance with a specific subject matter. However, Chapter 3 is dedicated to general requirements that are applicable to every chapter. The subject matters are divided as follows;

CHAPTERS	SUBJECTS
1	Administration
2	Definitions
3	General Regulations
4	Plumbing Fixtures and Fixture Fittings
5	Water Heaters
6	Water Supply and Distribution
7	Sanitary Drainage
8	Indirect Wastes
9	Vents
10	Traps and Interceptors
11	Storm Drainage
12	Fuel Gas Piping
13	Health Care Facilities and Medical Gas and Medical Vacuum Systems
14	Firestop Protection
15	Alternate Water Sources for Nonpotable Applications
16	Nonpotable Rainwater Catchment Systems
17	Referenced Standards
Appendix A	Recommended Rules for Sizing the Water Supply System
Appendix B	Explanatory Notes on Combination Waste and Vent Systems
Appendix C	Alternate Plumbing Systems
Appendix D	Sizing Storm Water Drainage Systems
Appendix E	Manufactured/Mobile Home Parks and Recreational Vehicle Parks
Appendix F	Firefighter Breathing Air Replenishment Systems
Appendix G	Sizing of Venting Systems
Appendix H	Private Sewage Disposal Systems
Appendix I	Installation Standards
Appendix J	Combination of Indoor and Outdoor Combustion and Ventilation Opening Design
Appendix K	Portable Rainwater Catchment Systems
Appendix L	Sustainable Practices
Appendix M	Peak Water Demand Calculator
Appendix N	Impact of Water Temperature on the Potential for Scalding and Legionella Growth
Appendix O	Non-Sewered Sanitation Systems
Appendix P	Professional Qualifications
Appendix Q	Indoor Horticultural Facilities
Appendix R	Tiny Houses
Appendix S	Onsite Stormwater Treatment Systems

Figure 4 A page from the introduction of the Uniform Plumbing Code®.
Source: Courtesy of International Association of Plumbing and Mechanical Officials (IAPMO).

Chapter 6 is titled *Water Supply and Distribution.* It covers the requirements for potable and nonpotable water supply systems, including prohibited and unlawful connections, approved devices and assemblies, suitable materials for fittings and connections, approved valves, trenching, and water hammer. This chapter provides guidance on calculating water supply fixture units (WSFUs) and drinking-water treatment units, which is a measure used in determining the water softening needs of a system.

Chapter 7 is titled *Sanitary Drainage.* This chapter is divided into two parts: Drainage Systems and Building Sewers. Drainage Systems covers materials and sizing requirements for drainage piping and fittings. It also covers aspects such as joints, cleanouts, drainage, and testing. Building Sewers addresses requirements related to public and private sewer systems, including materials, grading, support and protection, cleanouts, and disposal facilities.

Chapter 8 is titled *Indirect Wastes.* It defines indirect wastes and spells out the requirements for air gaps and air breaks; connections and traps; piping and receptors; condensate and cooking-water wastes; steam and hot-water drainage condensers and sumps; venting; clear-water wastes; and swimming-pool wastewater.

Chapter 9 is titled *Vents.* General requirements for the various types of vents are covered, as are instances where vents are not required. The chapter covers interceptors, vent grades and horizontal drainage pipe, combination waste-and-vent systems, and engineered vent systems.

Chapter 10 is titled *Traps and Interceptors.* This chapter goes into detail about when and where various types of traps and interceptors are required, and how to size and install them.

Chapter 11 is titled *Storm Drainage*. It identifies the conditions under which storm drainage systems are required and their approved uses. The various types of storm drainage systems are discussed, as are the requirements for system components, such as conductors, leaders, gutters, traps, and drains. The chapter also covers the testing of storm drainage systems.

Chapter 12 is titled *Fuel Gas Piping*. After introducing the scope and coverage of approved gas-piping systems, this chapter covers inspection requirements and the various authorities having jurisdiction over gas service and supply. Piping and the various types of approved fittings and valves are discussed in detail, as is testing. Gas supply requirements are also covered in this chapter.

Chapter 13 is titled *Health Care Facilities and Medical Gas and Medical Vacuum Systems*. This chapter is divided into two parts: Special Requirements for Health Care Facilities and Medical Gas and Vacuum Systems. The first part thoroughly covers the installation requirements for medical gas and vacuum systems in health care facilities, including proper brazing procedures, sterilization, and venting. The second part of the chapter provides comprehensive details on system requirements and sizing; flow rates; accepted engineering practices; cleaning and joining piping; approved valves, labels, and alarms; and the testing and inspection process required for medical gas and vacuum systems.

Chapter 14 is titled *Firestop Protection*. This chapter covers the proper procedures for firestopping when installing water supply system and DWV systems. It addresses acceptable fire-resistance ratings, connections, couplings, sleeves, and penetrations; it also specifies the inspection procedure to be followed to ensure that the firestopping meets or exceeds the code's minimum requirements.

Chapter 15 is titled *Alternate Water Sources for Nonpotable Applications*. It covers the design, maintenance, testing, and inspection of graywater systems, and establishes minimum water-quality requirements for these systems.

Chapter 16 is titled *Nonpotable Rainwater Catchment Systems*. This chapter discusses the process of installing rainwater catchment systems, and how to connect them safely to potable or reclaimed water systems. It also covers backflow prevention, and the coloring and marking of catchment systems.

Chapter 17 is titled *Referenced Standards*. As its name indicates, this chapter provides a list of all standards referenced in the code and that must also be followed to be in compliance with the code. The standards are provided in a table format by standard number, followed by title, application, and the sections of the code in which they are referenced.

The code contains numerous appendices, which supplement the requirements of the UPC, but which are not considered part of the code unless formally adopted as such. The 19 appendices cover such topics as the recommended rules for sizing water supply systems; combination waste-and-vent systems; sizing of alternate plumbing systems, such as engineered plumbing systems, water heat exchangers, and vacuum systems; stormwater drainage systems; parks for manufactured homes, mobile homes, and recreational vehicles; replenishment systems for firefighter breathing air; sizing of venting systems that serve appliances fitted with draft hoods and other special appliances; private sewage-disposal systems; installation standards; the design of combination indoor-outdoor combustion and ventilation systems; catchment systems for potable rainwater; sustainable practices; calculating peak water demand; the impact of water temperature on the potential for scalding and Legionella growth; non-sewered sanitation systems; professional qualifications; indoor horticultural facilities; tiny houses; and on-site stormwater treatment systems.

1.2.3 Using the IAPMO Model Plumbing Code

The UPC is available in print in loose-leaf and softcover formats with chapter tabs, on CD-ROM, and in e-book format. IAPMO also publishes an illustrated training manual, study guide, and guide to important code changes to accompany the UPC. These publications are available in print and e-book formats. IAPMO also offers printed sizing guides for natural gas, DWV, and water supply system piping.

The print version of the UPC can be searched using the detailed table of contents as well as an alphabetical index. The e-book is searchable by keyword, section number, figure numbering, and table numbering. Clicking on an item in the table of contents will take the reader to the appropriate section in the text. The e-book version also provides a hierarchical table of contents that can be expanded or collapsed by clicking on icons to the left of the chapter or section title.

1.0.0 Section Review

1. In the IPC, the design, installation, storage, handling, and use of medical gas and vacuum systems are covered in the chapter titled _____.
   a. Indirect/Special Waste
   b. Fixtures, Faucets and Fixture Fittings
   c. Health Care Facilities
   d. Special Piping and Storage Systems

2. IAPMO publishes model codes for each of the following categories except _____.
   a. swimming pools, spas, and hot tubs
   b. fire
   c. plumbing
   d. solar energy and hydronic systems

2.0.0 Plumbing-Code Development and Revision

Objective

Explain how plumbing codes are developed and revised.

a. Explain model code standards.
b. Explain how model codes are revised and adopted.
c. Explain typical code changes.

Performance Tasks

There are no Performance Tasks in this section.

Plumbing codes are intended to reflect the current state of the science, technology, and practice of the trade. However, codes must change over time to include new materials and technologies and to reflect changing standards and practices. Both model and local codes change, though not always at the same time. Changes to model codes can drive changes to local codes, and vice versa. The code change process is a constant cycle (and can vary from organization to organization). The ICC's model code cycle, which is representative of this process, is shown in *20406B*. The IAPMO has its own code-development cycle and maintains significant documentation providing more detail on this.

2.1.0 Explaining Model Code Standards

Several organizations oversee code standards for their specific industries. Because these standards organizations conduct detailed research and testing as part of the development of their standards, model code organizations such as the ICC and IAPMO can reference those standards in their model codes rather than duplicate that work. This section provides basic information about those organizations and the particular standards they set.

The Air-Conditioning, Heating, and Refrigeration Institute (AHRI) is a trade association with 70 established standards and guidelines. AHRI represents the manufacturers of more than 90 percent of US-produced central air conditioning, heating, and commercial refrigeration equipment.

The American National Standards Institute (ANSI) is a private, not-for-profit organization that administers and coordinates the US voluntary standardization and conformity assessment system. Its mission is to enhance global competitiveness by promoting and facilitating voluntary consensus standards.

The American Society of Mechanical Engineers (ASME) is a not-for-profit, educational, and technical organization serving 130,000 members in 158 countries. ASME sets many industrial and manufacturing standards to promote and enhance the technical competency and professional well-being of its members.

The American Society of Sanitary Engineering (ASSE) International is dedicated to continually improving the performance, reliability, and safety of plumbing systems through consensus product performance and professional qualification standards; active product seal control standards and credentialing; advocacy; and public awareness programs.

The American Welding Society (AWS) is a multifaceted, not-for-profit organization with more than 66,000 members, including engineers, scientists, educators, researchers, welders, and inspectors. AWS supports welding education and technology development in order to advance the science, technology, and application of welding.

The American Water Works Association (AWWA) is an international, not-for-profit scientific and educational organization that promotes healthy drinking water and high-quality water supply. AWWA has more than 50,000 members, including 4,000 utilities that supply water to 180 million people.

ASTM International, formerly the American Society for Testing and Materials International (ASTM), is a developer and provider of more than 12,000 voluntary consensus standards in a wide range of technical fields.

CSA Group, formerly the Canadian Standards Association (CSA), is a not-for-profit association serving business, industry, government, and consumers in Canada. CSA Group develops standards that address and enhance public safety and preserve the environment.

The Cast Iron Soil Pipe Institute (CISPI) seeks to advance the manufacture, use, and distribution of cast-iron soil pipe and fittings. CISPI strives to improve industry products, achieve standardization of cast-iron soil pipe and fittings, and provide a continuous program of product testing, evaluation, and development.

Federal Specifications (FS) are standards recognized by the General Services Administration. FS standards are available from the Superintendent of Documents, US Government Printing Office.

The International Code Council (ICC) is a not-for-profit organization dedicated to developing a single set of comprehensive and coordinated national model construction codes. *Appendix 02406B* outlines the ICC Code Development Process.

The National Fire Protection Association (NFPA) promotes fire and electrical safety by publishing and advocating more than 300 consensus codes and standards. NFPA's membership consists of more than 70,000 individuals from more than 100 countries, as well as national trade and professional organizations.

The NSF International, formerly the National Sanitation Foundation (NSF), is an independent, not-for-profit organization committed to public health, safety, and protection of the environment. NSF focuses on food, water, consumer products, and the environment; developing standards; and testing and certifying products and systems.

The Plumbing and Drainage Institute (PDI) includes manufacturers of engineered plumbing drainage products. PDI promotes the advancement of engineered plumbing products through outreach, advocacy, research, and standardization of product requirements.

Low-Flush Toilets

The Energy Policy Act of 1992 mandated the use of low-flush toilets as a water conservation measure. The law required newly installed toilets to use 1.6 gallons per flush, less than half the amount they used before the law was passed. The latest version of the law requires newly installed toilets to use just 1.28 gallons per flush. Although the technology has improved considerably since the law was introduced, low-flush toilets may cause plumbing problems in homes built before the new requirements existed, because their drainage plumbing is designed to work with higher flow rates. To address these problems, plumbers should ensure that the building's water pressure is appropriate for the model of toilet installed.

2.2.0 Explaining the Revision and Adoption of Model Codes

Model codes need to keep pace with changes in technology, practice, and safety standards that affect the industries they cover. As a result, model code organizations adopt a regular update cycle to prepare new editions that reflect those changes. This cyclical approach allows the model code organization to review changes and to decide whether, and how, to incorporate them, while also ensuring that the people who rely on their codes—from plumbers to manufacturers—have a chance to provide input based on their experience. States and cities that use model codes follow a similar cyclical process for adopting revised model codes, in whole or in part, for use in their jurisdictions. This section provides a summary overview of how these revision and adoption processes typically work.

2.2.1 Revision

Model code revision is traditionally a public process, involving members of the construction trades, the regulatory community, and the industry. In short, the process may include anyone with a professional interest in ensuring that the model codes are current and comprehensive. Model code organizations generally hold regular meetings to discuss proposed changes. Attendees may speak in favor of, or in opposition to, proposed changes, after which Technical Committee members vote for or against modifications. While change proposals were once submitted through printed forms, today they are almost always submitted online. Any trade professional, regulatory official, industry representative, or even a member of the general public may propose changes.

When all proposals have been submitted and all the arguments have been heard, the proposed changes are voted on. Some organizations allow a select committee of public officials to vote on the proposed code changes. Regardless of the method used, the approved changes are included in the next edition of the model code, which is usually published every three years.

The ICC Revision Process. Changes to the ICC codes, including the IPC, are reviewed by two groups of committees that meet in alternating years. The codes are divided between the two groups. The IPC, for example, is reviewed by the same committee that reviews the ICC's building, fuel gas, mechanical, and private sewage disposal codes. Changes may be submitted by code enforcement officials, representatives from industry, design professionals, and other interested parties.

The IAPMO Revision Process. Changes to the IAPMO codes, including the UPC, are open to the general public as well as to code officials, industry representatives, installers, and manufacturers. As discussed in the section titled The International Association of Plumbing and Mechanical Officials, IAPMO follows an ANSI-accredited open-consensus process that encourages members and nonmembers alike to vote on proposed changes. IAPMO has identified three main goals for its code-development process:

- Effective preservation of public health, safety, and welfare
- Prompt attention to technological changes that affect construction regulations
- Decisions reached through open discussion and consensus

2.2.2 Adoption

Model codes are comprehensive in scope, but they can offer only the minimum acceptable standards and the broadest guidelines. Therefore, the jurisdictions that adopt them are free to make changes and add amendments to accommodate their state and local regulations. Changes to model codes are typically determined by an appointed advisory commission of regulatory experts, the members of which are determined by law. The commission reviews the code that is current in the jurisdiction. It also reviews the most recent edition of various model codes, unless the jurisdiction has developed its own. The commission decides which model code to adopt and what, if any, changes or amendments are necessary. Adopted codes must be signed into law before they can take effect. Codes usually have to be submitted as an ordinance and approved by the regulatory officials of the jurisdiction, as well as by officials all the way up to the state government.

Adopted codes—for example, building, mechanical, electrical, plumbing, and fire-protection codes—are reviewed on a regular basis, usually every three to five years. Reviews ensure that the jurisdiction's codes remain up to date and comprehensive. Otherwise, the constant need to amend codes with variances and exceptions for each new technology and process would result in a patchwork code and bureaucratic headaches. A local code may not change the requirements that are specified in the model code on which it is based.

The review process also gives regulators an opportunity to adopt a different model code from the one they have used in the past. Regulators may choose to adopt a new model code to bring the jurisdiction into closer alignment with neighboring jurisdictions, or the model code from which a local code was adopted may no longer be updated.

Authority for code administration, inspection, and enforcement varies from jurisdiction to jurisdiction. The responsibility may lie with a building inspector, the health department, the city engineer, or a commissioner's office. Take the time to become familiar with how codes are regulated in your area so you will always know whom to call when you have questions or concerns about code compliance.

Did You Know?

Forbidden Fittings

Some codes prohibit certain fittings, while others may permit them. Plumbers are expected to know what fittings may or may not be used in plumbing installations, but consumers may be unaware of plumbing-code requirements and install prohibited fittings. You may come across such fittings from time to time. Learn from local plumbing experts the proper course of action to take if you discover something that is in violation of the local code.

One prohibited fitting is the saddle fitting, which is used to make a connection to a pre-installed pipe. Saddle fittings are often used in humidifiers and ice makers. These widely used and easily available fittings may or may not be acceptable in the local code. Homeowners run the greatest risk of violating a code that forbids saddle fittings because these fittings can be easily obtained at local hardware stores.

2.3.0 Explaining Typical Code Changes

Changes in model and local codes frequently result from developments in plumbing materials, the growing demand for energy and water conservation, and the availability of alternative energy sources. New materials are approved for inclusion in model and local codes only after they are proven consistently safe and reliable. For example, local codes increasingly accept the use of new technologies in plumbing installations that may have been considered unsuitable for use in certain plumbing installations years ago. In addition to the piping material itself, codes also approve the new fastening and joining methods that these pipes require.

To illustrate just some of the factors that can influence what is included in model codes, the 2012 UPC incorporates changes that were adopted to address the following developments:

- New industry guidance for using alternative water sources in nonpotable applications
- New methods of joining piping in water supply and DWV systems
- Changes in plumbing fixture requirements for greater ease of use by the elderly
- New research conducted by the American Society of Plumbing Engineers (ASPE) and other organizations into design methods that reflect the concept of "potty parity," or the standard ratio of women's restroom facilities to those for men (resulting in higher numbers of plumbing fixtures required for each occupancy classification)

Codes increasingly stress energy conservation measures, such as insulation for flow restrictors on hot- and cold-water outlets, pipes, and water heaters. Solar hot-water installations are a popular alternative source of hot water in residential and commercial installations. They are addressed in special model codes, such as IAPMO's Uniform Solar Hydronics and Geothermal Code™ or the ICC's International Energy Conservation Code®. Be aware that model solar-energy codes do not address specific technical issues, such as whether to use double-wall or single-wall heat exchangers, or whether it is more appropriate to use toxic or nontoxic circulating fluids in solar water-heater installations.

Code changes reflect the most recent consensus among the construction trades, the regulatory community, and the industry. As technology, practice, and the law change, so too will model codes and adopted codes. As a result, there will never be a perfect code or an ultimate code—just the latest code.

Did You Know?

House Traps and Regional Code Differences

Differences between codes may arise from a variety of factors, such as geography, climate, urban population density, or even the age of a city and its infrastructure. A good example of the variations between local codes and model codes is the case of house traps. Both the IPC and the UPC model codes prohibit house traps but provide an exception when the local authority requires them. This exception is necessary because many counties and municipalities in the northeastern part of the United States actually require them in every structure, especially in cities with older sections and high population density.

House-trap rules demonstrate the dramatic differences in codes that can be found in different regions and states, and how important it is to know the code governing your specific area.

2.0.0 Section Review

1. The international organization dedicated to promoting healthy drinking water and high-quality water supply is _____.
   a. AHRI
   b. ASME
   c. AWWA
   d. AWS

2. New materials are approved for inclusion in model and local codes _____.
   a. when the manufacturer demonstrates that they can be produced safely
   b. when they are considered affordable
   c. immediately upon their introduction
   d. when they are proven consistently safe and reliable

Module 02406 Review Questions

1. In addition to establishing and defining acceptable practices and materials, a model code also _____.
   a. lists prohibited installations
   b. regulates contractor/owner responsibilities
   c. outlines apprenticeship requirements
   d. establishes work hours and conditions

2. Developing a consistent and uniform set of international codes is a goal of _____.
   a. BOCA
   b. IAPMO
   c. ICC
   d. SBCCI

3. Requirements for systems that collect and dispose of water used in applications such as subsurface irrigation systems are outlined in the IPC chapter entitled _____.
   a. Storm Drainage
   b. Nonpotable Water Systems
   c. Water Supply and Distribution
   d. Indirect/Special Waste

4. The International Plumbing Code® is available in either loose-leaf or soft-cover versions as well as a(n) _____.
   a. audiobook version
   b. enhanced version
   c. password-protected chip version
   d. online version

5. The Uniform Plumbing Code® consists of 17 chapters and 19 _____.
   a. addendums
   b. accessory sections
   c. allied trades listings
   d. appendices

6. The Uniform Plumbing Code® is available in loose-leaf or softcover versions as well as _____.
   a. an audiobook version
   b. an enhanced version
   c. online and device app versions
   d. a password-protected chip version

7. The private, not-for-profit group that administers the voluntary standardization and conformity assessment system in the United States is _____.
   a. ANSI
   b. ASME
   c. AWS
   d. AWWA

8. Model code organizations prepare new editions based on _____.
   a. supply and demand
   b. government agency requirements
   c. a regular update cycle
   d. equipment and materials improvements

9. A jurisdiction that adopts a model code _____.
   a. can amend it to include local requirements
   b. must implement it exactly as written
   c. may purchase a version customized to its needs
   d. is required to get permission before making changes

Answers to odd-numbered review questions are found in the Module Review Answer Key at the back of this book.

Answers to Section Review Questions

Answer	Section	Objective
Section One		
1. d	1.1.2	1a
2. b	1.2.1	1b
Section Two		
1. c	2.1.0	2a
2. d	2.3.0	2c

MODULE 02408

Private Water Supply Well Systems

Objectives

Successful completion of this module prepares you to do the following:

1. Explain how to drill good wells.
    a. Explain how to locate wells.
    b. Explain how to size wells.
    c. Explain how to construct wells.
    d. Explain how to clean wells.
2. Explain the operation of various types of pumps and pump components.
    a. Explain the operation of a shallow-well jet pump.
    b. Explain the operation of a deep-well jet pump.
    c. Explain the operation of a submersible pump.
3. Explain how to select and install water supply and storage components.
    a. Explain how to select and install water supply lines.
    b. Explain how to select and install water storage tanks.

Performance Task

Under supervision, you should be able to do the following:

1. Assemble and disassemble given components of private water supply well systems.

Overview

Private water supply systems provide potable water in rural areas. Most private systems in the United States use wells with pumps that lift water from aquifers. The four most common well types are bored, drilled, driven, and dug. The common types of pumps used in private water supply systems are shallow- and deep-well jet pumps and submersible pumps. When selecting pipe, consider the cost, life expectancy, and the likelihood of corrosion and freezing. Codes require well water to be tested for turbidity, pH (potential hydrogen) value, hardness, and biochemical oxygen demand (BOD).

CODE NOTE

Codes vary among jurisdictions. Because of the variations in code, consult the applicable code whenever regulations are in question. Referring to an incorrect set of codes can cause as much trouble as failing to reference codes altogether. Obtain, review, and familiarize yourself with your local adopted code.

NCCER Industry-Recognized Credentials

If you are training through an NCCER-accredited sponsor, you may be eligible for credentials from NCCER. The ID number for this module is 02408. Note that this module may have been used in other NCCER curricula and may apply to other level completions. Contact NCCER at 1.888.622.3720 or go to **www.nccer.org** for more information.

You can also show off your industry-recognized credentials online with NCCER's digital credentials. Transform your knowledge, skills, and achievements into digital credentials that you can share across social media platforms, send to your network, and add to your resume. For more information, visit **www.nccer.org**.

Digital Resources for Plumbing

SCAN ME

Scan this code using the camera on your phone or mobile device to view the digital resources related to this craft.

1.0.0 Drilling Good Wells

Performance Tasks

There are no Performance Tasks in this section.

Objective

Explain how to drill good wells.

a. Explain how to locate wells.
b. Explain how to size wells.
c. Explain how to construct wells.
d. Explain how to clean wells.

Cistern: A water storage tank used to supply some private water supply well systems with potable water.

Well: A hole in the ground that connects with an aquifer. Wells are equipped with pumps to lift water from the aquifer into water storage tanks.

Aquifer: An underground reservoir containing water-saturated soil, sand, and rock.

Hydrologic cycle: The process through which water cycles through the environment as both a liquid and a vapor.

Many people and businesses in rural areas and in other remote locations do not have access to municipal water supplies. These people rely on private sources for their potable, or drinkable, water. Some private water supply systems get water from springs or from storage tanks called **cisterns**. However, most private systems in the United States get water from **wells**. A well is a hole in the ground that connects with an **aquifer**, or underground reservoir. Wells use pumps to lift water from the aquifer into pipes that connect to a home or business. A trouble-free private water supply well system begins with a good well. A good well is a reliable source of clean, potable water. It should also be a sound investment for the customer. Customers have a right to expect years of trouble-free service from a private water supply well system.

Plumbers should know how wells work. In many areas, plumbers install pumps, supply piping, and storage tanks. In some states, certified well drillers perform these tasks. Elsewhere, the duties are split between the plumber and the well driller. Local codes specify the plumber's responsibilities. As mentioned earlier, wells draw water from aquifers. Aquifers are naturally filled with water as part of the **hydrologic cycle**. The hydrologic cycle is the process by which water circulates through the environment. This process is illustrated in *Figure 1*.

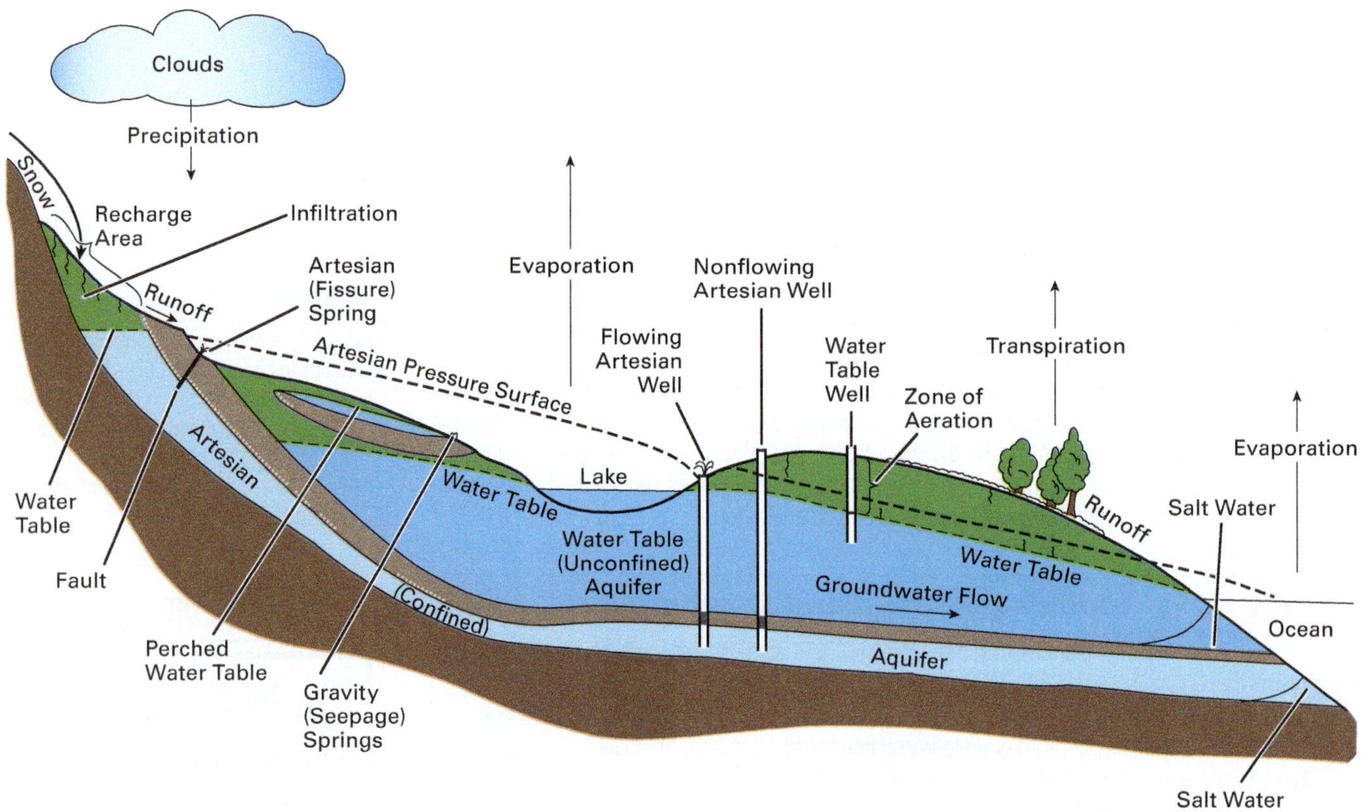

Figure 1 Hydrologic cycle.

Water vapor rises into the air, or evaporates, from the ground, bodies of water, and plants. Once in the air, water condenses into clouds and returns to Earth in the form of rain, snow, or hail. This process is called precipitation. Back on the ground, the water seeps, or percolates, into the soil, where it becomes part of an aquifer. Surface runoff and underground flow fill lakes, rivers, and oceans. Then the process begins again.

The sand, soil, and rocks in an aquifer are completely saturated with water. The upper limit of an aquifer is called the **water table**. The water table can be close to or far below the land surface. Whatever its depth, though, the water table usually follows the contours of the land. The two types of aquifers are **water table aquifers** and **artesian aquifers**.

Water table: The top of an aquifer, below which rock, sand, and soil are completely saturated with groundwater.

Water table aquifer: An aquifer in which water flows toward bodies of water. Also called an unconfined aquifer.

Artesian aquifer: A deep aquifer that does not lose water through runoff. Also called a confined aquifer.

Did You Know?

Private Water Supply Systems

The US Environmental Protection Agency (EPA) estimates that approximately 42 million people in the United States—16 percent of the population—obtain their potable water from private water supply systems.

Water table aquifers are refilled, or recharged, primarily by water percolating down from Earth's surface. Artesian aquifers are recharged only by precipitation that falls in the recharge area (usually a higher-elevation formation of rock and rock outcrops). These aquifers are restricted between impermeable layers of soil and rock and are therefore under pressure. Artesian pressure slopes downward from the recharge area, and the artesian-pressure surface is the level to which the water in an artesian aquifer would rise if allowed. Drilling a well into an artesian aquifer with the outlet positioned below the pressure surface will produce a flowing well.

Wells can be drilled into either water table aquifers or artesian aquifers. Artesian aquifers do not lose groundwater through runoff. They are often called confined aquifers for this reason. Water table aquifers are not as deep as artesian aquifers. The water in water table aquifers flows toward bodies of water. These aquifers are also called unconfined aquifers.

A **perched water table** is a smaller water table that is cut off from a larger, nearby aquifer by an impermeable layer of rock. If a perched water table is tapped by many wells, its water supply may eventually be exhausted, and the wells will run dry.

Perched water table: A small water table that is cut off from a larger reservoir.

1.1.0 Locating Wells

Wells must be correctly located over the water supply. An incorrectly located well will cause problems for the customer. In most cases, plumbers are not directly involved in the well-drilling process. However, customers often ask plumbers to recommend a reputable well driller. Take the time to learn the names of several dependable drillers in your area.

A driller should be able to determine which type of well would work best for each installation. The driller should review drilling logs to learn about other water wells in the area. Drilling logs list the following information for each well:

- The depth at which water was first found when the well was drilled
- The well's total depth
- The different types of rock, sand, and soil encountered during drilling
- The water quality

Finally, the driller should get all the required permits. In most jurisdictions, well permits are issued by the health department or a similar agency.

Work closely with the customer, driller, and code officials to find a good location for a well. Be sure to review state and local regulations that apply to wells. They will list what types of wells are approved for household and commercial use. Regulations will also specify the minimum allowable distances between a well and property lines, septic tanks, buried fuel tanks, and livestock fields. Most codes require at least a 100' distance between a well and a leach field, for example. Consider the distance from the well to the structure being serviced. The distance will influence the amount of pipe and the number of fittings needed. Locating a well on high ground will ensure that surface water drains away from, not into, the well opening.

There are always risks when locating a well. The well might not be able to replenish itself fast enough to meet demand. Worse, the water table could eventually fall below the well, causing the well to go dry. Study the drilling logs carefully. Talk to drillers who know the area. If you are concerned that the well may not meet the customer's needs, talk to the customer and the driller. Taking the time to locate the well correctly will save everyone time and money in the long run.

1.2.0 Sizing Wells

Seven-minute peak demand period: An estimate of the greatest possible water demand in a private water supply well system.

Sizing the well and selecting a pump can be difficult. One popular way to determine residential water needs is to calculate the **seven-minute peak demand period**. This is an estimate of the heaviest possible water demand. The estimate is usually presented in a table (*Table 1*). Local plumbing officials can provide a table for your area.

TABLE 1 Sample Table for Estimating the Seven-Minute Peak Demand Period

Outlets	Flow Rate(gpm)	Total Usage(gal.)	Bathrooms in House			
			1	$1^1/_2$	2 to $2^1/_2$	3 to 4
Shower or bathtub	5	35	35	35	53	70
Lavatory	4	2	2	4	6	8
Toilet	4	5	5	10	15	20
Kitchen sink	5	3	3	3	3	3
Automatic washer	5	35	—	18	18	18
Dishwasher	2	14	—	—	3	3
Normal seven-minute peak demand (gal.)			45	70	98	122
Minimum-size pump required to meet peak demand without supplemental supply			7 gpm (420 gph)	10 gpm (600 gph)	14 gpm (840 gph)	17 gpm (1,020 gph)

Notes: Values given are average and do not include extremes. Peak demand can occur several times during morning and evening hours. Farm, irrigation, and sprinkler requirements are not shown. These values must be added to the peak demand figures if usage will occur during normal demand periods.

Some fixtures are less likely than others to operate at their maximum output. For example, a kitchen sink is not likely to be run at its peak output for seven minutes at a time. Three gallons per minute (gpm) is a reasonable compromise. That is the figure used in the table. Because demand may exceed the sample minimum, use a flow rate higher than the minimum acceptable rate for the system.

1.3.0 Constructing Wells

There are four common types of wells. These four wells are classified by how they are created:

- Bored wells
- Drilled wells
- Driven wells
- Dug wells

Bored wells (*Figure 2*) are drilled out by an auger. They can be used where the ground is relatively soft. Once the hole has been completely drilled, a pipe section called a **well casing** is inserted. The well casing holds the shape of the well and prevents contamination of the water. The diameter of a bored well can range from 2" to 30". In many areas, bored wells are less common than other types of wells.

Drilled wells are constructed where the ground is too hard to be dug or driven and/or where the water table is deep underground. Chisels or rotary drills are used to construct drilled wells (*Figure 3*). The diameter of a drilled well depends on the size of the chisel or drill, but it is usually smaller than that of bored or **dug wells**. Drilled wells are currently the most common type of well.

Bored well: A type of well drilled by an auger and lined with a well casing.

Well casing: A pipe section inserted into a well that maintains the shape of the well and prevents contamination.

Drilled well: A type of well created by chisels or rotary drills.

Dug well: A type of well excavated by shovel and backhoe and lined with brick, stone, or concrete.

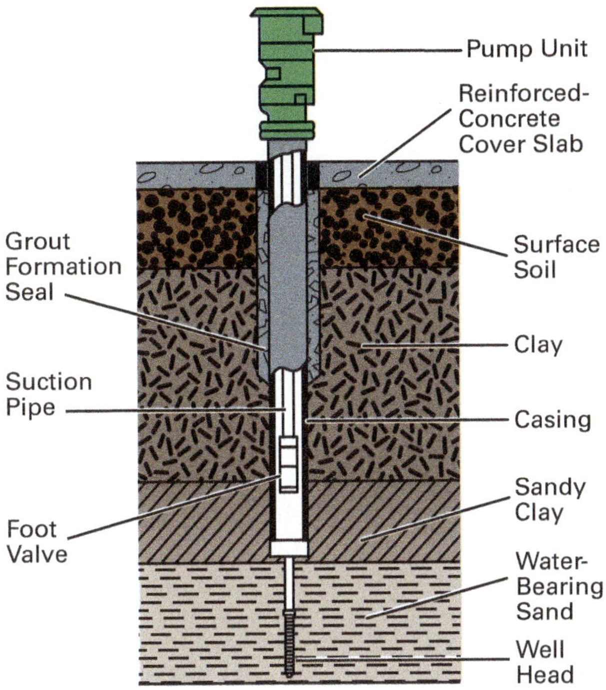

Figure 2 Cross section of a typical bored well.

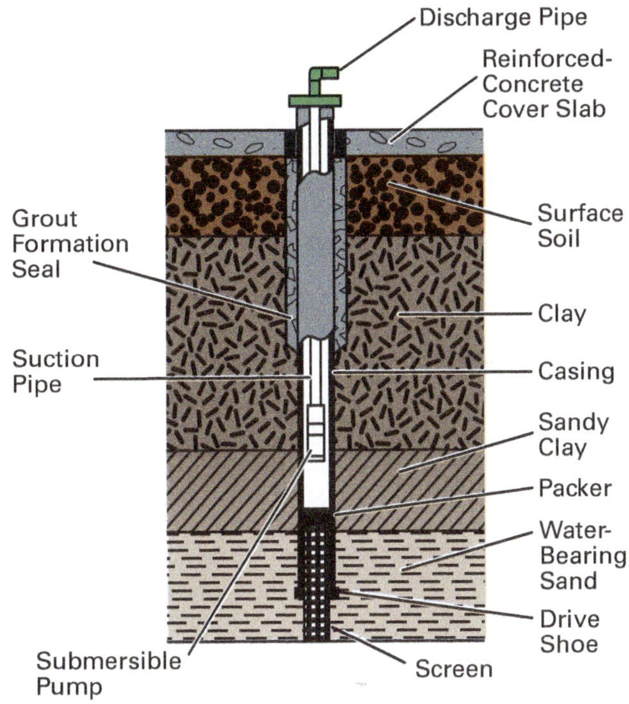

Figure 3 Cross section of a typical drilled well.

Driven wells are used where the soil is relatively free of rocks. A pointed **well head** forced into the ground with a hammer creates driven wells (*Figure 4*). Well heads are usually between $1^1/_2$" and 2" in diameter.

Dug wells are excavated with shovels and backhoes, and are lined with brick, stone, concrete, tile, or metal (*Figure 5*). Their diameters depend on the excavating tools used to dig them.

Wells that are 25' or less in depth are called **shallow wells**. Wells that are deeper than 25' are called **deep wells**. Bored, dug, and driven wells are usually shallow wells. Drilled wells are most often deep wells. Regardless of how the well is created, it is correct to say you are drilling a well.

Due to the enormous variety of geological conditions in different regions, the types and methods of well construction differ from region to region. Some types of wells may be common in some regions and rarely used, or even prohibited, in others. Be sure to check your local code when making decisions on what type of well to construct.

Driven well: A type of well drilled by a pointed well head driven by a hammer.

Well head: A pointed rod used to create a driven well by being forced into the ground with a hammer.

Shallow well: A well that is no more than 25' deep.

Deep well: A well that is drilled deeper than 25'.

WARNING!

A flooded well is a serious health hazard. Flooded wells may not be safe for months after a flood. Floodwater contains bacteria, wastewater, and chemicals. Repeated testing is usually required to ensure the safety of the water. The EPA does not recommend disinfecting or using flooded dug wells.

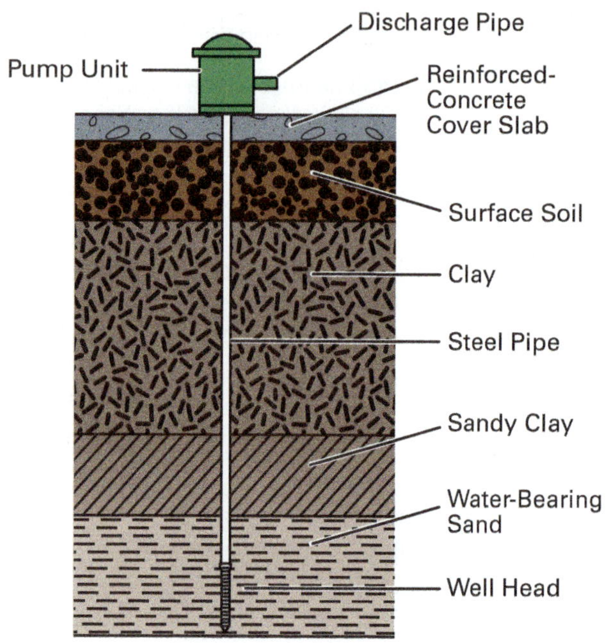

Figure 4 Cross section of a typical driven well.

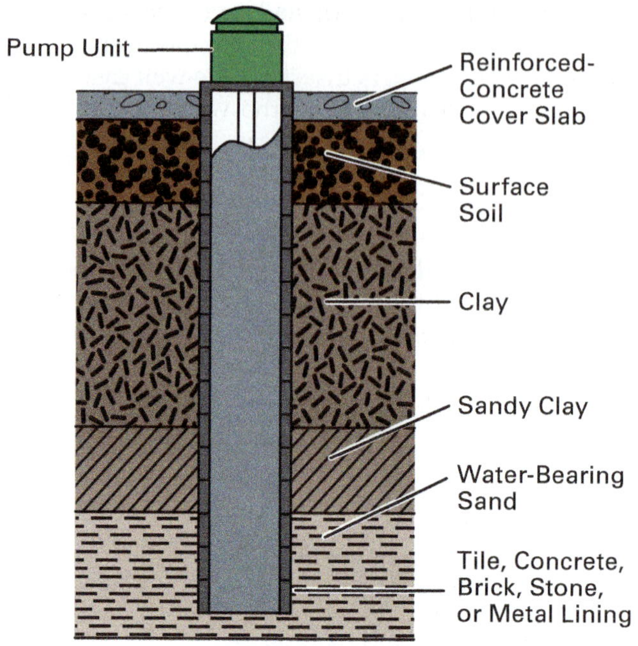

Figure 5 Cross section of a typical dug well.

1.4.0 Cleaning Wells

Chlorinate: To disinfect a well with a chlorine compound such as calcium hypochlorite or liquid bleach.

After all the sand and dirt left in the well pit from the drilling operation have been removed, **chlorinate**, or disinfect, the well. Add a chlorine compound, such as calcium hypochlorite or liquid household bleach, to the well water. Dilute the chlorine with water before adding. For calcium hypochlorite, use a 70 percent solution with water. For liquid bleach, use a 5.25 percent solution with water. Your local code will provide guidelines on how much disinfectant to use (*Table* 2). Health officials are responsible for testing the water quality of private water supplies. Most testing for contaminants occurs after installation, annually, or when a problem is suspected.

WARNING!

Disinfectants do not protect against pesticides, heavy metals, and other nonbiological contaminants. Consult local health officials if you suspect that a well contains nonbiological contaminants. Home water-treatment units, also called point-of-use or point-of-entry units, may be required to eliminate such contaminants.

Did You Know?

Arsenic Levels

Arsenic is a natural chemical that is released by rock, mineral, and soil erosion. Arsenic is also a by-product of industrial processes. Scientific studies have linked continued exposure to arsenic in drinking water to certain types of cancer. In 2006, the limit on the amount of arsenic in drinking water was lowered to 10 parts per billion (ppb).

Water-Quality Testing

Health officials test well water when private water supply installations are complete. However, tests for nitrate and coliform bacteria should be performed annually. Tests for radon and pesticides should be performed more frequently if a problem is suspected.

Private laboratories will test for nitrate and bacteria for a small fee. Tests for other types of contaminants could cost hundreds or thousands of dollars. Contact your local water-quality or certification officer for a list of testing laboratories in your area.

Laboratories usually mail test results within a few weeks. If the results are positive for contamination, contact the public health department for guidance and instructions.

TABLE 2 Sample Table for Determining Chlorination Levels in Well Water

Water Depth(ft)		Well Diameter (in) 2	3	4	5	6	8	10	12	16	20	24	28	32	36	42	48
5	A	1T	1T	1T	1T	1T	1T	2T	3T	5T	6T	3 oz	4 oz	5 oz	7 oz	9 oz	12 oz
	B	1C	1C	1C	1C	1C	1C	1C	1C	2C	4C	1Q	2Q	3Q	3Q	4Q	5Q
10	A	1T	1T	1T	1T	1T	2T	3T	5T	8T	4 oz	6 oz	8 oz	10 oz	13 oz	$1\frac{1}{2}$ lbs	$1\frac{1}{2}$ lbs
	B	1C	1C	1C	1C	1C	1C	2C	2C	1C	2Q	3Q	4Q	4Q	6Q	8Q	$2\frac{1}{2}$ G
15	A	1T	1T	1T	1T	2T	3T	5T	8T	4 oz	6 oz	9 oz	12 oz	1 lb	$1\frac{1}{2}$ lbs	$1\frac{1}{2}$ lbs	2 lbs
	B	1C	1C	1C	1C	1C	2C	3C	4C	2Q	$2\frac{1}{2}$ Q	4Q	5Q	6Q	2G	3G	4G
20	A	1T	1T	1T	2T	3T	4T	6T	3 oz	5 oz	8 oz	—	—	—	—	—	—
	B	1C	1C	1C	1C	1C	2C	4C	1Q	$2\frac{1}{2}$ Q	$3\frac{1}{2}$ Q						
30	A	1T	1T	2T	3T	4T	6T	3 oz	4 oz	8 oz	12 oz	—	—	—	—	—	—
	B	1C	1C	1C	1C	2C	4C	$1\frac{1}{2}$ Q	2Q	4Q	5Q						
40	A	1T	1T	2T	4T	6T	8T	4 oz	6 oz	10 oz	1 lb	—	—	—	—	—	—
	B	1C	1C	1C	2C	2C	1Q	2Q	$2\frac{1}{2}$ Q	$4\frac{1}{2}$ Q	7Q						
60	A	1T	2T	3T	5T	8T	4 oz	6 oz	9 oz	—	—	—	—	—	—	—	—
	B	1C	1C	2C	3C	4C	2Q	3Q	4Q								
80	A	1T	3T	4T	7T	9T	5 oz	8 oz	12 oz	—	—	—	—	—	—	—	—
	B	1C	1C	2C	4C	1Q	2Q	$3\frac{1}{2}$ Q	5Q								
100	A	2T	3T	5T	8T	4 oz	7 oz	10 oz	1 lb	—	—	—	—	—	—	—	—
	B	1C	2C	3C	1Q	$1\frac{1}{2}$ Q	$2\frac{1}{2}$ Q	4Q	6Q								
150	A	3T	5T	8T	4 oz	6 oz	10 oz	1 lb	$1\frac{1}{2}$ lbs	—	—	—	—	—	—	—	—
	B	2C	2C	4C	2Q	$2\frac{1}{2}$ Q	4Q	6Q	$2\frac{1}{2}$ G								

Notes: A = calcium hypochlorite solution; B = liquid household-bleach solution; T = tablespoon; oz = ounces (by weight); C = cup; lb = pound; Q = quart; G = gallon

1.0.0 Section Review

1. Each of the following types of information can be found in a drilling log except _____.
   a. the well's total depth
   b. the quality of the water in the well
   c. the type of pump used in the well
   d. the types of rock, sand, and soil encountered during drilling

2. When estimating fixture output, use a flow rate that is _____.
   a. at least 75 percent of the flow rate of the fixture with the greatest seven-minute peak demand
   b. the same as the minimum acceptable rate for the system
   c. lower than the minimum acceptable rate for the system
   d. higher than the minimum acceptable rate for the system

3. The type of well that can be used where the ground is relatively soft is the _____.
   a. bored well
   b. drilled well
   c. driven well
   d. dug well

4. Disinfect wells using a(n) _____.
   a. muriatic acid solution
   b. chlorine compound
   c. mixture of lime and water
   d. ultraviolet purifier

2.0.0 Operation of Pumps and Wells

Performance Tasks

There are no Performance Tasks in this section.

Objective

Explain the operation of various types of pumps and pump components.

a. Explain the operation of a shallow-well jet pump.
b. Explain the operation of a deep-well jet pump.
c. Explain the operation of a submersible pump.

After the well has been drilled and cleaned, it must be fitted with a suitable pump and the components that allow the pump to operate. In fact, pump selection may be the most critical decision of the entire project. In some jurisdictions, only licensed installers are allowed to install pumps. If you are allowed to install pumps, be sure to follow the manufacturer's specifications and the local codes. You learned how to install and troubleshoot sewage, sump, and recirculation pumps earlier in this curriculum. Well pumps have many mechanical similarities to these pumps, but well installations have their own requirements. You can use the same troubleshooting techniques you learned earlier.

Consider the following factors when selecting a well pump (*Figure 6*):

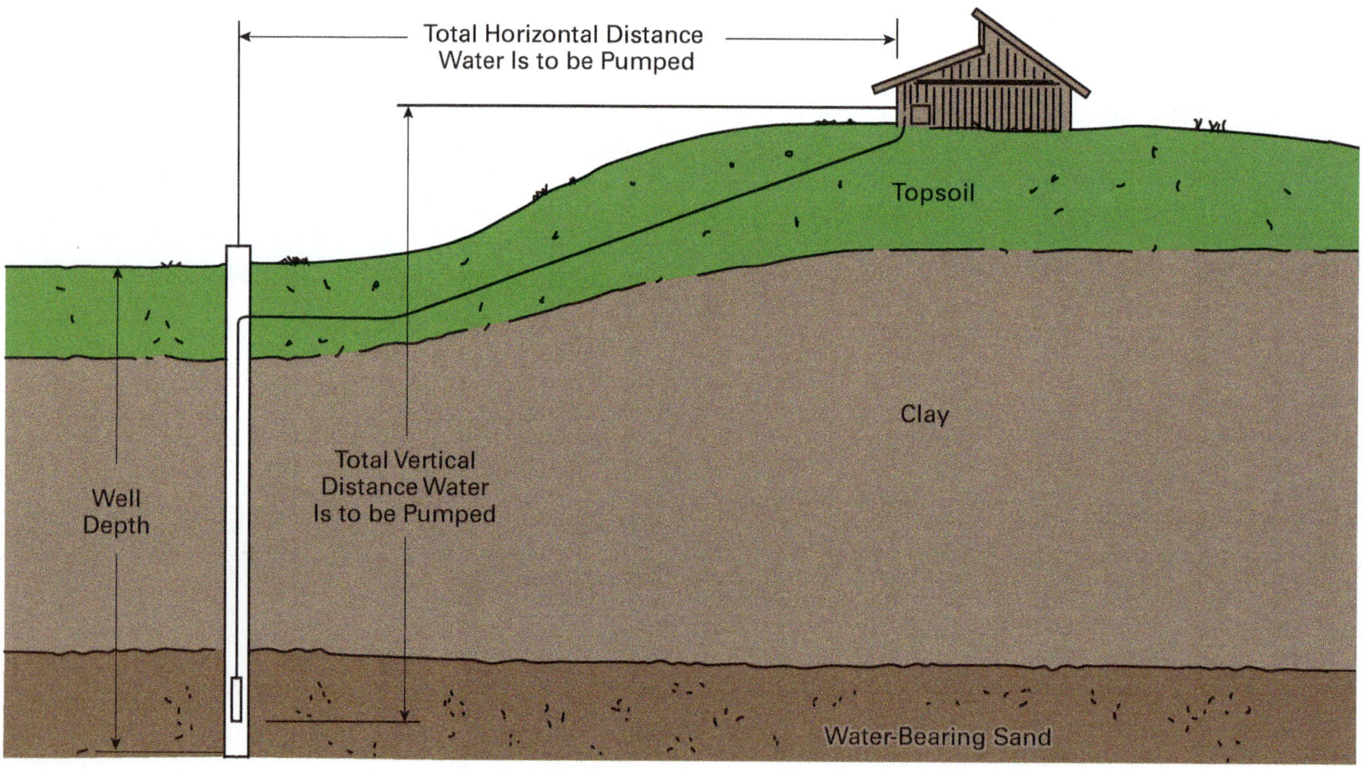

Figure 6 Factors to consider when selecting an appropriate pump.

- The total vertical height that the water must be lifted in the well
- The total vertical and horizontal distance that the water must be pumped to the structure
- The rate at which the water is pumped

Remember that 1 cubic foot of pure water weighs 62.4 pounds. Every vertical foot that water must be lifted, first in the well and then to the structure, adds to the weight that the pump must overcome. Long vertical and horizontal pipe runs require more pipe and fittings, resulting in friction loss. A high pumping rate will also add to friction loss. Select a pump that will be able to overcome these obstacles and still provide water at the correct pressure.

Ensure that the pump is set at the proper water depth within the well. If the pump is set at the bottom of the well, sediment could clog the pump and burn out the motor. If the pump is placed too high in the well, it may not be able to pump enough water. Refer to the manufacturer's specifications.

Three types of pumps are commonly used in wells:

- Shallow-well jet pumps
- Deep-well jet pumps
- Submersible pumps

Each type of pump is reviewed in the following sections. Take time to read the information closely. The safe and efficient operation of the well depends largely on selecting the proper pump.

CAUTION

Sediment and floodwater can damage pumps and motors. If a pump has been flooded, remove it from the well and clean it thoroughly. Ensure that all valves and moving parts are free of debris. Some codes require that licensed electricians or well contractors perform this work. Consult your local code.

WARNING!

Do not turn on a nonsubmersible pump in a well that has been flooded. There is a danger of electric shock and damage to the well or pump. Remove the pump and inspect it thoroughly before trying to use it again.

2.1.0 Understanding the Operation of Shallow-Well Jet Pumps

As the name suggests, **shallow-well jet pumps** are used in wells that are no more than 25' deep (*Figure 7*). The pump uses an impeller to draw water through a jet assembly. The jet assembly consists of a nozzle and a **venturi** attached to the pump A venturi, also called a diffuser, is a narrowing of the pipe that accelerates water flowing through it. The accelerated water lowers the pressure in the pump body so normal air pressure pushes the well water up the suction pipe and into the structure (*Figure 8*). Note that at high altitudes, the lower air pressure will reduce the depth at which shallow-well jet pumps can function.

Shallow-well jet pump: A pump used in a shallow well that uses a nozzle and venturi attached to the pump to accelerate the water flow.

Venturi: A device that accelerates water and lowers its pressure by passing it through a narrow pipe opening. It is also called a diffuser.

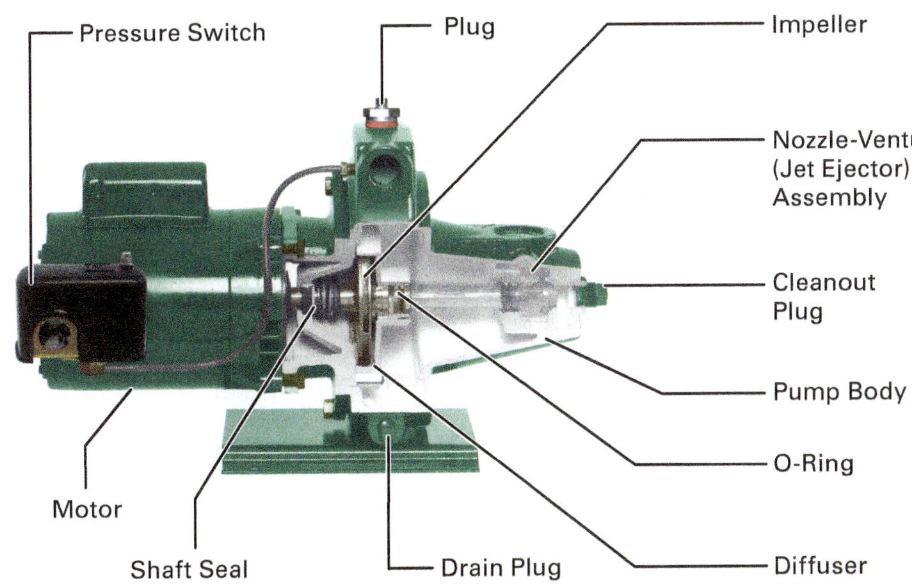

Figure 7 Cut-away of a shallow-well jet pump.
Source: Courtesy of A.Y. McDonald Mfg. Co., www.aymcdonald.com

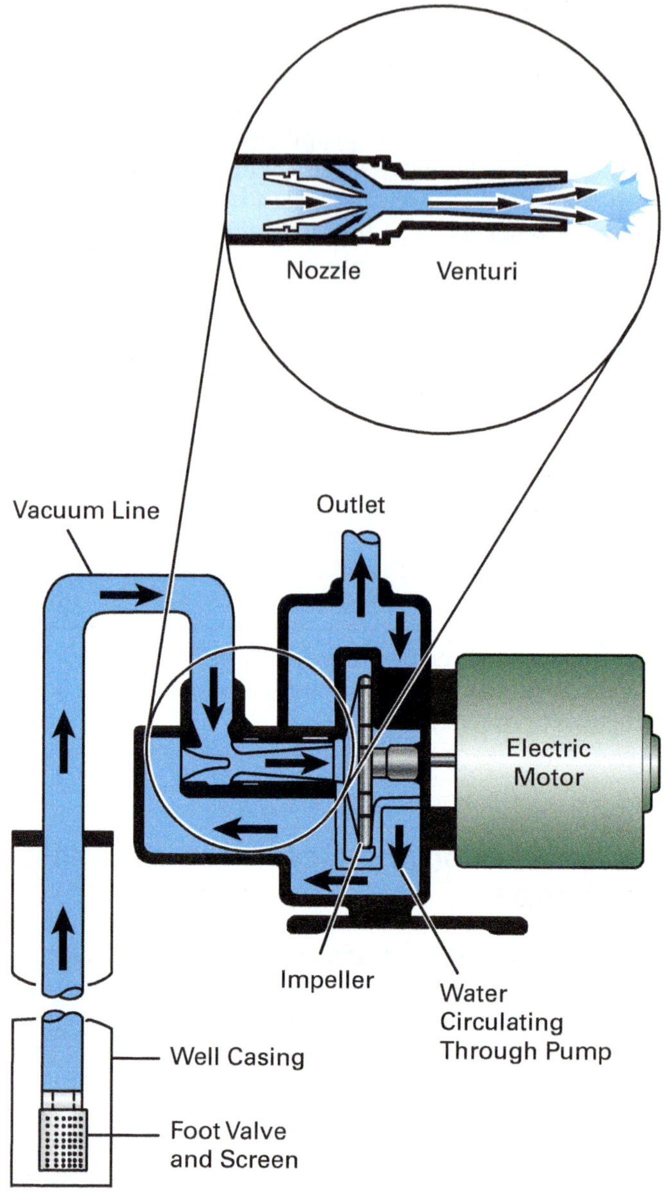

Figure 8 Water flow in a shallow-well jet pump.

Foot valve: A valve used to prime water-lubricated pumps in wells deeper than 32'.

Follow these steps when installing shallow-well jet pumps (*Figure 8*):

Step 1 Attach the **foot valve** to the end of the pipe. This is a self-priming valve for water-lubricated pumps in wells deeper than 32', which is the maximum height that a single pump can pump a column of water.

Step 2 Lower the assembly into the well, adding lengths of water line pipe as the assembly descends.

Step 3 When the assembly is properly positioned, place a well seal on top of the well. The seal will protect the well from contamination.

Step 4 Place a tee at the top of the vertical suction pipe, and plug the top opening. Attach the suction side of the pump to the horizontal opening of the tee.

Step 5 Connect the water storage tank to the discharge side of the pump.

Step 6 Install a shutoff valve and a union between the pump and the tank.

Did You Know

Hydraulic Fracturing

Most of the wells that plumbers help install are designed to remove water from the ground. Wells can also be used the other way, to inject water deep into the ground to cause rocks to fracture. This process is called hydraulic fracturing, or fracking. In hydraulic fracturing, water mixed with sand and trace amounts of chemicals is injected under high pressure into a well to force underground rock to split apart, causing thin cracks called fissures. When the pressure is removed, natural gas, liquid petroleum, and even uranium can then flow through the fissures into the well, where they can then be pumped out and processed for use by consumers and industry.

Hydraulic fracturing allows previously inaccessible deposits of gas, petroleum, and minerals to be tapped more easily than other mining or drilling efforts, which results in lower fuel costs for consumers. In recent years, concerns over the potential contamination of aquifers from the chemicals in the pressurized water have led to increased scrutiny by regulators to ensure that hydraulic fracturing does not accidentally contaminate potable water supplies.

2.2.0 Understanding the Operation of Deep-Well Jet Pumps

Jet pumps used in deep wells work the same way as those used in shallow wells. However, they differ slightly in the arrangement of the jet assembly. In a **deep-well jet pump**, the nozzle and venturi are located in the well, usually about 5' below the low waterline. Some water is diverted from the discharge line back into the jet assembly (*Figure 9*). This allows the jet cycle to continue.

Deep-well jet pump: A pump used in a deep well that uses a nozzle and venturi, usually located separately from the pump about 5' below the water line within the well, to accelerate the water flow.

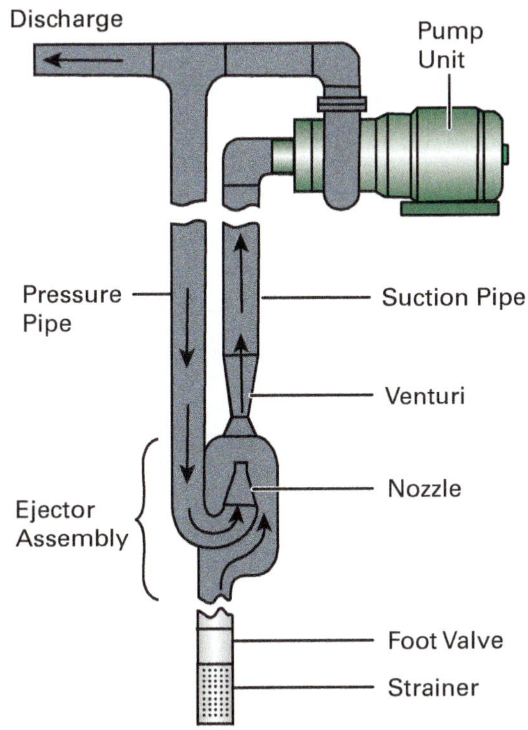

Figure 9 Water flow in a deep-well jet pump.

The installation procedure for deep-well jet pumps is like that for shallow-well jet pumps. Many deep-well jet pumps use a **two-pipe system** to pump water from a well (*Figure 9*). One pipe (the pressure pipe) feeds pressurized water to the jet assembly. Well water is drawn up the other pipe (the suction pipe).

Two-pipe system: For jet pumps, an alternative to the packer-type ejector discharge method. One pipe draws well water through an ejector, and another pipe feeds the pressurized water to the jet.

An ejector on the bottom of the pipe assembly allows well water to enter the suction pipe. Install the ejector after installing the foot valve. (Do not confuse the two-pipe system with the hydronic heating system of the same name. This other type of two-pipe system is discussed in the Hydronic and Solar Heating Systems module.)

If the well is too narrow for a two-pipe installation, install a **packer-type ejector** (*Figure 10*). In a packer-type ejector, the space between the water supply pipe and the well casing doubles as the pressure pipe. Though packer-type ejectors take up less space, their steel casings rust through over time. Two-pipe systems are less costly to install and easier to replace.

To protect the pump if the well runs dry, install a **low-water cutoff device** on the pump (*Figure 11*). Low-water cutoff devices turn off the pump when the water level in the well falls to a predetermined low point. The device turns the pump on when the water level recovers. Interestingly, while this equipment

Packer-type ejector: For jet pumps, an alternative discharge method to the two-pipe system in which the space between the water supply pipe and the well casing acts as a pressure pipe.

Low-water cutoff device: A device that deactivates a pump when the water level in a well falls to a preset low point. See also tail pipe assembly.

> ### Did You Know?
> **The United States Geological Survey**
>
> The Water Resources Division of the United States Geological Survey (USGS) is responsible for providing up to date hydrologic information and resources. USGS maintains a database of more than 3.5 million water-quality analyses. It also operates measuring instruments that continuously record and transmit water-quality data.
>
> USGS also maintains the Ground Water Site Inventory, with more than 850,000 records of wells, springs, test holes, tunnels, drains, and excavations throughout the country. This database is available through the USGS website at *waterdata.usgs.gov/nwis/gw*.

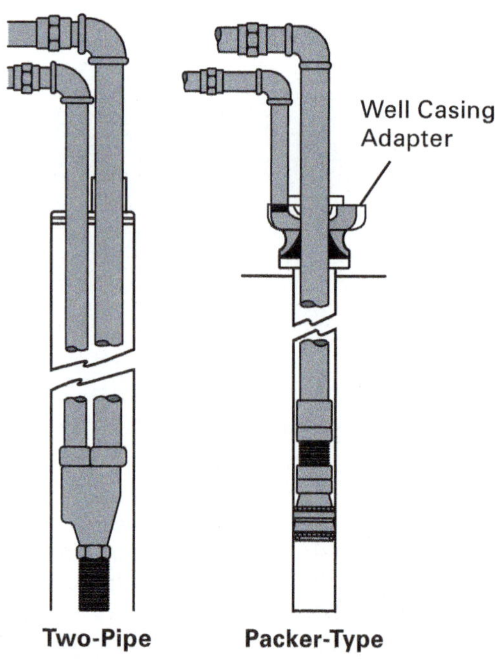

Figure 10 Comparison of a two-pipe system and a packer-type ejector in a deep-well jet pump installation.

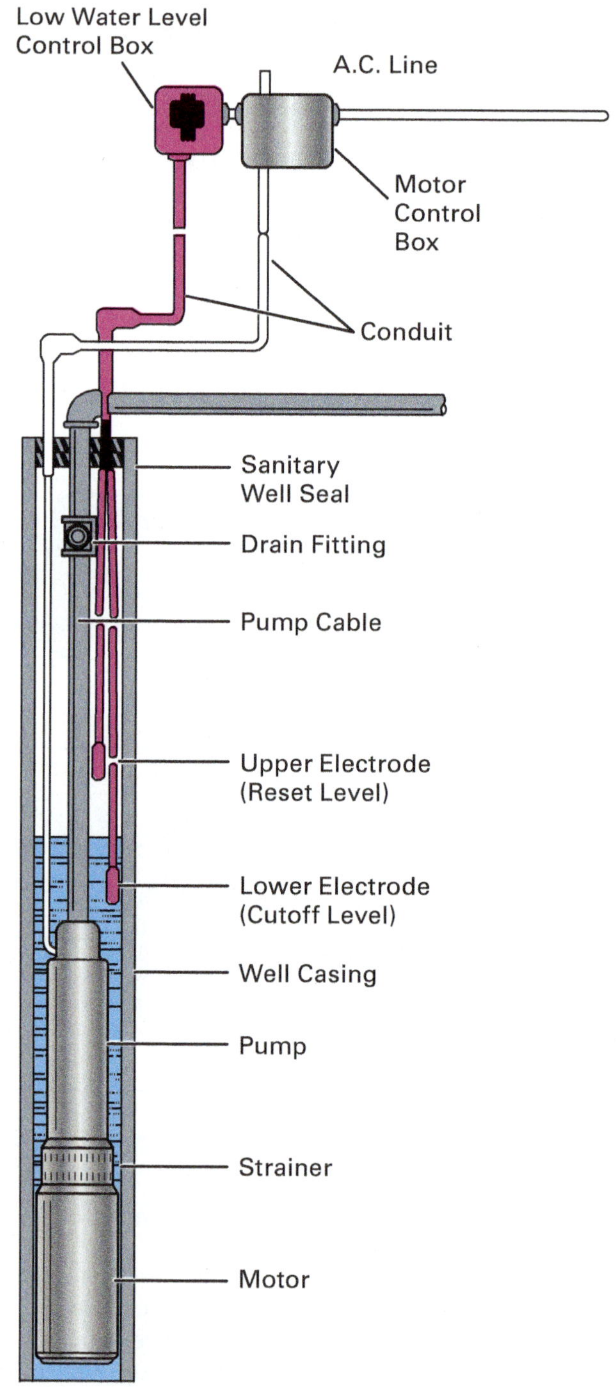

Figure 11 Typical low-water cutoff device.

operates on the same general principle as low-water cutoff devices in boilers, they are different devices. Those used in boilers are safety devices intended to prevent dangerous conditions, while the devices used in well pumps are meant to protect the pump itself from damage caused by running dry.

Another option on two-pipe systems is to install a **tail pipe assembly** below the jet assembly and above the foot valve (*Figure 12*). The length of the tail pipe assembly adjusts the pumping rate to balance with the well's recovery rate. If the recovery slows, the pumping rate also slows.

Install **pressure control valves** on all deep-well jet pumps. These valves divert some of the pump's output down the ejector. This ensures that there is enough pressure to operate the ejector, even during periods of minimum discharge. Finally, install a **well-casing adapter** to seal off the top of the well casing from air leaks.

Tail pipe assembly: A device that matches a pump's discharge rate to a well's recovery rate. See also low-water cutoff device.

Pressure control valve: A valve that diverts a portion of a pump's discharge through the ejector, thereby maintaining adequate pressure in the ejector.

Well-casing adapter: A well cover that protects the well casing from air leaks.

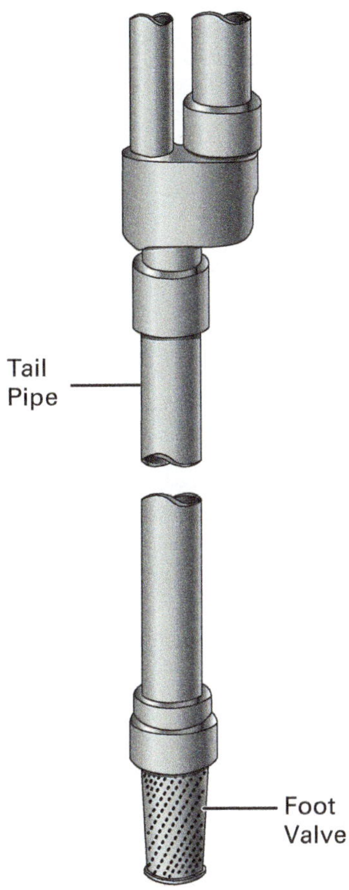

Figure 12 Tail pipe assembly.

2.3.0 Understanding the Operation of Submersible Pumps

Submersible pumps are mostly used in deep wells. They are more efficient and easier to install than jet pumps. Submersible pumps are installed at the bottom of the well, below the water level.

Submersible pumps have two major components (*Figure 13*):

- A multistage centrifugal pump
- A submersible electric motor

Well water enters the pump through the intake screen near the bottom of the pump. The impellers push the water up the pump body through the discharge pipe at the top of the pump. A check valve at the top of the pump prevents water in the discharge pipe from running back into the well when the pump shuts off. Note that in *Figure 13*, the motor-shaft extension at the top of the motor fits into the splined coupling at the bottom of the pump.

Submersible pump: A pump consisting of a multistage centrifugal pump and a submersible electric motor. Submersible pumps are usually installed at the bottom of deep wells.

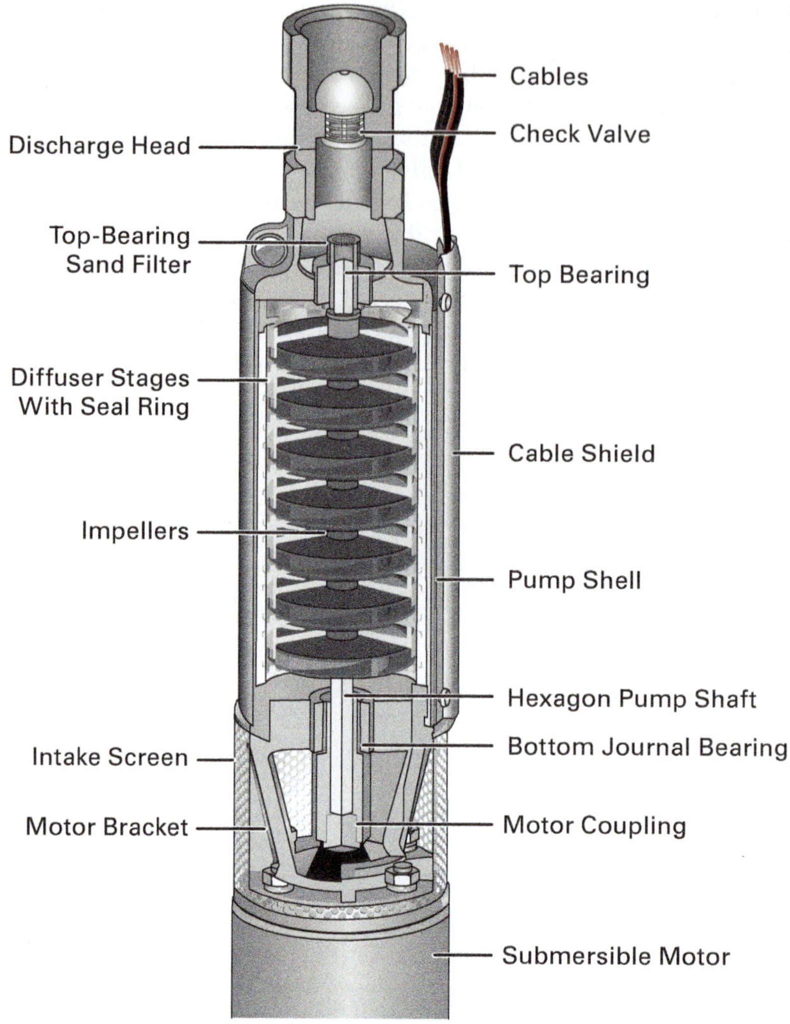

Figure 13 Components of a typical submersible pump.

WARNING!

In some jurisdictions, electricians must install and test pump electrical systems. Review your local codes and standards before installing and testing pump electrical systems. Never try to do electrical work by yourself if you have not had the proper training.

Test the pump before installing it, following the manufacturer's instructions. Use a collar clamp to support the pump and pipe assembly. Remember that submersible pumps are water lubricated. Submerge the pump in water before testing. Otherwise, the pump will be damaged. Test the check valve according to the manufacturer's instructions.

WARNING!

Always turn off the pump motor before attempting to repair or replace the pump. When wet, pump motors can cause injury and death from electric shock. Dry the pump motor thoroughly before testing. Wear appropriate personal protective equipment. Do not attempt to repair a pump or pump motor without proper training.

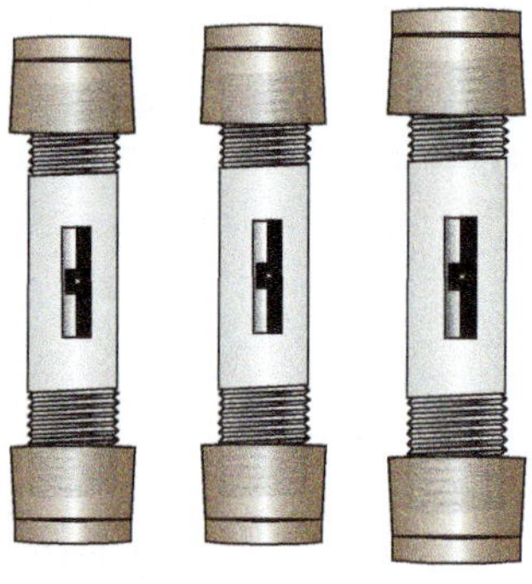

Figure 14 Waterproof mechanical splice kits.

When the pump passes all tests, it is ready to be installed. Splice the pump's electrical cable using a waterproof mechanical splice kit (*Figure 14*). Always be sure to use proper wiring techniques. Consult with local electrical officials if you have any questions about the electrical system. Tape the electrical line to the water pipe at intervals of no more than 10'. Install a secondary lightning arrester if required for safety.

Lower the pump assembly into the well using a hoist. Use the collar clamp to keep the pump assembly from falling out of control. As the pump descends, couplings on the end of each length of pipe will stop against the collar clamp. Attach the next length of pipe at that time. Use pipe dope or Teflon® tape on the fitting joints to prevent leaks. In many jurisdictions, plastic pipe is permitted for well installations. If you use plastic pipe, be careful when extracting the pump. Careless handling will break the pipe.

Install a **pitless adapter** in the well below the frost line. Pitless adapters let the discharge pipe, also called the **drop pipe**, leave the well and connect to the water storage tank (*Figure 15*). They are called pitless adapters because they eliminate the need to dig a pit to access the water line. Otherwise, contaminated water could enter the well through the pit. Because pitless adapters are installed below the frost line, they also protect the discharge line from freezing.

Pitless adapter: A device that permits a discharge pipe to exit a well and connect to a water storage tank. Pitless adapters protect well water from contamination and freezing.

Drop pipe: Another name for the discharge pipe that connects the well to the water storage tank.

(A)

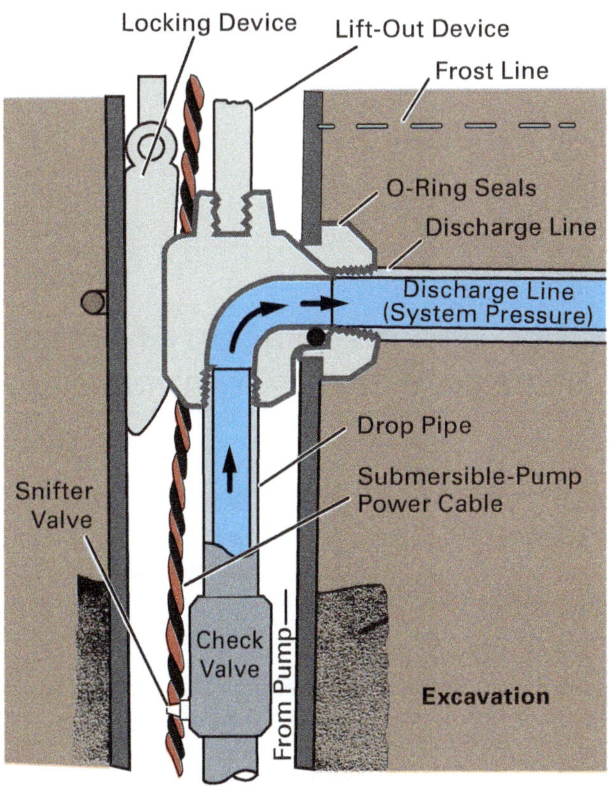

(B)

Figure 15 Pitless adapter installed in a well.
Source: (A) Courtesy of Merrill Manufacturing, Storm Lake, IA

Connect the electrical cable to the building's electrical system only after installing the water supply line. Ensure that the building's electrical system has been de-energized before connecting the cable. The electrical supply line can be buried in the same trench as the water supply piping. However, it must be at a separate level. In some areas, qualified electricians must perform this task. Refer to your local code. If you are allowed to perform the task, follow the local electrical code.

Most residential pump motors use 220 to 230 volts. Ensure that the electrical wire can handle the voltage for the length to be run. Select wire that is suitable for underground use. Bed the wire in soil or sand. This will protect the wire from cuts by sharp rocks. The wire should be laid loose or snaked. Otherwise, settling and freeze-thaw cycles may stress the wire. In most cases, bury the wire about 18" below the surface of the soil. Consult local electrical officials for requirements in your area.

2.0.0 Section Review

1. The maximum well depth for using a shallow-well jet pump is _____.
   a. 15'
   b. 25'
   c. 35'
   d. 45'

2. In a deep-well jet pump, the nozzle and venturi are located in the well at a typical depth of around _____.
   a. 5' above the low waterline
   b. 5' above the high waterline
   c. 5' below the low waterline
   d. 5' below the high waterline

3. In a submersible pump, water in the discharge pipe is prevented from running back into the well when the pump shuts off by the installation of a _____.
   a. vacuum breaker
   b. butterfly valve
   c. check valve
   d. backwater valve

3.0.0 Selection and Installation of Water Supply and Storage System Components

Performance Task

1. Assemble and disassemble given components of private water supply well systems.

Objective

Explain how to select and install water supply and storage components.

a. Explain how to select and install water supply lines.
b. Explain how to select and install water storage tanks.

In addition to pumps and their components, wells require water supply lines and storage tanks to operate. Like pumps, supply lines and storage tanks must be selected based on the needs of the particular installation. The local applicable code will specify the requirements for lines and tanks, including approved materials and sizing requirements. The following sections cover the general considerations for selecting and installing water supply lines and storage tanks. Always refer to the local applicable code for the selection and installation requirements that are applicable to your jurisdiction and for the needs of individual well installations.

3.1.0 Selecting and Installing Water Supply Lines

Your local plumbing code will specify approved piping materials in your area. The most popular piping materials are steel, copper, and plastic. Galvanized steel pipe corrodes in acidic soils. Copper usually lasts more than twice as long as galvanized steel pipe. However, copper pipe corrodes in soil with high carbon dioxide or hydrogen sulfide content. Plastic water supply pipe lasts the longest. Consult your local code and plumbing officials.

Ensure that the pipe is properly sized for the installation. When selecting water supply pipe, consider the following factors:

- Cost
- Life expectancy
- Likelihood of corrosion and freezing
- System pressure
- Local plumbing codes
- Ease of installation

With experience, you will learn how to select the proper pipe for the types of installations you will encounter.

Bed water supply lines in soft, clean soil or sand. Ensure that there are no large rocks or boulders in the material used to backfill the pipe trench. Falling rocks and boulders can damage water pipes during an improper backfilling procedure. Install dielectric fittings on the supply lines to protect them from galvanic corrosion. Talk to local plumbing officials to determine the risk of galvanic corrosion in your area.

Most codes require that a water sample be drawn from the well and tested. Water tests are usually done after the well has been purged but before the supply line has been connected to the building. Local water-quality officials perform the tests. Water is tested for four characteristics:

- **Turbidity** – The presence of suspended solids in the water. Turbidity is less of a concern in groundwater than in surface water.
- **pH value** – The number of hydrogen ions in the water. The term pH is an abbreviation for "potential hydrogen."
- Hardness – The presence of mineral salts in the water.
- **Biochemical oxygen demand (BOD)** – The amount of oxygen in the water available to support waterborne organisms.

Turbidity: A measure of suspended solids in well water.

pH value: A measure of the hydrogen ions in water. The term is an abbreviation for "potential hydrogen."

Biochemical oxygen demand (BOD): A measure of life-sustaining oxygen content in water.

Take the time to learn about the water quality of aquifers in your area. Your knowledge will help ensure that water supply line installations go smoothly.

Clean Pipes

Ensure that the water supply piping is as clean as possible before installation. Sand, gravel, or mud trapped in the pipes will flow through the piping system, lodge in the valves and faucets, and damage the washers and O-rings. The faucets will then leak. Store pipe in a clean, dry area, and cap all pipe ends at the end of each workday. Be sure to flush the line as thoroughly as possible when installation is complete to remove any sand, mud, or gravel from the system.

3.2.0 Selecting and Installing Water Storage Tanks

Water storage tanks serve three main functions:

- They store water for intermittent use.
- They prevent the short cycling of the pump.
- They provide water under constant pressure throughout the building.

Pressurized air-bladder storage tanks are widely used in private water supply well systems (*Figure 16*). In this type of tank, water is circulated at the proper system pressure by air pressure. The tank is equipped with an internal air bladder. The pump fills the tank to a preset pressure.

When the customer operates a fixture, air pressure in the tank pushes down on the bladder, allowing water to flow through the system at a constant pressure. The pump then refills the tank when it reaches a preset lower air pressure (*Figure 17*).

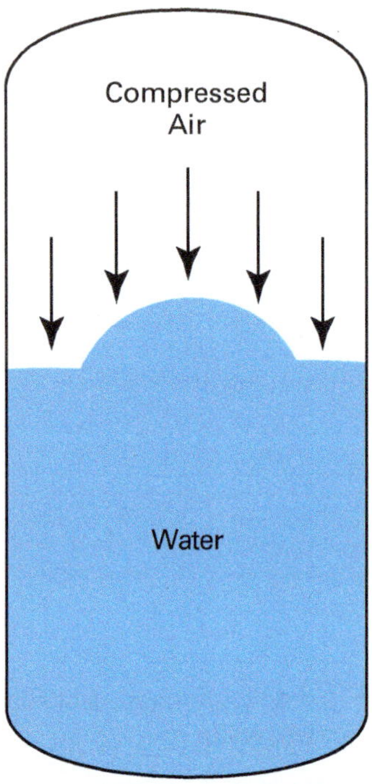

Figure 16 Typical pressurized air-bladder storage tank.

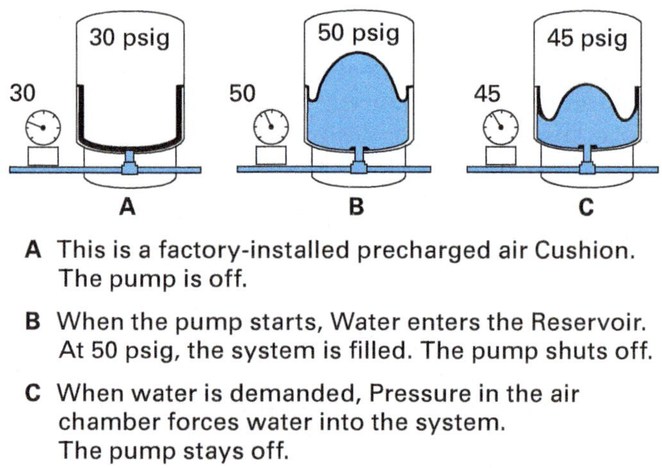

Figure 17 Operation of a typical pressurized air-bladder storage tank.

The main function of the pressure tank is to reduce pump cycling. Therefore, size the water tank according to the desired running time of the pump. Refer to the tank manufacturer's specifications to determine the proper-size tank. For example, some manufacturers recommend a running time of one minute for a submersible pump that has a flow rate of 10 gpm and an operating pressure range of 30 to 50 pounds per square inch (psi).

Locate the water storage tank before connecting the supply lines to the building. Install a pressure switch and a pressure gauge as close as possible to the tank itself. The pressure switch controls the amount of water and the range of pressure inside the tank, and turns the pump on and off based on whether a preset pressure differential has been reached. The pressure gauge shows the pressure in both the tank and the piping system. Many tanks also require the installation of a drain and a relief valve. The drain is used to empty the tank when necessary. If the tank exceeds the maximum preset pressure because the pressure switch fails, the relief valve can drain the tank and relieve the pressure. *Figure 18* displays the typical installation of a pressure switch, pressure gauge, relief valve, and drain on a pressurized air-bladder storage tank.

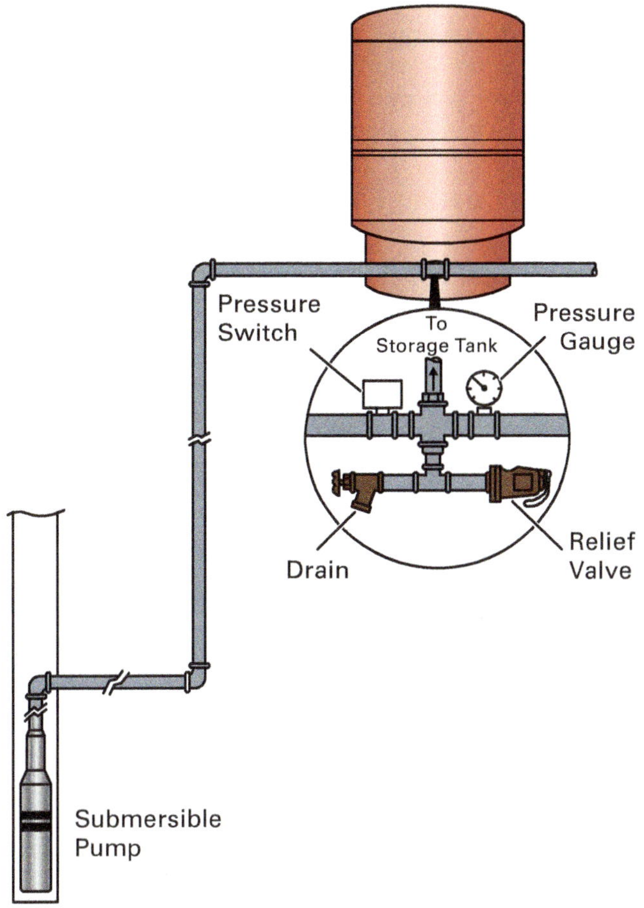

Figure 18 Typical installation of a pressure switch, pressure gauge, relief valve, and drain on a pressurized air-bladder storage tank.

After installing the pressure tank and controls, purge the entire water system. Flush the potable water lines thoroughly with water. This will wash out the chlorine used to disinfect the well. It will also remove sediment, joint compound, and other foreign matter still in the lines. Following the purge, perform final checks on the tank's pressure-switch settings. At this point, a water-quality authority should test the water.

Following the water-quality test, the installation of the private water supply well system is complete. From that point on, the installation of stacks, branches, and fixtures is just like in any other building. Install drain, waste, and vent (DWV) and water supply systems using the techniques you have already learned. Refer to your local code for other conditions that apply to private water supply well systems.

Think About It

Friction

The greater the number of fittings in the system (valves, tees, elbows, and so on), the greater the amount of friction, which causes a pressure drop in the system.

Did You Know?

Protecting Private Water Supplies

The EPA recommends several steps for protecting private water supply well systems. Periodically, one should inspect exposed parts of wells for cracked, corroded, or damaged well casings, broken or missing well caps, and settling or cracking of surface seals. The area around the well should be sloped to drain surface runoff away from the well. A well cap or sanitary seal should be in place to prevent unauthorized use or entry, while wells themselves should be tested once a year for coliform bacteria, nitrates, and other concerning contaminants.

When servicing wells, maintain accurate records. These should include the time and nature of disinfection, sediment removal, and any other chemicals used in the well.

Only certified well drillers should perform new well construction, modifications on existing wells, or abandonment and closure of wells.

Do not mix or use pesticide, fertilizers, herbicides, degreasers, fuels, or other pollutants near wells. Likewise, do not dispose of wastes or hazardous materials in dry or abandoned wells or septic systems. Never cut off the well casing below the land surface.

3.0.0 Section Review

1. Each of the following is a popular material for water supply piping in a private water supply well system except _____.
   a. copper
   b. steel
   c. plastic
   d. cast iron

2. The main function of the pressure tank is to _____.
   a. store more water in a smaller space
   b. reduce pump cycling
   c. use less energy to heat the water
   d. equalize service pressure across all fixtures

Module 02408 Review Questions

1. People depend on wells to provide reliable water that is _____.
   a. diluted
   b. artesian
   c. filtered
   d. potable

2. For any given installation, the person who decides which type of well works best is the _____.
   a. plumber
   b. driller
   c. customer
   d. code official

3. Driven wells are typically used where the soil is relatively free of _____.
   a. rocks
   b. roots
   c. metals
   d. air pockets

4. The excavating tools used to drill a dug well will affect its _____.
   a. liner
   b. well head
   c. diameter
   d. well casing

5. Testing the water quality of a well is the responsibility of _____.
   a. health officials
   b. code officials
   c. plumbers
   d. customers

6. A well that is no more than 25' deep uses a _____.
   a. venturi-well jet pump
   b. shallow-well jet pump
   c. diffuser-well jet pump
   d. pump-well jet pump

7. One disadvantage to using packer-type ejectors is _____.
   a. they take up a lot of space
   b. they are difficult to install
   c. their steel casings eventually rust
   d. their use is prohibited in many jurisdictions

8. A low-water cutoff device is used to protect the pump if the well _____.
   a. runs dry
   b. overflows
   c. becomes turbid
   d. becomes contaminated

9. A submersible pump has two major components: a submersible electric motor and a _____.
   a. multistage impeller pump
   b. multistage centrifugal pump
   c. motor shaft pump
   d. water-lubricated pump

10. Most residential pump motors use _____.
   a. 180 to 190 volts
   b. 200 to 210 volts
   c. 220 to 230 volts
   d. 240 to 250 volts

11. The type of water supply piping that lasts the longest when used in private water supply well systems is _____.
   a. galvanized steel pipe
   b. plastic pipe
   c. copper pipe
   d. lead pipe

12. Local water-quality officials test water for four characteristics: hardness, biochemical oxygen demand, pH value, and _____.
   a. hydrogen ions
   b. mineral salts
   c. turbidity
   d. galvanic corrosion

13. Air pressure is used to maintain water pressure at the proper system pressure in a pressurized _____.
   a. air-bladder storage tank
   b. high-pressure storage tank
   c. submersible-pump tank
   d. supply-line tank

14. Draining the tank and relieving the pressure is the purpose of a _____.
   a. gauge valve
   b. drain valve
   c. pressure valve
   d. relief valve

15. After installing the pressure tank and controls, the next step is to _____.
   a. perform final checks on the settings
   b. test the water for impurities
   c. purge the entire water system
   d. install a vent system

Answers to odd-numbered review questions are found in the Module Review Answer Key found at the back of this book.

Answers to Section Review Questions

Answer	Section	Objective
Section One		
1. c	1.1.0	1a
2. d	1.2.0	1b
3. a	1.3.0	1c
4. b	1.4.0	1d
Section Two		
1. b	2.1.0	2a
2. c	2.2.0	2b
3. c	2.3.0	2c
Section Three		
1. d	3.1.0	3a
2. b	3.2.0	3b

MODULE 02409

Private Waste-Disposal Systems

Source: Lost_in_the_Midwest/Shutterstock

Objectives

Successful completion of this module prepares you to do the following:

1. Describe the types of private waste-disposal systems.
   a. Describe soil absorption systems.
   b. Describe organic systems.
   c. Describe closed systems.
2. Explain how to locate and size private waste-disposal systems.
   a. Explain how to locate and size septic tanks.
   b. Explain how to locate and size leach fields.
3. Explain how to install private waste-disposal systems.
   a. Explain how to install septic tanks.
   b. Explain how to install leach-field distribution piping.
4. Explain how to clean and service private waste-disposal systems.

Performance Tasks

This is a knowledge-based module. There are no performance tasks.

Overview

Rural areas are often too far away from a city to connect to the municipal waste system. Farms and parks in such areas require private waste-disposal systems. The three categories of private waste-disposal systems (also called private sewage treatment systems) are soil absorption systems, organic systems, and closed systems. Soil absorption systems are the most common. The three types of soil absorption system are septic systems, aeration systems, and pressure distribution systems. Septic systems are the most common type of soil absorption system. They work by breaking down solid wastes and allowing liquid wastes, or effluent, to leach into the ground.

CODE NOTE

Codes vary among jurisdictions. Because of the variations in code, consult the applicable code whenever regulations are in question. Referring to an incorrect set of codes can cause as much trouble as failing to reference codes altogether. Obtain, review, and familiarize yourself with your local adopted code. On-site private sewage treatment systems are commonly regulated by health departments, so be sure to check with the local health department and building departments for regulations governing these systems.

NCCER Industry-Recognized Credentials

If you are training through an NCCER-accredited sponsor, you may be eligible for credentials from NCCER. The ID number for this module is 02409. Note that this module may have been used in other NCCER curricula and may apply to other level completions. Contact NCCER at 1.888.622.3720 or go to **www.nccer.org** for more information.

You can also show off your industry-recognized credentials online with NCCER's digital credentials. Transform your knowledge, skills, and achievements into digital credentials that you can share across social media platforms, send to your network, and add to your resume. For more information, visit **www.nccer.org**.

Digital Resources for Plumbing

SCAN ME

Scan this code using the camera on your phone or mobile device to view the digital resources related to this craft.

1.0.0 Types of Private Waste-Disposal Systems

Performance Tasks

This is a knowledge-based module. There are no Performance Tasks in this section.

Objective

Describe the types of private waste-disposal systems.

a. Describe soil absorption systems.
b. Describe organic systems.
c. Describe closed systems.

Municipal waste systems serve urban homes, offices, and businesses. Rural areas, such as farms and parks, are often too far away from a municipal system to be connected. Just as they rely on private water supply systems, people who live and work in such areas also use private waste-disposal systems. These systems are installations that treat and dispose of the sewage wastes from a single building. In this module, you will learn about the different kinds of private waste-disposal systems and the way those systems treat and dispose of wastes.

Like other plumbing installations, private waste-disposal systems must meet local code standards. The system must protect the health of the users and the safety of the environment. Plumbers working in rural and semirural areas must be familiar with the local code requirements that affect private waste-disposal systems. They must also know how to choose the right system for the customer's needs, as well as how to install the different types of systems. In this module, you will learn how to lay out, install, and maintain the common types of private waste-disposal systems.

The three main categories of private waste-disposal systems are:

- Soil absorption systems
- Organic systems
- Closed systems

Soil absorption system: A private waste-disposal system that uses tanks to separate out effluent, then allows it to drain into a leach field.

Organic system: A private waste-disposal system that allows waste to decompose through exposure to aerobic bacteria. The resulting dried waste is called humus.

Closed system: A private waste-disposal system that stores wastes until they can be disposed of.

Soil absorption systems store solid wastes and allow wastewater to drain into the ground. **Organic systems** allow wastes to decompose through contact with air. **Closed systems** store wastes until they can be disposed of permanently. Each of these systems is described in detail in the following sections. While the code definitions for these systems can be quite detailed, refer to the definitions here for purposes of completing this module.

Local codes, the environment, and the customer's needs will determine the most appropriate system. Take time to learn about each system. Ask experienced plumbers and code officials for their advice on which systems work best under which conditions. The more you learn, the easier it will be for you to select the system that meets each customer's needs.

1.1.0 Identifying Soil Absorption Systems

Effluent: Liquid waste that separates from solid waste. In soil absorption systems and cesspools, effluent is allowed to percolate into the soil.

Soil absorption systems are the most common type of private waste-disposal system. These systems separate solid waste from liquid waste. The liquid waste, called **effluent**, is then allowed to seep into the ground, where it is absorbed.

There are three kinds of soil absorption systems:

- Septic systems
- Aeration systems
- Pressure distribution systems

Soil absorption systems can be installed only where the soil is capable of absorbing effluent. Each location will have its own requirements.

Think About It

The International Private Sewage Disposal Code®

The International Code Council (ICC) publishes the International Private Sewage Disposal Code® (IPSDC) as part of a series of model codes that are collectively nicknamed the I-codes. Other I-codes cover plumbing, building construction, fire prevention, residential buildings, mechanical installations, fuel gas, energy conservation, property maintenance, electrical installations, and zoning. Many jurisdictions are adopting I-codes as their current codes become obsolete. The IPSDC covers the following:

- Regulations governing the installation of a private waste-disposal system
- Methods for evaluating a site
- Guidelines for layout and design
- Materials permitted for use in a private waste-disposal system
- Requirements for each type of private waste-disposal system
- Installation and maintenance of system components
- Inspection of installed systems

If your local private waste-disposal system code is based on the IPSDC, take the time to familiarize yourself with its requirements. Ask a local code expert to explain differences between the model code and your local code. The IPSDC code, as well as information on the EPA's SepticSmart program, are available publicly online. Develop a good working knowledge of your private waste-disposal system codes. The quality of your work will improve as a result.

1.1.1 Septic Systems

Septic systems are the most common type of soil absorption system. Septic systems consist of a **septic tank**, a **distribution box**, a **leach field** (sometimes called an absorption field or a drain field), and the piping that connects these parts (*Figure 1* and *Figure 2*). Septic systems work by breaking down solid wastes and allowing liquid wastes to percolate, or **leach**, into the ground. This is the most common system used today.

Septic system: A soil absorption system that uses a septic tank to separate out sludge.

Septic tank: A tank used in a soil absorption system to settle out sludge and scum and to decompose them using anaerobic bacteria.

Distribution box: A watertight container that directs effluent to different parts of a leach field.

Leach field: In a soil absorption system, an area of soil that is designed to accept effluent from distribution pipes. Leach fields are also called absorption fields or drain fields.

Leach: To percolate into soil and be absorbed.

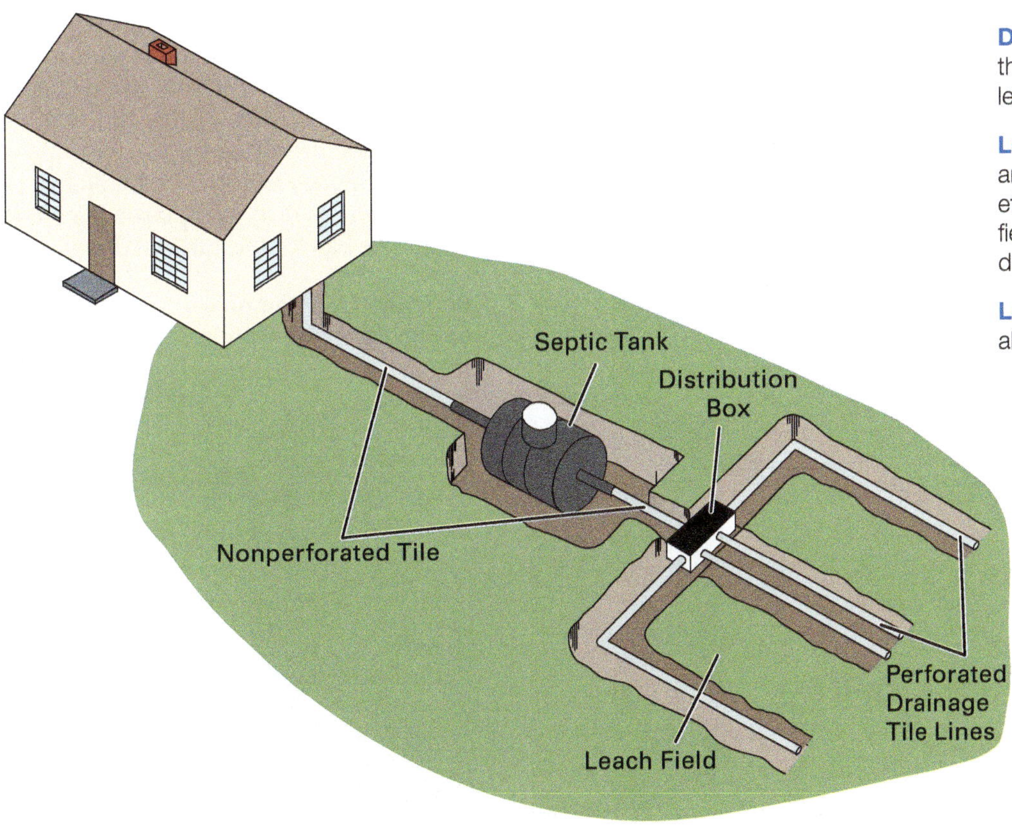

Figure 1 Typical layout for a septic system.

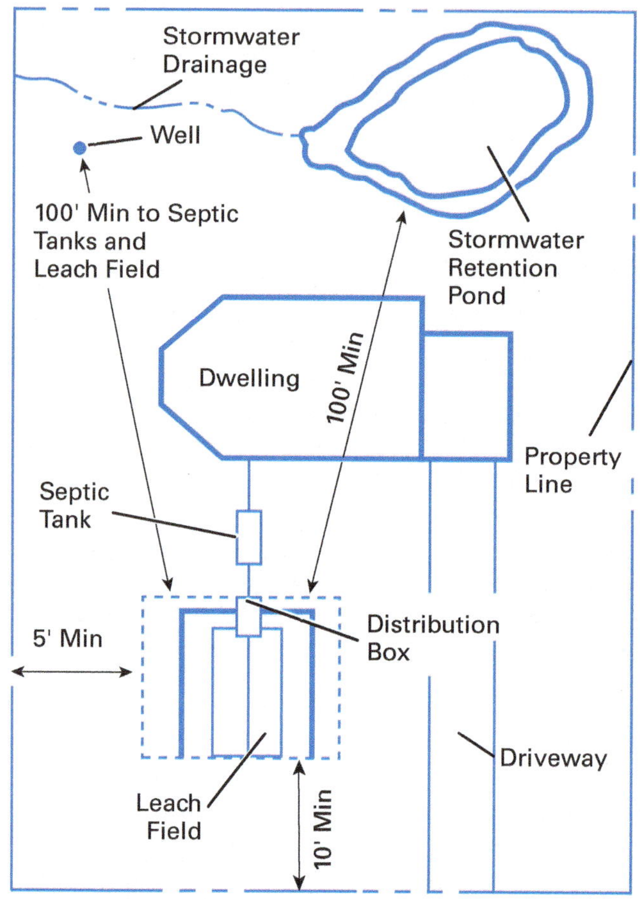

Figure 2 Typical layout for a septic system.

Anaerobic bacteria: Bacteria that do not need oxygen to live. They are used in septic tanks and aeration tanks to decompose solid wastes.

In septic systems, wastewater is first discharged into a septic tank directly from the building's waste system. A septic tank is a container that allows liquid wastes to separate from solid wastes. Wastewater remains in the tank for at least a day, allowing large solids to settle out as sludge. **Anaerobic bacteria**, or bacteria that do not need oxygen to live, decompose much of the sludge. Grease, oils, and floating particles rise to the top. Methane and carbon dioxide produced by the decomposition are vented from the effluent side of the tank. Septic tanks can be made from a variety of materials, including fiberglass, concrete (*Figure 3*), and plastic (*Figure 4*). On-site septic tanks can also be built with concrete blocks and tar-based sealers.

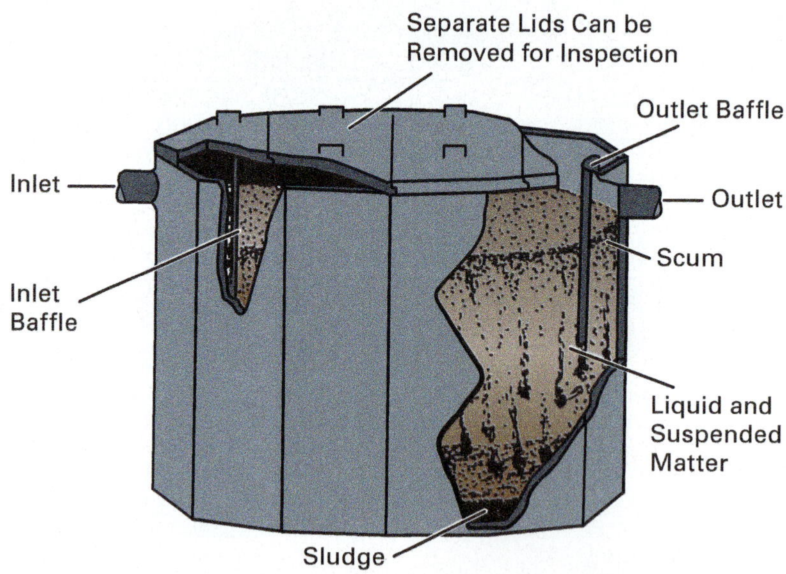

Figure 3 Typical concrete septic tank.

Figure 4 Typical plastic septic tank.
Source: pryzmat/Shutterstock

From the septic tank, the effluent flows through a sealed pipe to a distribution box. A distribution box (*Figure 5*) is a watertight container with one inlet and many outlets. It directs the effluent to different parts of a leach field or distributes it among several leach fields. A distribution box can also be used to direct effluent flow to a second or adjacent field. Most distribution boxes have lids for inspection and cleaning. Often, the outlets have knockout plugs, which allow the installer to use only the number of outlets required. Like septic tanks, distribution boxes can be made from different types of materials.

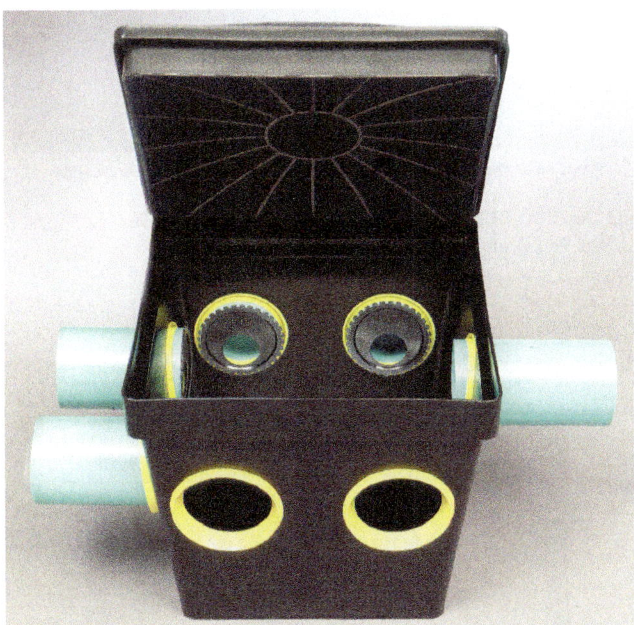

Figure 5 Plastic distribution box.
Source: Courtesy of Tuf-Tite Inc. 2023

From the distribution box, effluent drains through distribution pipes. Distribution pipes have holes or gaps that allow effluent to drain out of them. Distribution pipes are laid underground in a leach field.

WARNING!

Sewage contains bacteria and other organisms that can cause illness and death. Follow local codes when disposing of sewage. Never allow waste to drain into soil that cannot absorb it. The result could be contaminated groundwater and ruined soil.

1.1.2 Aeration Systems

Aeration system: A type of soil absorption system that uses an aeration tank to clarify effluent.

Aeration tank: A waste storage and separation tank used in an aeration system. In an aeration tank, anaerobic bacteria decompose sludge and then aerobic bacteria clarify the effluent.

The second kind of soil absorption system is an **aeration system**. Aeration systems are similar to septic systems. The major difference is that they use an underground **aeration tank** instead of a septic tank. Aeration tanks clarify wastewater more completely than septic tanks do. However, they are more mechanically complex. Because of this, some codes prohibit aeration systems. Consult your local code before installing an aeration system.

Aeration tanks are designed with two chambers (*Figure 6*) or three (*Figure 7*). In a two-chamber aeration tank, solid waste first enters the tank, where sludge settles out and is decomposed by anaerobic bacteria. The liquid waste then

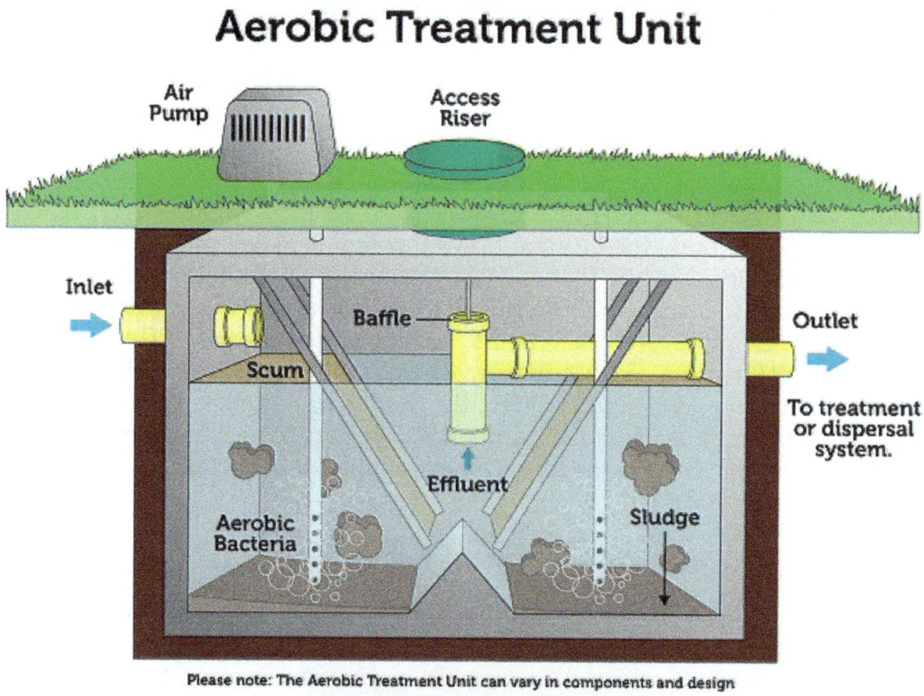

Figure 6 Cross section of a two-chamber aeration tank.
Source: Courtesy of US Environmental Protection Agency

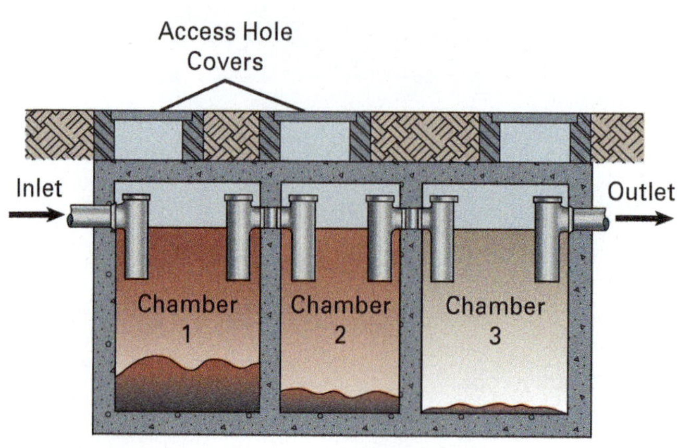

Figure 7 Cross section of a three-chamber aeration tank.

Licensed Septic Tank Installers

Some jurisdictions require plumbers to have a separate license to install septic tanks. If your job will involve septic system installations, find out what your local requirements are. Do not attempt to install a system if you are not authorized by your local code. If you need a license, ask your local code official how and where you can get one.

flows into a second chamber, into which air is injected. The air contains **aerobic bacteria**, bacteria that need oxygen to live. Aerobic bacteria consume the organic matter in the wastewater, and then the clarified wastewater flows out to the distribution box and into the leach field.

Aerobic bacteria: Bacteria that need oxygen to live. They are used in aeration tanks to clarify effluent.

The wastewater that comes out of aeration tanks is mostly clear and odorless. Because of this, aeration systems often use smaller leach fields than septic systems. Aeration systems are often used in places where codes limit the types of wastes that can be disposed of into the ground. Some jurisdictions allow wastewater to be used for irrigation and in flush toilets. Check with your local code to determine how best to handle aeration-system wastewater in your area.

1.1.3 Pressure Distribution Systems

The third type of soil absorption system is the **pressure distribution system**. Like other soil absorption systems, pressure distribution systems separate out effluent and allow it to drain into the ground. The main difference is that they use pumps to move the effluent. The effluent is pumped down a main to a manifold, where it is channeled into distribution pipes (*Figure 8*). Pressure distribution systems are more efficient than septic and aeration systems and require smaller pipe.

Pressure distribution system: A soil absorption system in which effluent is pumped through the distribution pipes under pressure.

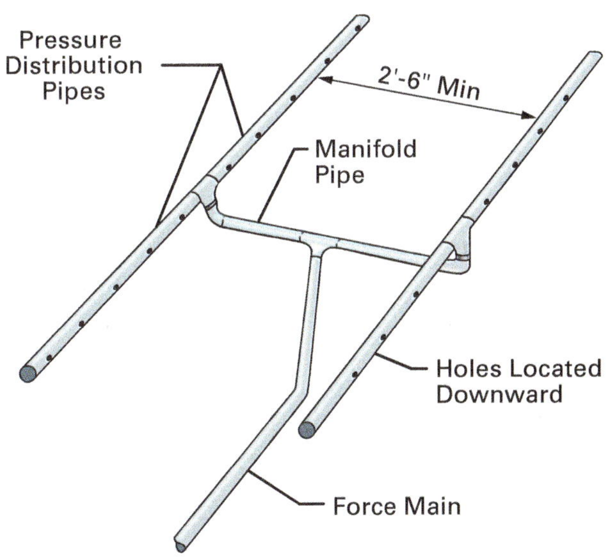

Figure 8 Typical design of a pressure distribution system.

Source: 2021 International Private Sewage Disposal Code. Copyright © 2024. International Code Council, Inc. All rights reserved. Reprinted with permission.

1.2.0 Identifying Organic Systems

Septic and aeration systems use large amounts of water to drain, separate, and decompose wastes. Organic systems break down wastes without using water (*Figure 9*). Bathroom and kitchen wastes fall into a sealed composting tank. In the tank are peat and topsoil that contain bacteria. The bacteria decompose the waste. Air flows through the tank to assist in the decomposition. As the waste decomposes, it emits carbon dioxide and water vapor. These are vented out of the tank. When the cycle is complete, the dried waste that remains is called **humus**. Humus can eventually be added to soil to fertilize it.

Humus: In an organic system, the dried solid waste that results from decomposition. Humus can eventually be added to soil to fertilize it.

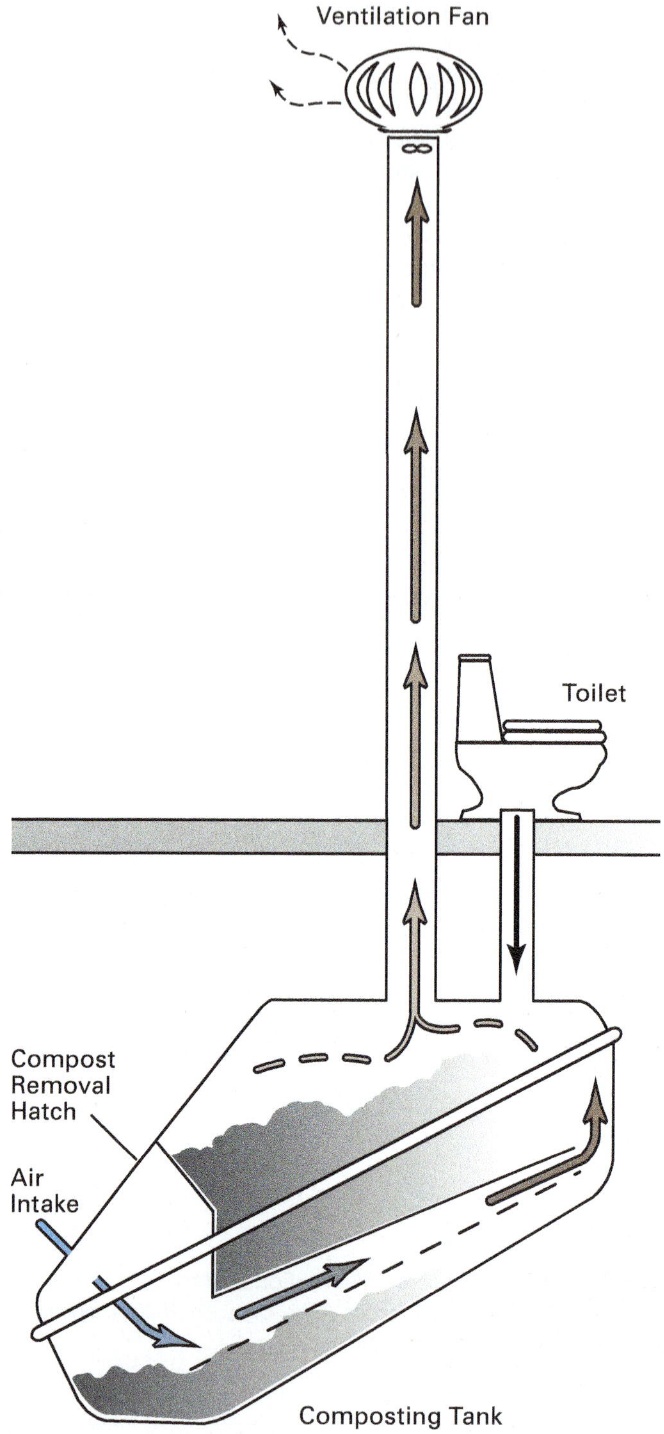

Figure 9 Typical residential installation for an organic waste-disposal system.

Organic systems are small, self-contained units. They can be located in basements or crawl spaces. They are often used in vacation houses. Some jurisdictions do not permit organic systems. Refer to your local code before installing an organic system.

1.3.0 Identifying Closed Systems

If you have worked on site at a construction job, you have probably used a portable toilet (*Figure 10*). If you look at how a portable toilet is designed, you will see that the entire waste cycle is self-contained. A portable toilet is the simplest form of a closed waste-disposal system. Closed systems are used where either the terrain or the soil is not suitable for waste drainage. The process is illustrated in *Figure 11*.

Figure 10 Portable toilet.
Source: g215/Shutterstock

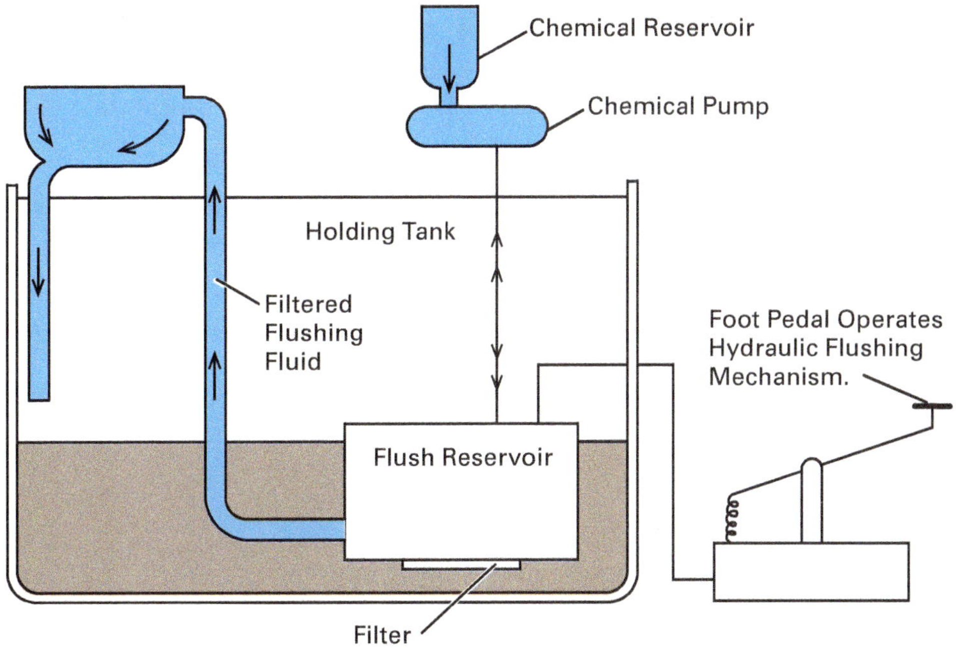

Figure 11 Closed waste-disposal system.

Waste settles in a tank filled with water. The water is pumped through a reservoir, where it is mixed with chemicals that deodorize and purify it. This clean water is then used to flush new waste. Closed systems require little or no plumbing. However, they must be cleaned and recharged regularly.

1.3.1 Drywells and Cesspools

Some private systems may include **drywells**. These are covered pits that accept roof and basement drainage. Drywells are filled with loose sand and gravel. They collect drainage wastewater and allow it to leach into the surrounding soil. Do not allow drywells to drain into private waste-disposal systems.

A **cesspool** is a lined, covered pit that receives solid and liquid wastes directly from the main waste system. Cesspools do not require a septic tank or aeration tank. Cesspools are designed to hold solid wastes and allow effluent to seep into the surrounding soil. Cesspools are not considered as sanitary as the other systems described in this module. They are also considered obsolete technology; when they fail, you will not install another cesspool. Instead, the failed cesspool will be replaced with newer, better technology.

Drywell: A covered pit filled with loose sand and gravel that collects roof and basement drainage and allows it to percolate into the surrounding soil.

Cesspool: A lined, covered pit for storing solid and liquid wastes. Cesspools allow effluent to seep into the soil.

WARNING!

A variety of materials should never be disposed of in a private waste-disposal system. These include ashes, cinders, and rags; flammable, poisonous, or explosive liquids and gases; oil, grease, and other material that can obstruct or damage the system; and surface water, rainwater, or other clear water.

Did You Know?

Connections to Public Sewers

In many parts of the country, new suburbs and developments are reaching areas that were once far from municipal services. Most codes require that buildings be connected to a public sewer system as soon as such a system is available. Codes usually set a time limit for the changeover. The IPSDC, for example, requires that it take place within one year.

When a private system is abandoned, all pipes must be plugged or capped. The remaining wastes should be disposed of properly. Remove or fill the septic tanks.

1.0.0 Section Review

1. The type of soil absorption system that uses pumps to move the effluent is called a(n) _____.
   a. septic system
   b. pressure distribution system
   c. aeration system
   d. closed system

2. Covered pits that do not require a septic tank or aeration tank are called _____.
   a. distribution boxes
   b. organic systems
   c. cesspools
   d. drywells

2.0.0 Locating and Sizing Private Waste-Disposal Systems

Performance Tasks

This is a knowledge-based module. There are no Performance Tasks in this section.

Objective

Explain how to locate and size private waste-disposal systems.

a. Explain how to locate and size septic tanks.
b. Explain how to locate and size leach fields.

Organic and closed systems are self-contained. This means that they work without affecting the outside environment. Soil absorption systems distribute effluent into the soil. Therefore, they must be safely located away from houses, property lines, wells, and trees. Tanks and leach fields must be big enough to handle the wastes. In this section, you will learn how to locate and size a private soil absorption system. Note that these are general guidelines only. Refer to your local code to determine whether effluent can safely be discharged on the property. Consult local health officials for specific details.

A permit is required before a private system can be installed. Contact your local code official to find out how to apply for permits. Applications must include the following information:

- Property survey
- Two copies of the system's design drawings
- A site plan
- Soil-quality data
- Permit fees

Code officials usually inspect the system while it is being installed and again when it is completed. Anything that will be concealed (underground piping, pumps, etc.) must be inspected before it is concealed. Follow the local code when installing a private waste-disposal system. Doing so will help ensure that the system passes inspection.

Refer to *Figure 1*. Most codes specify the minimum distances from specific structures that septic tanks can be installed. The local applicable code will also specify the minimum distance from wells that leach fields can be located. If the field is located nearer to a well than permitted by code, the effluent from the field could contaminate the water supply. Install leach-field drainpipes at least 10' away from property lines and plants with large root structures. If the system is located on an uphill slope, the building drain will be below the level of the septic tank; therefore, install a sewage pump to raise the waste to the level of the tank.

Ensure that the distribution pipes have sufficient grade to allow the effluent to seep slowly by the force of gravity. Most codes suggest a grade of between 2" and 4" per 100′. This grade will allow effluent to distribute equally throughout the leach field.

Do not locate leach fields in swampy areas. The soil cannot absorb effluent, so the effluent would contaminate the area. Instead, build a raised mound system and locate the leach field there. Because leach fields are near the surface of the ground, heavy traffic areas are not suitable for use as leach fields. The leach-field ground must have a slow percolation rate; clay soil is ideal. Prevent erosion on slopes by excavating and filling (*Figure 12*). Many codes require two separate fields for drain, waste, and vent (DWV) systems that produce more than 5,000 gallons per day.

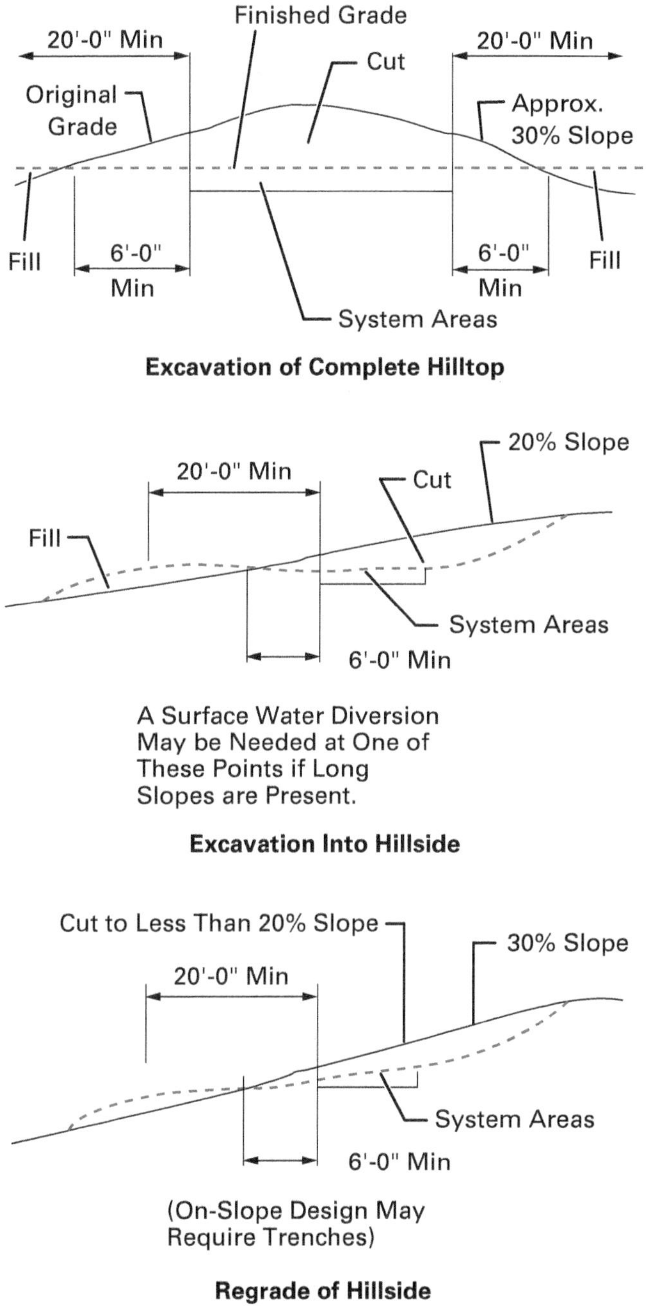

Figure 12 Preventing erosion in hillside leach fields by excavating and filling.
Source: 2021 International Private Sewage Disposal Code.

Some installations may require a grease interceptor (see *Figure 13*). Install the interceptor on a different line from the septic tank. The two waste lines can be connected downstream from the septic tank (see *Figure 14*).

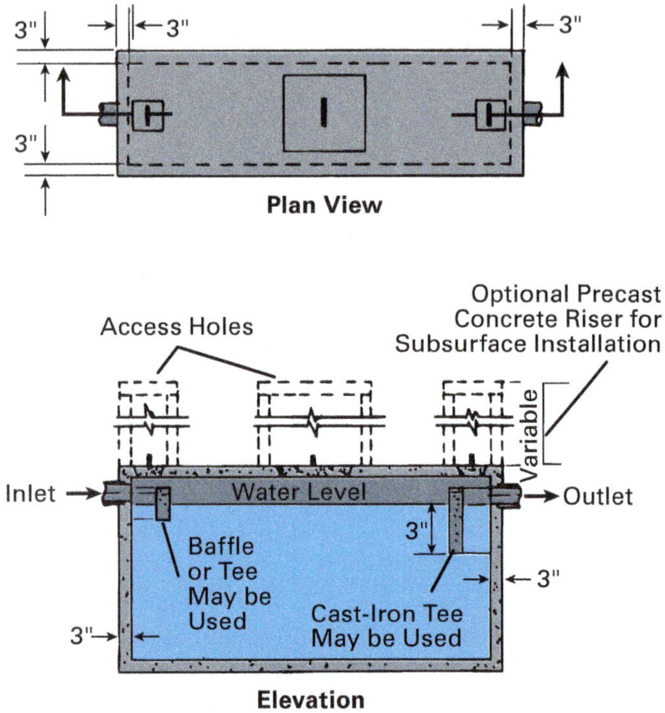

Figure 13 Typical site-built grease interceptor.

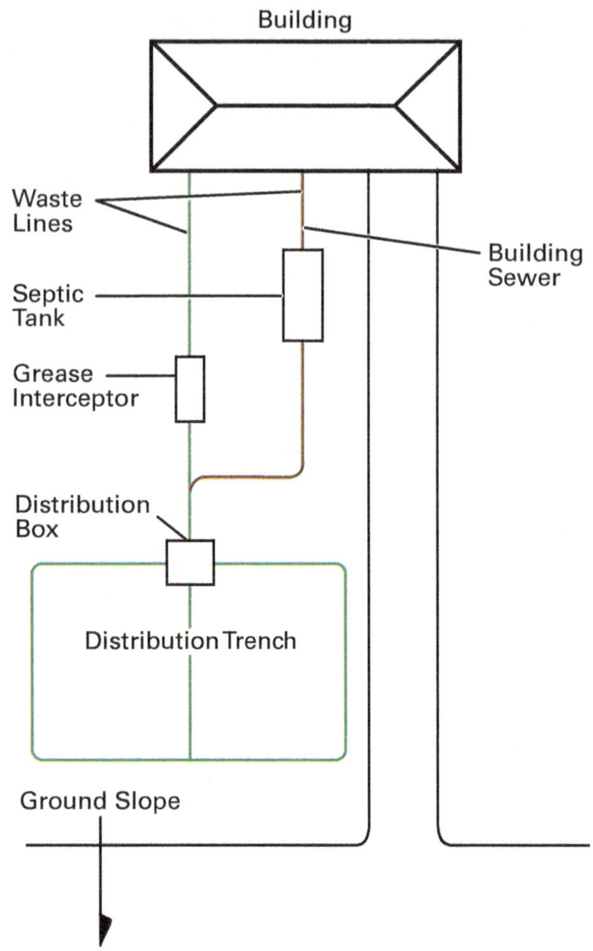

Figure 14 Typical grease interceptor installation in a septic system.

2.1.0 Locating and Sizing Septic Tanks

Local codes specify the minimum distances that a septic tank can be located from other objects (*Table 1*). Size the septic tank based on the number of bedrooms in the building (*Table 2*). Tanks should not be less than 750 gallons. Many codes allow up to four septic tanks to be connected in series in place of a larger single tank.

TABLE 1 Sample Table for Determining Minimum Separation of Septic Tanks from Other Objects

Element	Distance (feet)
Building	5
Cistern	25
Foundation wall	5
Lake, high-water mark	25
Lot line	2
Pond	25
Reservoir	25
Spring	50
Stream or watercourse	25
Swimming pool	15
Water service	5
Well	25

TABLE 2 Sample Table for Determining Minimum Septic Tank Size in Gallons Based on Number of Bedrooms

Number of Bedrooms	Septic Tank (gallons)
1	750
2	750
3	1,000
4	1,200
5	1,425
6	1,650
7	1,875
8	2,100

2.2.0 Locating and Sizing Leach Fields

The requirements for locating and sizing leach fields are specified in the local applicable code. Be sure to follow these requirements closely, as failure to do so can result in contamination and health issues. Codes require evaluation and testing of soil in the area selected for the leach field to establish the percolation rate (which is determined through a percolation test). The percolation rate is then used to determine the size of the leach field. The following sections describe these procedures in more detail. Refer to the local applicable code for guidelines that are specific to your area.

2.2.1 Soil Evaluation and Percolation Testing

Before installing a leach field, determine the characteristics of the soil in the area. Bore at least three observation holes. Codes require each hole to be at least 3' deep and wide enough for a person to observe the characteristics of the soil. Holes can be dug by backhoe, auger, or hand.

Write a **soil profile description** for each bore hole. This is a report that indicates the depths and thicknesses of the different soil layers encountered. Also note the different types of **soil mottle**, or spots and streaks that indicate saturated soil, in each bore hole. In some cases, you may have to determine the depth of the bedrock. This information will allow the code official to determine whether the area is suitable to serve as a leach field.

Soil profile description: A report that specifies the depths and thicknesses of soil layers in a proposed leach field.

Soil mottle: Spots and streaks in soil that indicate water saturation.

When the soil test indicates that the area can be used as a leach field, the next step is to determine the rate at which water percolates into the soil. This will help you determine the size of the leach field. Refer to your local code for specific guidelines on how to conduct a percolation test.

Local health departments observe soil tests (which are performed by the contractor). You should know how these tests are performed. Use the following steps to test the soil's average percolation rate in a leach field:

Step 1 Dig at least six evenly spaced holes in the area of the proposed leach field. Dig the holes at least 3' deep and 4" to 12" in diameter.

Step 2 Roughen the sides of the holes. This will allow them to absorb water.

Step 3 Remove all loose dirt from each hole, and then add 2" of sand or gravel to each hole.

Step 4 Using a hose, fill each hole with about 1' of water. Add water to the hole as necessary to maintain this level for at least 4 hours. This is called the swelling period. Some codes require a swelling period of between 16 and 30 hours.

Step 5 At the end of the swelling period, allow each hole to retain 6" of water, and then measure the drop in water in each hole over 30 minutes. Multiply the result by two. This will be the rate of drop, in inches per hour, for each hole.

Step 6 Add the rate of drop for each hole. Divide the sum by the total number of holes (in this example, by 6). This will be the average rate of drop.

Step 7 Convert the average rate of drop per hour to the average percolation rate in minutes. Divide the average rate of drop per hour by 60. The result is the soil's average percolation rate in minutes per inch.

Next, convert the average percolation rate into the allowable rate of effluent application. This is a measure of how much effluent can be discharged into the leach field. It is measured in gallons per square foot (*Figure 15*). Consult your local code or contact your local code official. The sample chart in *Figure 15* uses the following common formula:

$$\text{Rate of application} = 5 \div \sqrt{\text{(average percolation rate)}}$$

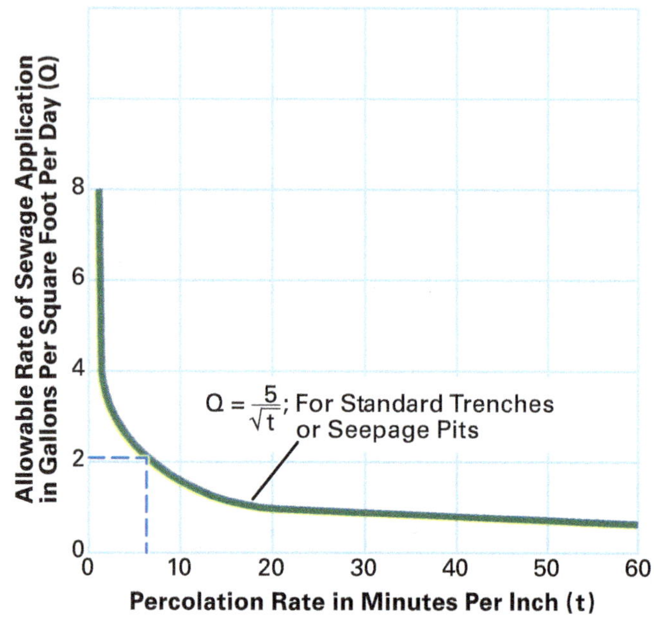

Figure 15 Sample chart for determining the rate of sewage application.

Inspections

Local code officials are required to inspect all private waste-disposal systems as they are being built and upon completion. Do not backfill over any part of the installation until the inspector has approved the entire installation. Any modifications, additions, or alterations must also be inspected and approved.

In other words, if the average percolation rate is 6 minutes per inch, the allowable rate of discharge is 2.04 gallons per square foot per day, or $5 \div \sqrt{(6)}$.

Percolation rates affect the size of the leach field. Slower percolation rates require larger fields. If the soil is extremely sandy, local codes may require a raised leach field. Raised leach fields are made of large piles of sod. Sod provides an adequate percolation rate for effluent.

If water seeps into the holes during the percolation test, the water table is too high to permit a septic system. If the percolation rate is less than 1" per hour, the soil is too dense or too saturated for use as a leach field. If the holes will not hold water because the soil is too sandy, shorten the test. Hold the water levels at 6" for 50 minutes, and then measure the drop in water over a period of 10 minutes.

2.2.2 Sizing the Leach Field

Once the percolation test has been completed, size the leach field using the test information. Your local code will provide information on how to determine the leach-field size based on the percolation rate (*Table 3*). Codes may provide different tables for sizing leach fields for residential and non-residential installations.

TABLE 3 Sample Table for Calculating Size of Leach Field Using Percolation Rate Data

Percolation Class	Percolation Rate (minutes required for water to fall 1")	Seepage Trenches or Pits (square feet per bedroom)	Seepage Beds (square feet per bedroom)
1	0 to less than 10	165	205
2	10 to less than 30	250	315
3	30 to less than 45	300	375
4	45 to 60	330	415

Did You Know?

George E. Waring, Jr., Sanitation Pioneer

Colonel George E. Waring, Jr., was an educator and administrator who helped Americans live healthier, cleaner lives. He served as New York City's street-cleaning commissioner from 1895 until his death three years later. During that time, his department became famous for its army of street sweepers. They dressed in white uniforms and were known as the White Wings. In three years, the city went from one of the nation's dirtiest cities to one of its cleanest.

Waring was a retired Civil War officer as well as a sanitation engineer. Before he became a New York commissioner, he made his mark in the plumbing field. In 1876, he wrote a book called *The Sanitary Drainage of Houses and Towns.* It became a model for many cities throughout the country. It was only the first of many plumbing books that he wrote. His books made it easy for the average reader to understand the importance of sanitary plumbing. Historians credit him with helping to usher in the era of safe and sanitary plumbing.

2.0.0 Section Review

1. Size septic tanks based on the number of a building's _____.
   a. bedrooms
   b. occupants
   c. water closets
   d. floors

2. When conducting a percolation test in extremely sandy soil that will not hold water, the water levels should be held at _____.
   a. 6" for 30 minutes
   b. 6" for 50 minutes
   c. 3" for 30 minutes
   d. 3" for 50 minutes

3.0.0 Installing Private Waste-Disposal Systems

Objective

Explain how to install private waste-disposal systems.

a. Explain how to install septic tanks.
b. Explain how to install leach-field distribution piping.

Performance Tasks

This is a knowledge-based module. There are no Performance Tasks in this section.

Most private waste-disposal systems use a tank to store solid wastes and a leach field to drain liquid wastes. The size and arrangement of each system is unique. However, the steps required to install a private waste system are fairly standard no matter how large or small the system. You have already learned how to join pipes and connect them to building drains. In this section, you will learn how to install septic tanks, distribution pipe, and leach fields.

When doing so, ensure that all components are the proper size and that they are made of approved materials. Your local code will provide sizing guidelines. It will also list materials that can be used safely in a private waste-disposal system. All pipes from the building that drain to the entrance to the leach field should be sloped downward to allow the waste to flow by gravity. The standard slope is $^1/_4$" per foot of pipe.

3.1.0 Installing Septic Tanks

Refer to the local applicable code for the minimum distance that septic tanks may be located away from buildings. Install the tank so that its top is at least 1' below ground level. In colder climates, the top should be 2' or 3' below ground. Refer to your local code for the appropriate depth. Provide access to the tank's manholes to allow for inspection, cleaning, and repair. Ensure that the tank has an inspection port above either the inlet baffle or the outlet baffle.

Pipes leading from the building drain should be watertight. Common pipe materials include cast-iron soil pipe, plastic pipe, and sealed clay pipe. Consult your local code for approved pipe materials. Use approved watertight pipe to connect the septic tank to the distribution box. Most codes specify 4" pipe for the connection between septic tanks and distribution boxes.

The decomposition process creates three layers of waste in a septic tank. Solids form a layer of sludge on the bottom of the tank. Greasy scum floats to the top of the tank. Between these two layers is the effluent that drains into the leach field. Locate inlets and outlets to allow the wastes to separate while allowing effluent to drain.

On the inlet, install a sanitary tee or ell that extends at least 15" into the effluent (*Figure 16*). This will deposit new sewage below the scum layer. Another option is to direct the flow to a baffle (*Figure 17*). Effluent entering the tank through the inlet will hit the baffle and follow it deep into the tank. Outlets should extend 18" into the tank. Install a sanitary tee on the outlet. Leave the top of the tee open to act as a vent (*Figure 18*). On prefabricated tanks, ensure that the existing inlets and outlets meet local code standards.

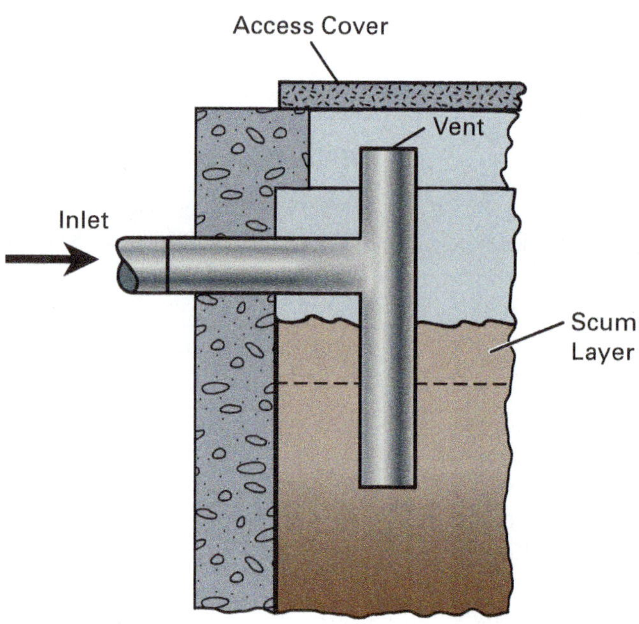

Figure 16 Septic tank inlet extending below the scum layer.

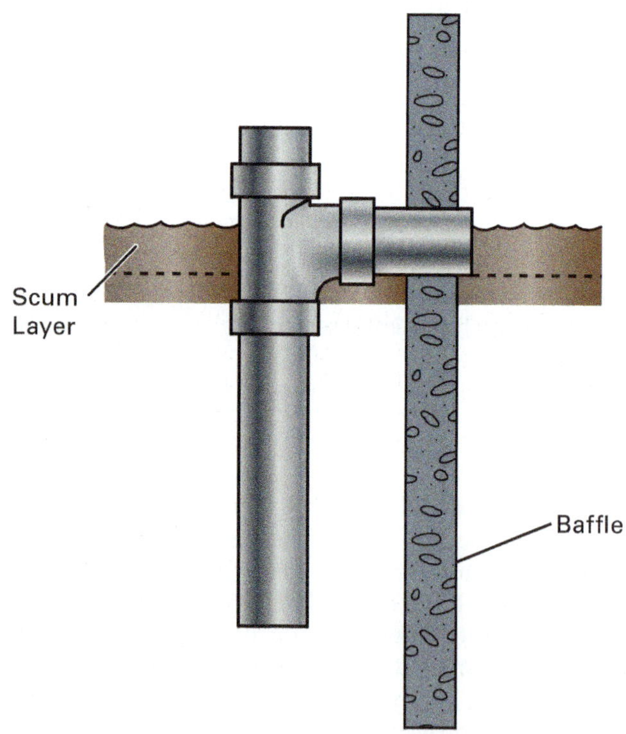

Figure 17 Septic tank inlet with inlet baffle.

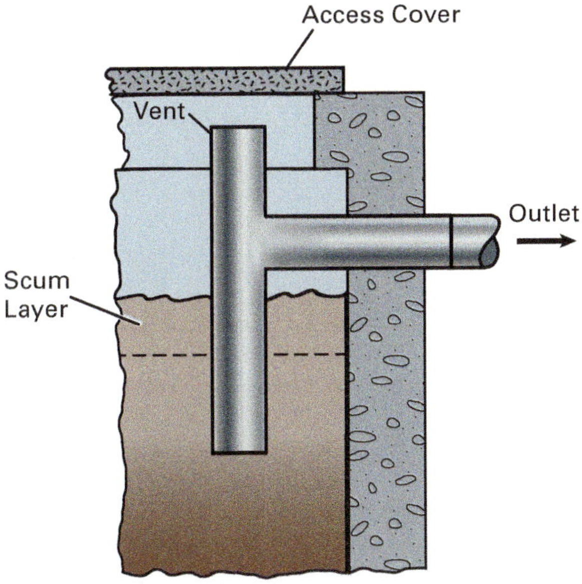

Figure 18 Septic tank outlet with sanitary tee vent.

WARNING!

Do not attempt to join different types of plastic pipe with solvent cement. If the bond fails, sewage waste can leak and contaminate the groundwater and soil.

3.2.0 Installing Leach-Field Distribution Piping

Distribution pipe is similar to regular DWV pipe, but it has drain holes (*Figure 19*). Some codes require covering the pipe with a nylon screen to prevent silt from entering the pipe. Lay the pipe with the perforations facing down. Join plastic perforated pipes using solvent weld or friction fit. Rigid-wall plastic pipe may be used instead of specially made distribution pipe. Clay pipe is no longer installed for use as leach-field distribution piping, but you may encounter it on service calls. When repairing clay pipe, do not seal the butt joints. This will allow the effluent to leach through the butts (*Figure 20*).

Flexible plastic distribution pipe bundled in synthetic aggregate (*Figure 21*) has become one of the most widely used methods of wastewater drainage into leach fields in the United States and Canada. Synthetic aggregate systems (often referred to by the trade name EZflow®) use aggregate material that is typically manufactured from recycled plastic. They can be installed in areas that have a smaller footprint than trenches or bed systems constructed solely using perforated pipe laid in natural aggregate material.

Figure 19 Perforated plastic distribution pipe.

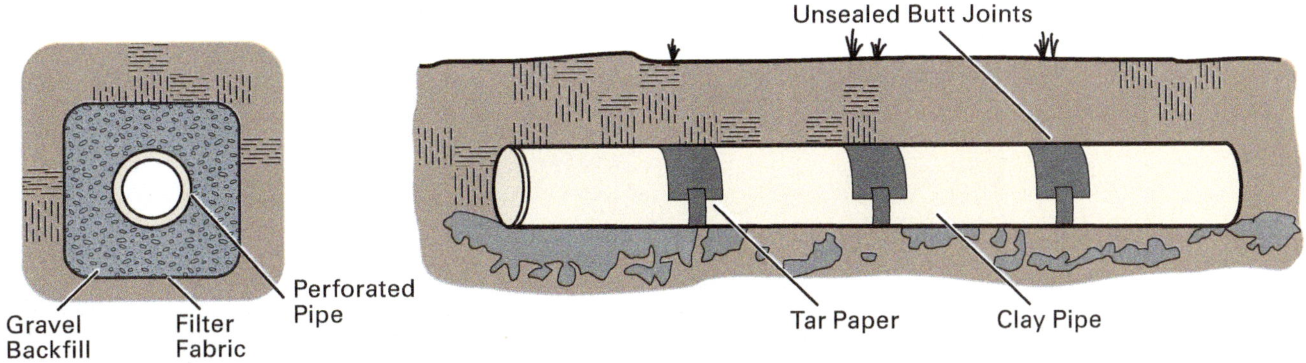

Figure 20 Typical installation of clay pipe in a leach field.

Figure 21 Distribution piping in bundled synthetic aggregate.
Source: Pat Canova/Alamy Images

Synthetic aggregate has been shown to improve leach field efficiency and capacity because it eliminates the use of stones that can clog drain perforations or the soil's surface, both of which can significantly reduce the infiltration of effluent. Synthetic aggregate systems are available in standard lengths and can typically be installed in less time, and with less site disruption, than a leach field constructed using mineral aggregate.

Install distribution pipe in trenches lined with washed gravel, crushed stone, or slag ranging in size from $^1/_2$" to $2^1/_2$" (*Figure 22*). This material should be at least 6" deep below the pipe, extend from side-to-side of the trench, and cover the pipe. Your local code provides allowable dimensions for leach-field trenches (*Table 4*).

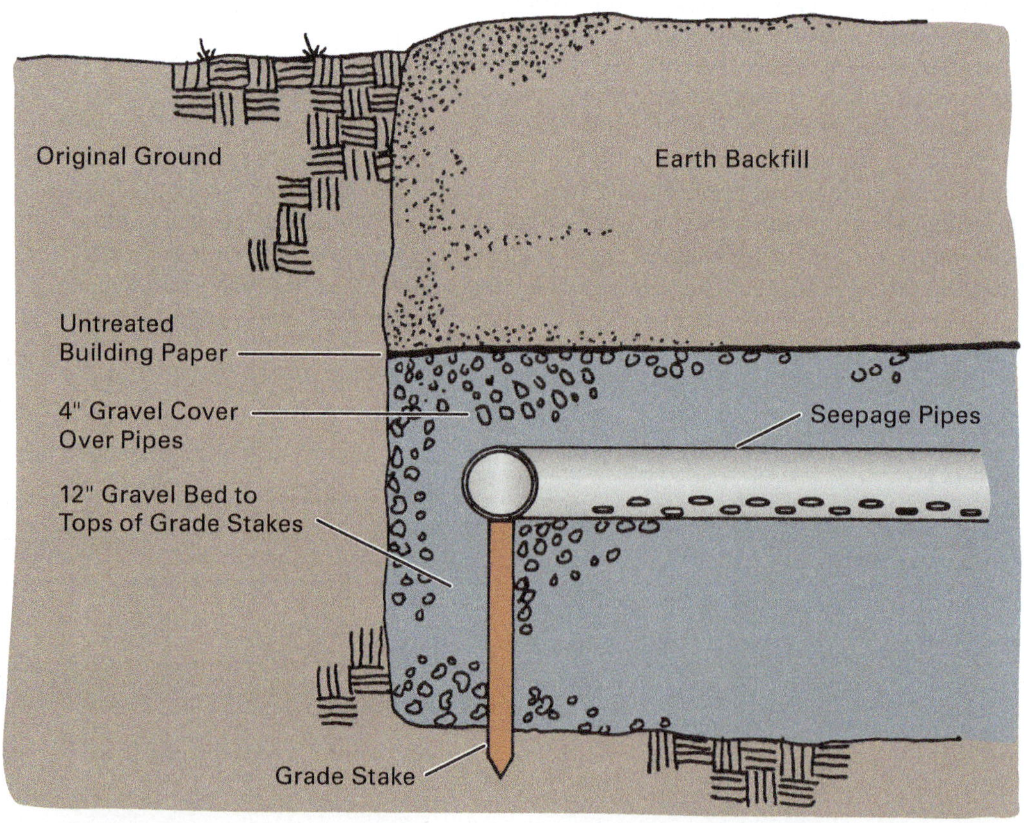

Figure 22 Typical trench for a leach-field distribution pipe.

TABLE 4 Sample Table for Determining Leach Field Trench Sizes by Width of Trench at Bottom

Width of Trench at Bottom (in)	Recommended Trench Depth (in)	Effective Absorption Area (in/sq ft)
18	18–30	1.5
24	18–30	2.0
30	18–30	2.5
36	24–36	3.0

Protect distribution pipe from being crushed under the weight of soil or aboveground traffic by installing heavy-duty nylon covers over the pipe once it has been laid (*Figure 23*). Also called chambers or crush protectors, these covers are designed with holes to allow effluent to flow into the surrounding leach field. They are available in a variety of standard lengths and can be joined together with snap fittings that allow for limited deflection. End caps are fitted with one or more knockout ports through which distribution piping can be fed. Covers eliminate the need to use gravel and other fine natural aggregate materials in leach-field trenches.

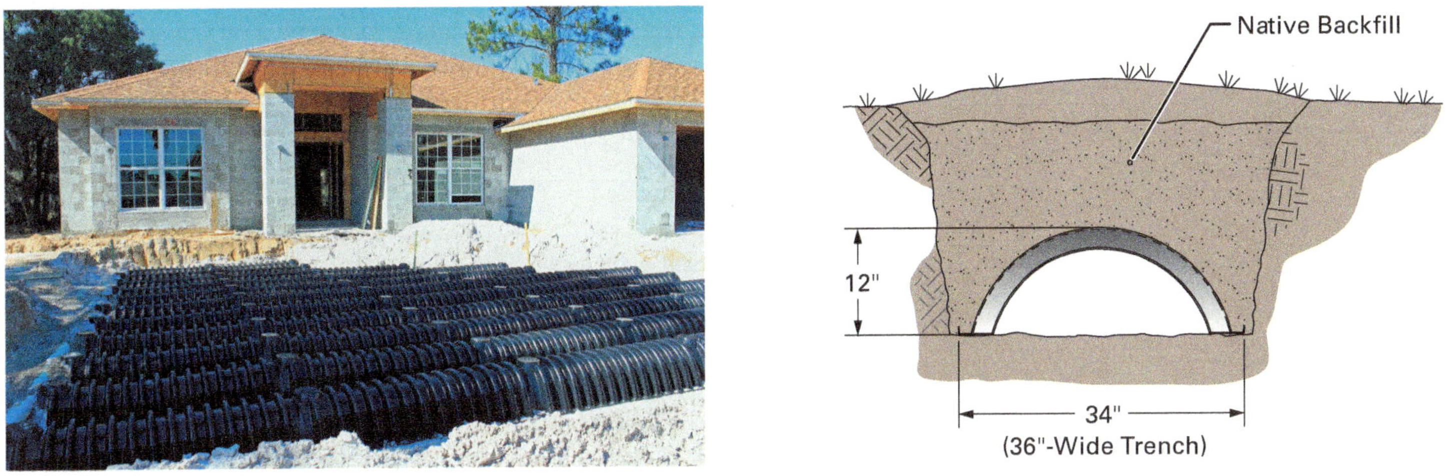

Figure 23 Nylon cover for protecting distribution piping from being crushed.
Source: (Left) KGERWX/Alamy Images

Most codes require that perforated pipe be buried 3' below ground level. Install at least two lines of pipe per leach field. Each line should not be longer than 100'. Separate lines by at least 6'. A 10' separation is ideal. Lay the pipe so that the terrain's natural grade will provide steady flow.

3.0.0 Section Review

1. In most codes, the size of pipe specified for the connection between septic tanks and distribution boxes is _____.
   a. 2"
   b. 3"
   c. 4"
   d. 5"

2. Flexible plastic distribution pipe that has been bundled with loose polystyrene aggregate material bound in mesh is often called _____.
   a. SharkBite®
   b. EZflow®
   c. Big Hawg®
   d. FlowRite®

4.0.0 Cleaning and Servicing Private Waste-Disposal Systems

Performance Tasks

This is a knowledge-based module. There are no Performance Tasks.

Objective

Explain how to clean and service private waste-disposal systems.

You learned that both septic and aeration tanks use anaerobic bacteria to break down sludge. These bacteria can reduce the volume of sludge by only about 40 percent. Over time, tanks will gradually fill with sludge. As a rule of thumb, sludge should be pumped out and disposed of when it fills about one-third of the tank's volume. In most tanks, this will take two or three years. Codes specify when to clean tanks, depending on the tank's capacity and the amount of waste generated. For manufactured tanks, consult the instructions for cleaning requirements.

If sludge is allowed to overfill the tank, the solids will not be able to settle out. That means that the effluent will contain large amounts of solid waste. If these wastes are allowed to percolate into the leach field, they will clog the soil. If enough solid wastes contaminate the leach field, the field will fail. If a field fails, sewage could back up into the structure or contaminate groundwater.

WARNING!

Septic tanks contain gases, such as methane, carbon dioxide, and radon, which are toxic and flammable. Septic tanks are confined spaces. Confined-space safety procedures must be followed, because lack of oxygen in the tank can suffocate unprotected workers.

Take proper precautions when working in or near septic tanks. Wear appropriate personal protective equipment. Do not attempt to work near a septic tank without prior emergency-rescue training. Never use an open flame near a septic tank. If someone falls into an unventilated tank, do not enter the tank without proper respiratory equipment. Install a fan over the manhole to provide fresh air, and call for emergency assistance.

Antibacterial soaps and other cleaning substances that are designed to kill bacteria should never be disposed of in private waste-disposal systems. They will also kill the anaerobic bacteria that break down the waste in the tank. The undigested waste will be unable to flow out of the tank into the field, causing the tank to overflow. Furthermore, the entire drain field will have to be dug up and replaced to remove the antibacterial contamination.

Stress to the customer the importance of regular tank cleaning. Schedule the first cleaning when you finish installing the system. Then, at every cleaning, schedule another appointment. Remind customers a few days before each appointment. Failure to perform regular cleaning could be both harmful and expensive.

Clean the tank according to local code requirements. Professional contractors usually service tanks using pumper trucks. Before attempting to clean a tank, allow the tank to ventilate for several days. Do this by opening the manhole covers. Install signs and barriers around the tank to prevent people from falling into it.

After ventilating the tank, begin the cleaning process by breaking up the scum layer. Then mix the sludge with the liquid in the tank. To mix the sludge, pump out some liquid from the tank. Then pump it back into the bottom of the tank. Repeat the cycle until the sludge and liquid are completely mixed. Then, pump out the contents of the tank completely. Always pump out the tanks through the manhole. Do not attempt to pump the tank through the inspection ports. Doing so could damage the baffles.

Remember to take all necessary safety precautions when cleaning tanks. Do not enter a septic or aeration tank to clean it or perform maintenance. Provide adequate ventilation when working on a tank. Use electric lights only. Never use open flames around a septic or aeration tank. Wear appropriate personal protective equipment.

When the cleaning is complete, inspect the inlet and outlet baffles. If they have deteriorated or are missing, replace them. Secure the manhole covers to ensure a tight seal. Finally, dispose of the septic tank waste according to your local code.

WARNING!

Septic tank sludge is a biohazard. Dispose of sludge according to approved disposal procedures. Check with local health officials for the procedures in your area. Never dump septic tank sludge into a city sewage system without first verifying that this method is permitted by code.

Cold-Weather Installations

Codes require that private waste-disposal systems be installed only when the weather conditions are suitable. Do not try to test and install a leach field when the ground is frozen. If there is snow on the ground, remove it from the area of the leach field. Always protect the backfill material from freezing. To prevent freezing, use loose, unfrozen soil for the first foot of backfill.

4.0.0 Section Review

1. Anaerobic bacteria can reduce the volume of sludge only by approximately _____.
   a. 40 percent
   b. 30 percent
   c. 20 percent
   d. 10 percent

Module 02409 Review Questions

1. A private waste-disposal system that allows wastes to decompose through contact with air is called a(n) _____.
   a. decomposition system
   b. anaerobic system
   c. aeration system
   d. organic system

2. In a septic system, liquid effluent from the septic tank is directed into the leach field by a _____.
   a. distribution box
   b. gate valve
   c. diverter manifold
   d. wye connector

3. Decomposition of organic matter in a septic tank produces the gases carbon dioxide and _____.
   a. hydrogen
   b. propane
   c. methane
   d. nitrous oxide

4. In some jurisdictions, codes allow wastewater from aeration systems to be used for _____.
   a. washing vehicles
   b. irrigation
   c. concrete mixing
   d. food processing

5. In a portable toilet, wastes are deodorized and purified by _____.
   a. anaerobic bacteria
   b. filtration and aeration
   c. chemicals
   d. aerobic bacteria

6. Local codes may require two separate leach fields if the daily output of a DWV system is greater than _____.
   a. 10,000 gallons
   b. 5,000 gallons
   c. 3,000 gallons
   d. 1,000 gallons

7. The size of a leach field is determined by the soil's _____.
   a. percolation rate
   b. chemical composition
   c. water-retention index
   d. average particle size

8. Observation holes bored to permit examination of soil characteristics must have a depth of at least _____.
   a. 24"
   b. 36"
   c. 48"
   d. 60"

9. The preliminary step in percolation testing that involves holding a 1' water level in holes for at least four hours is referred to as the _____.
   a. saturation phase
   b. settlement period
   c. swelling period
   d. sealing phase

10. Soil is too dense or saturated for use as a leach field if it has a per-hour percolation rate of less than _____.
   a. 1"
   b. 2"
   c. 3"
   d. 4"

11. The standard slope used to permit gravity flow of waste through a pipe is _____.
   a. 1" per foot
   b. $^3/_4$" per foot
   c. $^1/_2$" per foot
   d. $^1/_4$" per foot

12. The minimum distance below ground level for a septic tank is _____.
   a. 1'
   b. 5'
   c. 6'
   d. 10'

13. The bottom layer of solid in a septic tank is referred to as the _____.
   a. settlement layer
   b. decomposition layer
   c. debris layer
   d. sludge layer

14. A material no longer installed as distribution pipe in leach fields is _____.
   a. clay pipe
   b. perforated plastic pipe
   c. rigid-wall plastic pipe
   d. cast-iron pipe

15. To properly clean a septic tank, pumping should always be done through _____.
   a. an inspection port
   b. the distribution box
   c. the manhole
   d. a tank vent

Answers to odd-numbered review questions are found in the Module Review Answer Key at the back of this book.

Answers to Section Review Questions

Answer	Section	Objective
Section One		
1. b	1.1.3	1a
2. c	1.3.1	1c
Section Two		
1. a	2.1.0	2a
2. b	2.2.1	2b
Section Three		
1. c	3.1.0	3a
2. b	3.2.0	3b
Section Four		
1. a	4.0.0	4

MODULE 02410

Swimming Pools and Hot Tubs

Source: V Matsuk/Shutterstock

Objectives

Successful completion of this module prepares you to do the following:

1. Explain how to size and install swimming pool systems and components.
   a. Explain how to size a swimming pool.
   b. Explain how to install water supply and recirculation systems.
   c. Explain how to install gutters and drains.
   d. Explain how to maintain water quality.
2. Identify hot tub and spa systems and their components.
   a. Identify hot tub and spa components.
   b. Explain how to install hot tubs and spas.

Performance Tasks

Under supervision, you should be able to do the following:

1. Calculate the volume of a pool.
2. Identify the components of piping for a spa.

Overview

Swimming pools, hot tubs, and spas are popular installations for recreational use. Local code requirements for these fixtures vary widely. Swimming pools use a recirculation system to clean and circulate the pool water. Pools also use pumps, filters, gutters, drains, and chlorinators. Use the pool's capacity to establish the turnover rate, expected bathing load, and pump size. Acidity and alkalinity levels in pool water must be correctly balanced and maintained. When installing hot tubs or spas, follow local code and manufacturers' instructions. Proper pH (potential hydrogen) and chlorine levels are even more important in these installations than in pools.

CODE NOTE

Codes vary among jurisdictions. Because of the variations in code, consult the applicable code whenever regulations are in question. Referring to an incorrect set of codes can cause as much trouble as failing to reference codes altogether. Obtain, review, and familiarize yourself with your local adopted code. Free online access to the 2024 International Swimming Pool and Spa Code is available online.

NCCER **Industry-Recognized Credentials**

If you are training through an NCCER-accredited sponsor, you may be eligible for credentials from NCCER. The ID number for this module is 02410. Note that this module may have been used in other NCCER curricula and may apply to other level completions. Contact NCCER at 1.888.622.3720 or go to **www.nccer.org** for more information.

You can also show off your industry-recognized credentials online with NCCER's digital credentials. Transform your knowledge, skills, and achievements into credentials that you can share across social media platforms, send to your network, and add to your resume. For more information, visit **www.nccer.org**.

Digital Resources for Plumbing

NCCER SCAN ME

Scan this code using the camera on your phone or mobile device to view the digital resources related to this craft.

1.0.0 Swimming Pool Systems and Components

Performance Task

1. Calculate the volume of a pool.

Objective

Explain how to size and install swimming pool systems and components.

a. Explain how to size a swimming pool.
b. Explain how to install water supply and recirculation systems.
c. Explain how to install gutters and drains.
d. Explain how to maintain water quality.

Pools designed for public use adhere to very strict codes and regulations, the classification of which varies throughout the industry (hotel, competition, etc.). Officials from the public board of health inspect these pools regularly. As with all plumbing specializations, some companies choose not to install public pools, while others may choose to install public pools exclusively.

Codes distinguish between permanent and storable pools, and govern the locations where these pools may be (in ground, partially in ground, elevated, or on-ground storable). Permanent pools are located in the ground, partially above ground, or fully above ground. Semipermanent pools are considered "on-ground storable," but not marketable to public pools (only as residential pools). Both types of pool require easy access to water supply and drainage systems. In most cases, the pipes and fittings can be the same as those used in other water supply and drainage systems. They also require recirculation systems to sanitize the water. Use the installation techniques that you have already learned. Refer to your local code before installing swimming pools.

According to the 2024 *International Swimming Pool and Spa Code* (*ISPSC*®), private residential pools apply to detached one- and two-family dwellings and townhouses not more than three stories in height.

1.1.0 Sizing a Swimming Pool

Capacity: The volume of a pool in gallons.

Turnover rate: The time required for all the water in a pool to cycle through the filter.

Bathing load: The maximum number of people allowed in a pool per hour, based on capacity and turnover rate.

Begin by calculating the pool's volume in gallons, also called its **capacity**. Refer to the design drawings for the pool's dimensions. For a rectangular pool, multiply the length, width, and height. For a circular pool, multiply the height by the area of the circular base. Many pools are of a complex shape. Break the volume down into smaller, simpler shapes. Calculate the volumes of the smaller shapes and add them together to get the total area in cubic feet (*Figure 1*). See *Appendix 02410A* for additional practice questions related to pool volumes. Once you have calculated the capacity, convert the cubic-foot measurement into gallons. Remember that there are 7.48 gallons in a cubic foot.

Once you have calculated a pool's capacity, you can determine its **turnover rate** and **bathing load**. The turnover rate is the time that it takes to cycle all the water in the pool through the filter. The bathing load is the maximum number of people allowed in the pool per hour. The turnover rate affects the bathing load (*Table 1*). Shorter turnover rates mean that the water will be cleaned more often, allowing a higher bathing load. If the bathing load is exceeded, people could become ill from contaminants in the water. Some local codes determine the bathing load based on the square footage of pool water surface area per person. For example, the shallow area may require 10 ft^2 per person and the deep area 25 ft^2.

For example, suppose you are installing a pool with a capacity of 55,000 gallons. If you install a pump that can handle 92 gallons per minute (gpm), a complete turnover will take 10 hours. At that turnover rate, the pool can safely handle no more than nine people per hour. If, on the other hand, you install a pump that is rated at 155 gpm, the pump can cycle the water in just six hours. The pool could then handle 40 people per hour.

Calculate the customer's needs when sizing the pool. High turnover rates are not always best. Larger pumps also mean larger pipes and filters to handle the increased load. The result is a greater expense for the customer. In the example above, the bathing load might never exceed 15 people. In that case, you can

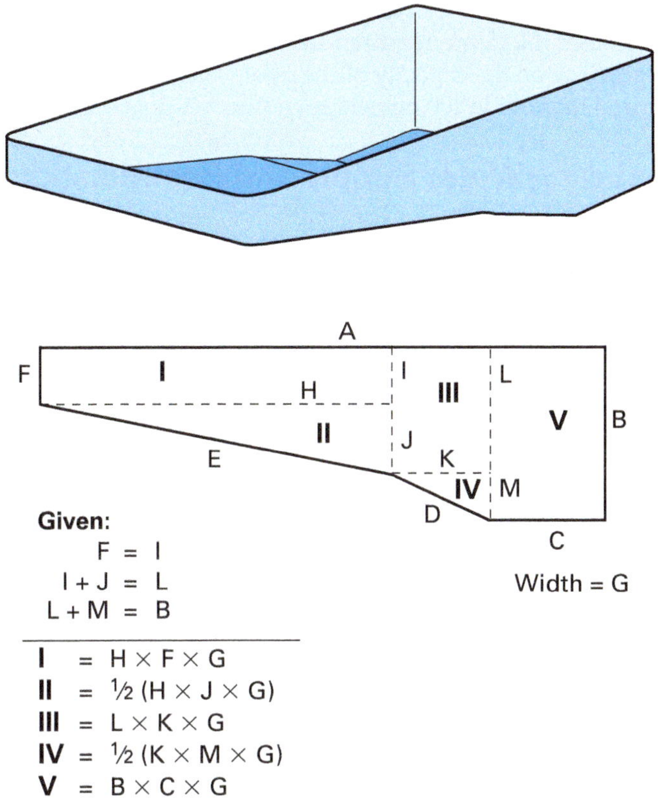

Figure 1 Calculating the area of a swimming pool with a complex shape.

compromise on a 115-gpm pump that will allow up to 17 people in the pool per hour (*Table 1*). Proper sizing ensures that the water supply, filtering, and drainage systems will be adequate and safe.

The *International Swimming Pool and Spa Code®* (ISPSC®) specifies turnover rates by the category of the pool. For example, the turnover time for swimming pools at hotels and residential complexes is equal to $1^1/_2$ times the average depth of the pool

TABLE 1 Calculating Turnover and Bathing Load Based on Capacity

Capacity to Waterline (gal)	Turnover (gpm)				Bathing Load per Hour at Turnover (Refiltration Rate)			
	6 hr	8 hr	10 hr	12 hr	6 hr (225 gal)	8 hr (400 gal)	10 hr (625 gal)	12 hr (900 gal)
31,600	88	66	53	44	23	10	5	3
38,400	105	80	64	54	23	12	6	4
55,000	155	115	92	77	40	17	9	5
80,900	255	170	135	115	60	25	13	7
119,500	335	250	200	165	89	37	19	11
158,700	440	330	265	220	117	50	25	15
207,600	575	435	345	290	154	65	33	19
254,300	710	530	425	355	188	80	41	24
305,700	850	640	510	425	226	96	49	28
422,400	1,170	880	705	585	314	133	68	39
557,600	1,550	1,160	930	775	413	179	89	52
632,100	1,750	1,320	1,050	880	468	197	101	58
703,900	1,950	1,460	1,170	975	521	220	112	65
883,000	2,450	1,840	1,470	1,220	654	276	142	82
1,074,900	2,990	2,240	1,790	1,490	796	336	172	99

in feet, for a maximum of six hours. Wave-action pools installed in public recreation facilities, on the other hand, are required to have turnover times between one and two hours, depending on the capacity of the pool. Permanent in-ground residential pools are required to provide 100 percent turnover within a minimum of 12 hours.

1.2.0 Installing Water Supply and Recirculation Systems

Once the pool has been properly sized and built, install a suitable water supply system. Pools can draw from a public water supply or a private well. In either case, install a backflow preventer on the supply line to prevent an accidental cross-connection while the pool is being filled. If a garden hose is used, install a vacuum breaker on the hose bibb. If the fill line is connected directly to the water supply system, install a backflow preventer designed to prevent back siphonage. If a permanent fill spout is used, ensure that there is an air gap of at least 6" between the outlet and the water overflow line.

Well water must be free of turbidity, hardness, and harmful organisms. Refer to your local code for guidelines. Use only approved pipe and fittings on a pool's fresh-water supply line. In most cases, pipe that is suitable for use in other water supply systems is permitted.

A recirculation system cleans the pool water. It also makes up for water lost through splashing, seepage, and evaporation. Make-up water helps keep the system's pumps primed. Recirculation systems use inlet and outlet fittings, pumps, and filters. The requirements for each of these are discussed in more detail in the following sections. Requirements for recirculation systems vary. Refer to your local code before installing a recirculation system.

1.2.1 Inlet and Outlet Fittings

The following fittings are used to direct water into and out of a recirculation system:

- Suction outlet fitting assemblies
- Skimmers
- Water supply fittings

While main drains and suction outlet fitting assemblies can be omitted, skimmers must always be present. Size and locate these inlets and outlets to allow clean water to circulate uniformly through the pool. A good rule of thumb is one inlet for every 350 ft^2 of pool surface. The ISPSC specifies the installation of one outlet for each 300 ft^2 of pool surface.

Suction outlet fitting assembly: A pool drain that serves as a connection for cleaning hoses.

Suction outlet fitting assemblies (*Figure 2*) are used as connections for cleaning hoses. When water and debris are vacuumed from the pool bottom, they are pumped through the filter. These fittings may be installed 18" below the water surface and should be easily accessible.

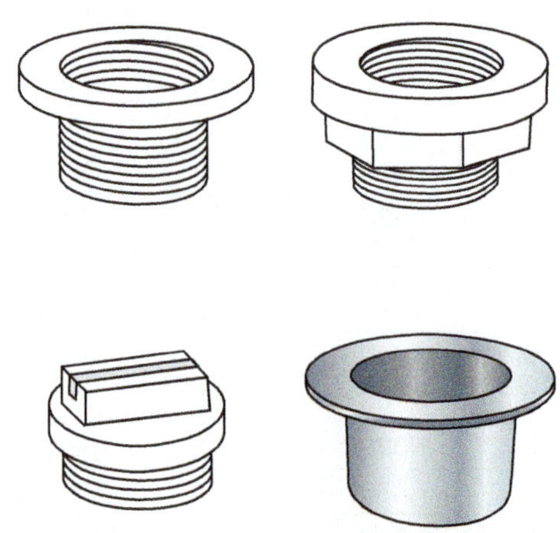

Figure 2 Plastic and chrome vacuum fittings (with plastic plug).

Sidewall drains (*Figure 3*) are installed on the pool wall below the waterline. This location prevents leaves or other objects from blocking the drain as the water enters.

Skimmers trap contaminants in a basket that can be removed for cleaning. The skimmer can be located in a gutter or behind an inlet on the pool wall (*Figure 4*). This allows easy access to the strainer or a vacuum hose connection. Skimmers can also be located on the pool deck to catch overflow (*Figure 5*). The ISPSC requires one skimmer per 500 ft^2 of pool surface area in a public pool and one per 800 ft^2 in a residential pool.

Sidewall drain: A drain installed on a swimming pool wall below the waterline.

Skimmer: A device that traps pool debris in a removable basket.

Figure 3 Sidewall drain.

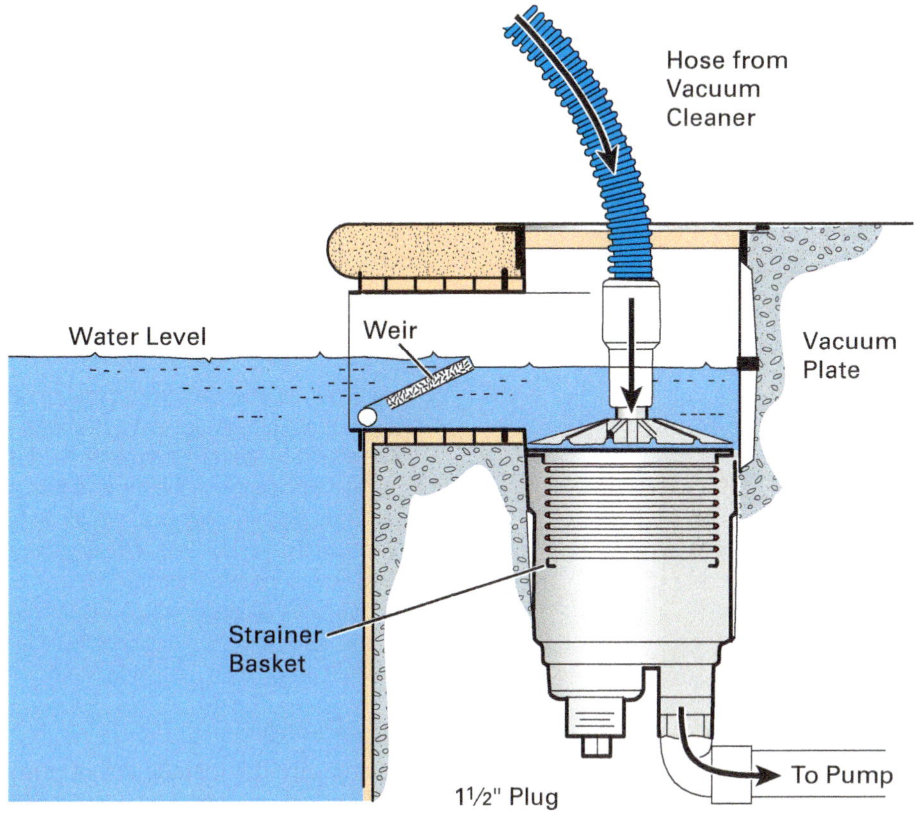

Figure 4 Skimmer installation in a pool wall.

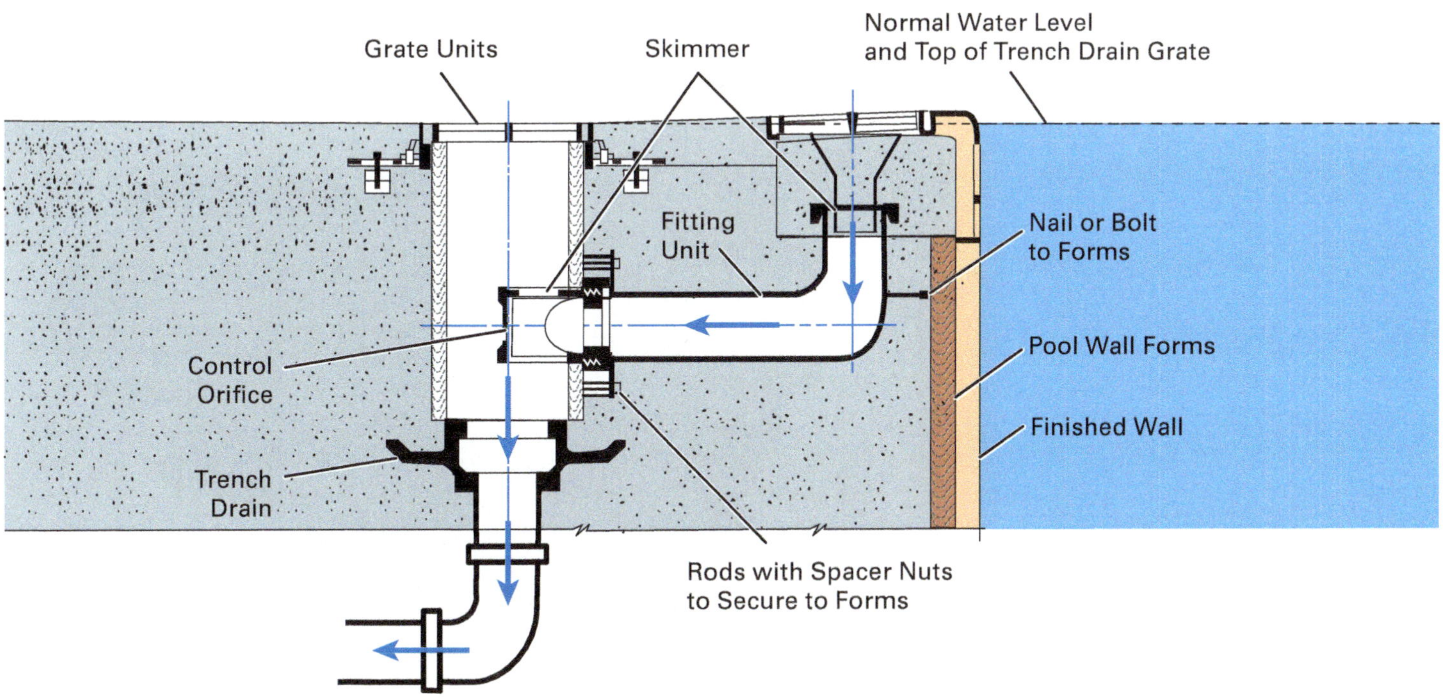

Figure 5 Skimmer installation on a pool deck.

Water supply fitting: A fixture that allows water to flow from a recirculation system back into a swimming pool.

Once the water has been filtered, it returns to the pool through **water supply fittings** (*Figure 6*). Water supply fittings are adjustable and come in a variety of sizes. Install water supply fittings 15' apart in the pool walls. Install at least one fitting within 5' of each corner of the pool. For placement in pools with irregular shapes, refer to your local code.

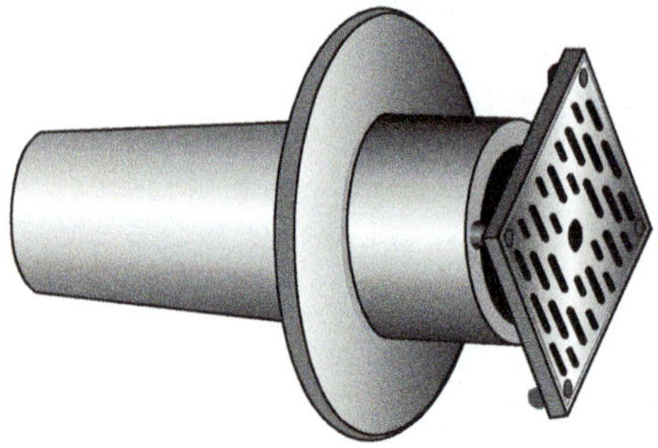

Figure 6 Water supply fitting.

Using Pumps and Skimmers to Ground a Pool

Local codes require protection against electrical shorts and accidental electrocution beyond the standard installation of a ground fault circuit interrupter (GFCI) to service pool pumps and skimmers. Pumps and skimmers can be used to ground plastic and fiberglass in in-ground, aboveground, and prefabricated pools.

Run solid copper grounding wire from at least four points on the metal frame of the pool to the designated grounding lug or connector on the pump and the skimmer, and ensure that all the wire is connected using slip bolts or other approved metallic connectors. If either the pump or the skimmer is double insulated, the grounding wire connected to the pump's or skimmer's electrical plug will then safely ground the pool. Otherwise, run another copper grounding wire from the pump or skimmer safely into the ground according to the local applicable code.

1.2.2 Pumps

Codes require the installation of pumps to circulate the pool's water through filters and chemical feeders. Once the water has been filtered and treated, it is pumped back into the pool. Many recirculation systems use water-lubricated centrifugal pumps (*Figure 7*). These are made of stainless steel or plastic so that they will not rust. Pool pumps are designed to run continuously. They should be fitted with removable protective screens to trap debris and hair before they can enter and clog the pump.

Figure 7 Water-lubricated centrifugal pump.
Source: Hans Verburg/iStock/Getty Images

Install the pump, piping, and electrical systems where they can be reached easily for service. Secure the pump firmly to its base according to the installation instructions. Union connections allow the pump to be removed easily for repair or replacement. Install an interceptor ahead of the pump on the suction line to trap hair and sediment. The interceptor should have a screen that has at least five times the area of the suction pipe.

Most pool packages include a pre-sized pump. For a site-built pool, you will need to size the pump to match the pool's capacity and turnover rate. Carefully set the pump's operating pressure. If the pressure is too high, water will move through the filter too fast. If the pressure is too low, the pump will not be able to move a sufficient volume of water. Filters, heaters, and other items will increase friction loss in the system. Size the pump accordingly. Install easily accessible shutoff valves on the pump's suction and discharge lines below the waterline. An emergency shutoff switch should be installed a minimum of 5' from the pool. Exceptions apply to on-ground storable pools, permanent in-ground residential swimming pools, residential spas, and residential water features.

1.2.3 Filters

During normal use, pool water collects a wide variety of materials that degrade its quality and increase turbidity. The following types of materials are common:

- Hair
- Dirt
- Leaves
- Insects
- Suntan oil

Pool recirculation systems use filters to reduce turbidity. Most private pools use cartridge filters (*Figure 8*). As the filter collects impurities, the

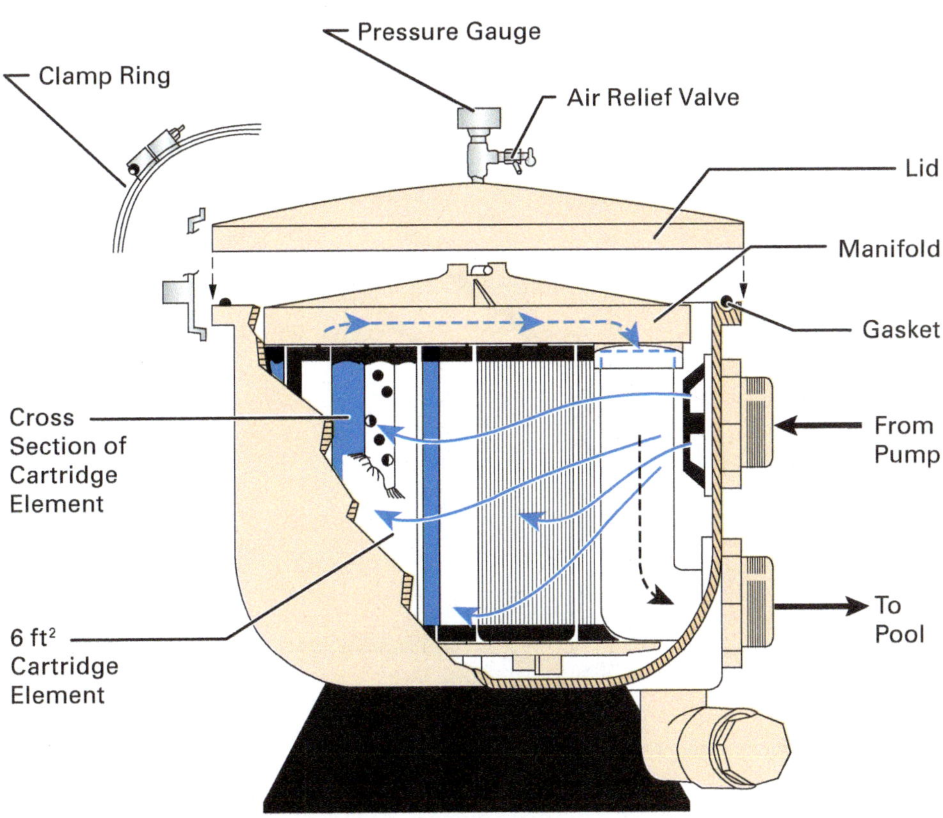

Figure 8 Cartridge-filter installation.

pump has to work harder to circulate water through it. As a result, the system pressure increases. When the pressure reaches 10 to 12 pounds above the normal pressure, change the filters. Follow manufacturers' instructions when replacing filters. Ensure that the replacement cartridge filters are the correct size.

Hair can clog a recirculation system. Install a hair strainer upstream of the cartridge filter. The strainer will trap hair before it can clog the filter, pump, or other fittings. Hair strainers have a basket that can be removed for cleaning.

1.3.0 Installing Gutters and Drains

Backwater valve: A valve that prevents the backflow of sewage into a pool drain.

Gutters (*Figure 9*) are an effective way to drain water from the pool into the recirculation and drainage systems. Gutters are installed around the entire perimeter of the pool. Drains are located along the gutter at intervals. The drains connect to a main gutter line, which drains water to the filter. If the gutter line drains into the sewer system, install a **backwater valve** on the line (*Figure 10*). Backwater valves prevent sewage backflow into the pool's drainage system.

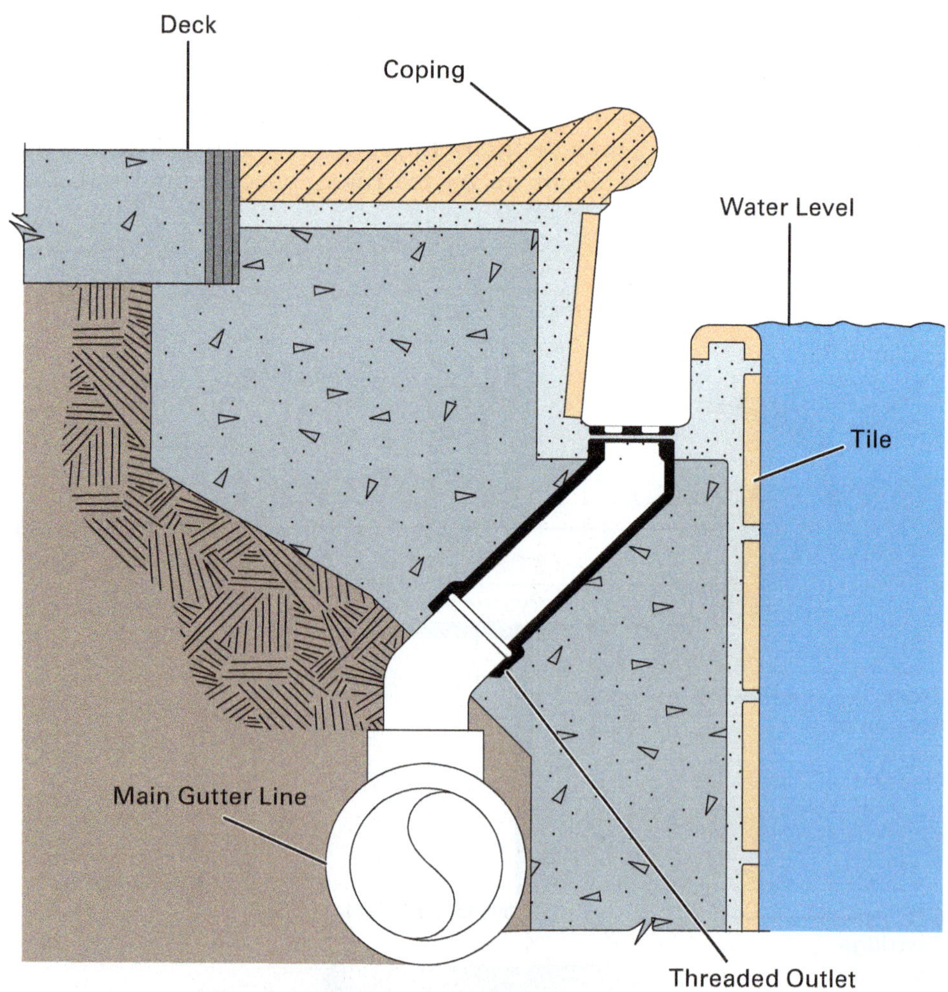

Figure 9 Cross section of a typical gutter drain.

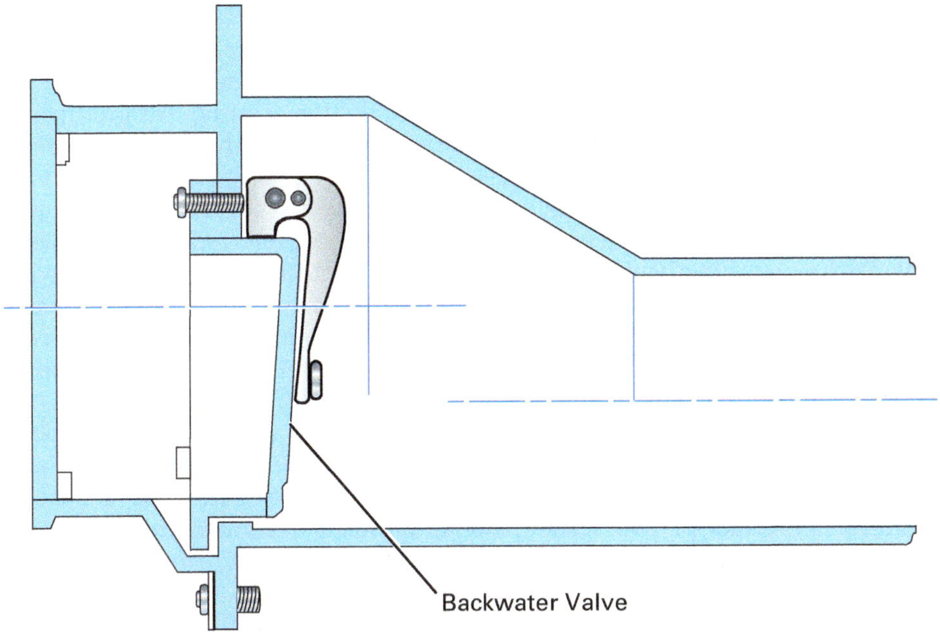

Figure 10 Backwater valve.

Deck drains catch the overflow and channel it to the filter. Ensure that the deck is sloped away from the pool. The slope prevents contaminated water from draining back into the pool. Ensure that the drainpipes connected to gutters and deck drains are pitched to provide easy discharge.

The main drain is designed to allow the entire pool to drain in no more than four hours (*Figure 11*). Locate the main drain at the pool's lowest point. The main

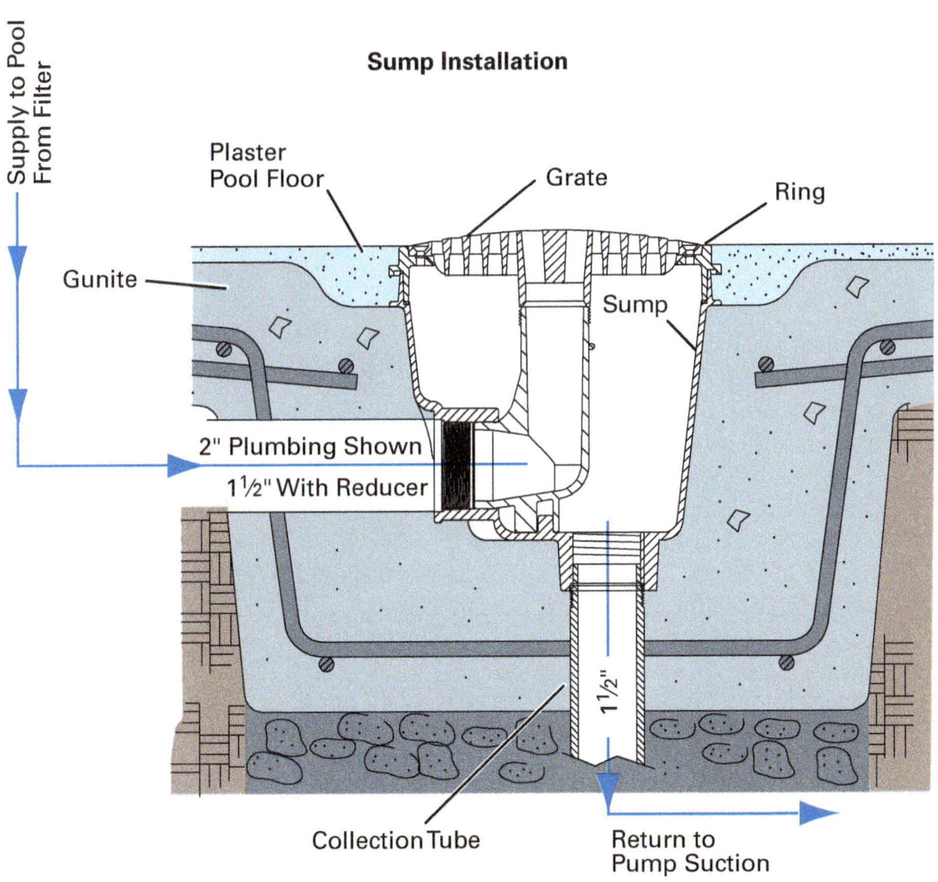

Figure 11 Main drain with a return line connection.

CAUTION

If filters are not maintained regularly, they could cause the recirculation system to become damaged. Remove cartridge filters, and soak them for at least 24 hours; then, re-install them. Replace sand filters every three to five years. Backwash and clean filters according to manufacturers' instructions.

drain must have a grate with an opening that is at least four times the area of the drainage pipe. Attach the grate in such a way that tools are required to remove it. The grate will prevent objects and children from getting caught by the drain's suction. Many codes specify the installation of antivortex gratings on the main drain to prevent the formation of a vortex during the draining process by limiting the velocity of the draining water.

Once drained, the pool water can be disposed of in several ways. Because pool water is wastewater, refer to your local code for preferred drainage methods. Pool water, filter backwash, and pool-deck drain water can be directed to the building drainage system. Connect the main drain to the drainage system, using an indirect waste pipe with an air gap. The main drain can also be connected directly to the sanitary sewer. Install a backwater valve to protect against contamination caused by a sewage backup.

Water can be channeled into a drain field or allowed to puddle on a properly sized and graded field. If the puddling method is used, the water must be able to percolate into the ground in less than 60 minutes. Fields must be at least 50' from wells used for potable water. Another option is to connect the drain to an irrigation sprinkler system. Codes prohibit backwash waste from entering an irrigation system.

A main drain may also serve the recirculation system. If so, ensure that the drain's flow rate to the recirculation system is less than 2' per second. A higher flow rate would create dangerously high suction. Install a gate valve on the main drain in an accessible location. Use the valve to close off the drain from the recirculation system when draining the pool.

Did You Know?

Overflow Standpipes

Overflow standpipes prevent spillover in fountains and decorative pools. The pipe has a dome to keep out leaves and other floating debris. The dome can be removed to provide access to the standpipe interior. The number and diameter of the pipes depends on the design of the fountain or pool. Consult the design drawings before installing standpipes.

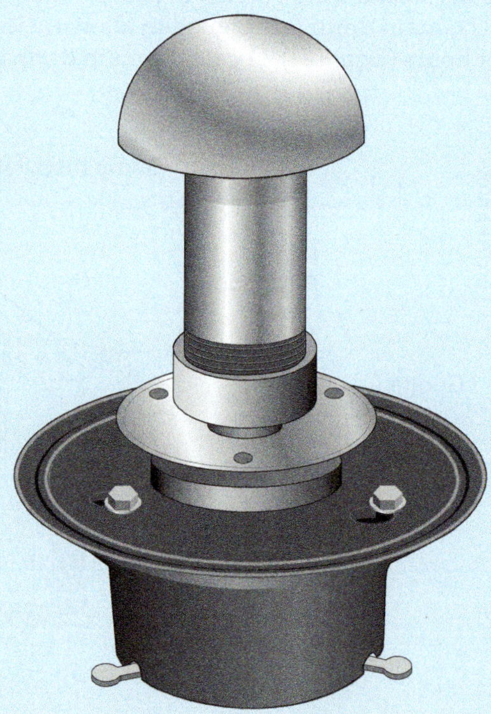

Hydrostatic pressure: The force applied to the bottom of a pool, spa, or hot tub by high groundwater levels beneath it.

In some places, high groundwater levels can actually float an empty pool out of the ground. This situation is caused by **hydrostatic pressure**. Protect the pool against hydrostatic pressure by installing a relief valve in the main drain (*Figure 12*). The relief valve allows water to flow into the empty pool until the

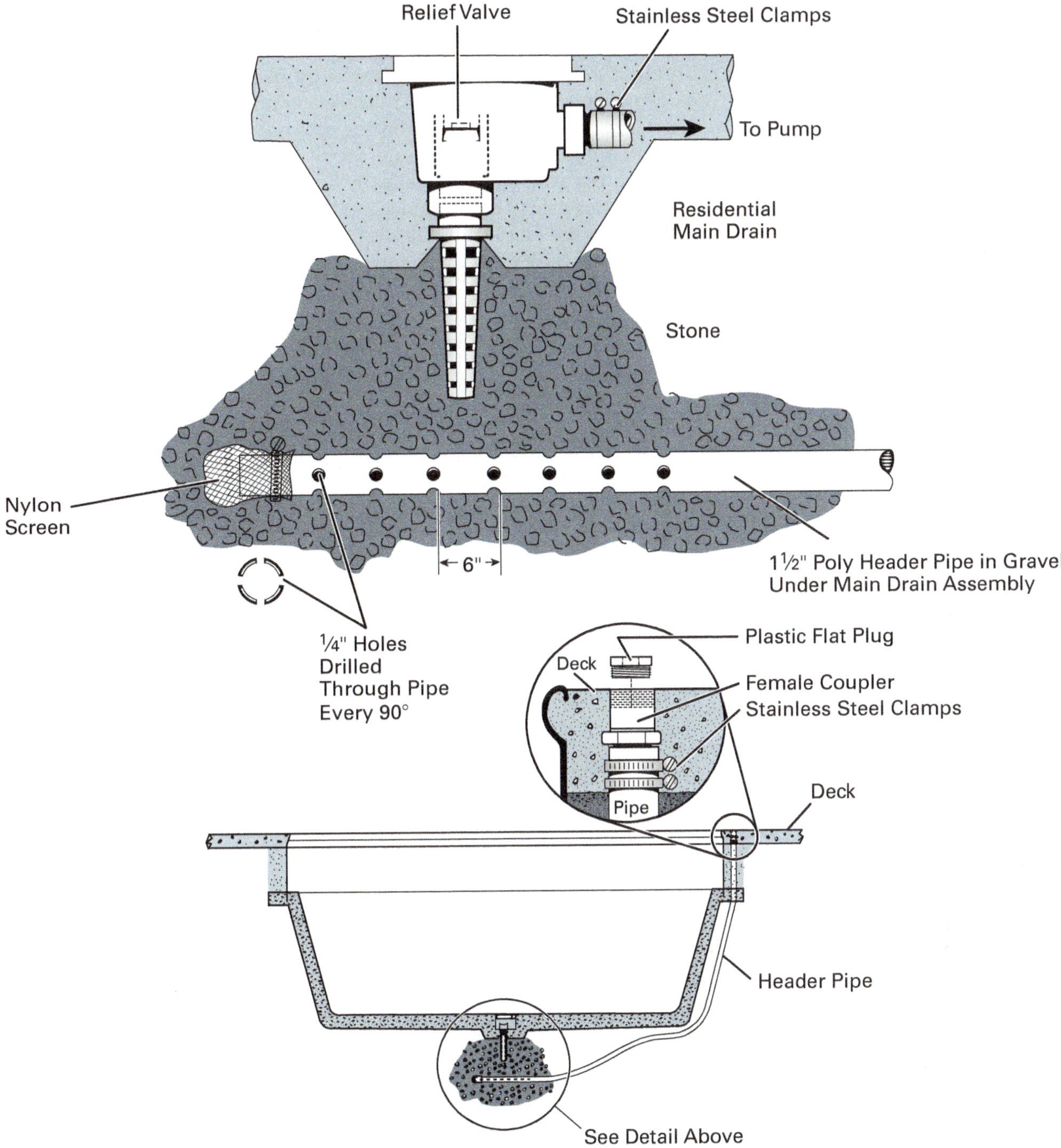

Figure 12 Relief-valve and header-pipe installations.

pressure is equalized. Another way to relieve hydrostatic pressure is to install a **header pipe** (*Figure 12*). The header pipe runs from underneath the pool to a connection on the pool deck. A pump connected to the pipe can siphon out the groundwater until the pressure is relieved.

Header pipe: A pipe used to relieve hydrostatic pressure. The pipe runs underneath the pool to a pump connection on the pool deck.

CAUTION

Never discharge swimming pool water into a septic system. The discharge could damage the tank. Chlorine in the water can contaminate the leach field and kill the bacteria in the soil. A sudden drainage from the pool could result in a sewage backflow into the house.

1.4.0 Maintaining Water Quality

Swimming pools can be contaminated by bacteria and other harmful organisms. Small amounts of chlorine in the water are enough to kill most organisms, so install an in-line chlorinator on the recirculation line. Chlorinators release a measured amount of chlorine into the water as it is pumped to the filter. The chlorine can be in tablet or stick form. Some large pools use chlorine gas under pressure. Another option is to place chlorine tablets in a **floater**. This is a device that floats in the pool and releases chlorine through vents on its underside (*Figure 13*). (These types of chemical treatments are not permitted in public pools.) Test chlorine levels regularly. Replace tablets or sticks according to manufacturers' instructions.

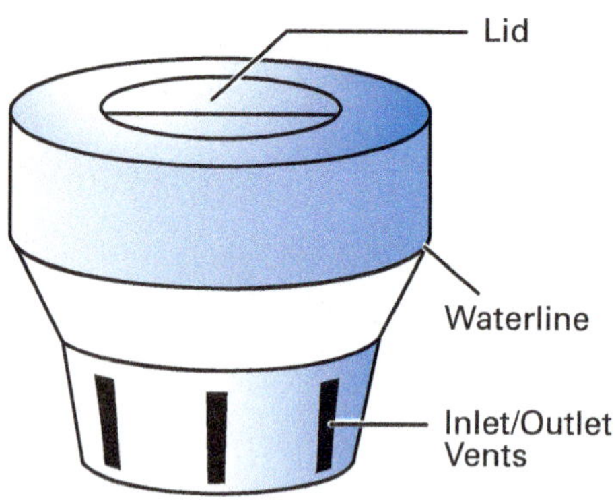

Figure 13 Floater chlorine dispenser.

Floater: A device that releases chlorine through vents while floating in a pool. These types of chemical treatments are not permitted in public pools.

Acidity and alkalinity levels in the pool water must be correctly balanced. High acid levels can cause the following problems:

- Corrosion of iron and copper pipes
- Damage to pumps and other recirculation components
- Irritation of swimmers' skin and eyes
- Inability of the water to hold chlorine

Algae

Alga is a type of seaweed. Algae (the plural of alga) can form in a pool that has not been used for a long time. Algae cause the water to turn green and have an unpleasant odor. They can leave a floating scum on the water surface or the pool walls. Algae can grow if the chlorine level is below normal. Adding chlorine is often enough to kill algae. If it is not, you may need to call in a pool service professional.

CAUTION

Do not store pool chemicals where they will be exposed to direct sunlight, because exposure to ultraviolet radiation can cause the chemicals to break down. Adding cyanuric acid to the pool water will help prevent breakdown of chlorine in the pool.

On the other hand, high alkaline levels can cause the following problems:

- Formation of scale on plumbing components
- Clouding of pool water
- Prevention of chlorine from killing harmful organisms

The measure of the acidity or alkalinity levels is called the pH (potential hydrogen) value. High pH values indicate high alkaline content. Low pH values indicate high acid content. A value of 7.0 indicates a neutral solution. The proper pH range for a pool is from 7.4 to 7.8. Use muriatic acid to lower the pH value. Refer to your local code. Consult a local pool expert for advice on how to maintain a correct pH value.

Disinfecting Pools Using Methods Other Than Chlorine

Chlorine is a popular and widely used disinfectant in swimming pools, but it has drawbacks. Chlorine can cause allergic skin reactions, and its disinfecting properties can be lessened by imbalances in the pH level of the water. Pool manufacturers offer a variety of other methods for disinfecting pool water. Saltwater systems, for example, are less expensive than chlorine and eliminate the strong chemical odor associated with chlorine. However, saltwater systems require regular cleaning to prevent calcium buildup. Furthermore, saltwater systems are currently more expensive than chlorine systems and must run all day, which results in greater energy consumption.

In copper-ion systems, a low-voltage electric current flows over copper or copper-alloy electrodes in the water. This releases electrically charged copper atoms, called ions, into the water. Copper ions are harmless to humans, but they prevent the buildup of algae and other organic materials in the water. Silver-ion systems use silver and silver-alloy electrodes to achieve the same results. Both copper-ion and silver-ion systems require little maintenance, although the electrodes have to be replaced every three to five years.

Ultraviolet (UV) light, which is used to disinfect potable water supplies, is also used to disinfect swimming pools. Ultraviolet systems are generally less expensive than chlorine systems to install and operate. They are, however, less effective than chlorine systems. If using UV in a residential system, a secondary system (typically chlorine) is necessary to assist in disinfection.

Swimming pool water should be tested regularly. Test kits can be purchased that will give you a quick reference to check the chlorine level and pH value. More thorough tests are required to identify other health hazards. For example, most over-the-counter test kits do not test for fecal coliform, a dangerous type of bacteria. The public health department usually conducts such tests. However, customers often contact the pool installer first. Take the time to help the customer obtain more information about testing. Ensure that your level of service matches the quality of the components and workmanship so that the swimming pool will remain a source of recreation, not headaches.

WARNING!

Pool chemicals are dangerous if swallowed, so store chemicals out of the reach of children. Ensure that the containers are sealed tightly. Contact emergency services immediately if someone swallows pool chemicals.

1.0.0 Section Review

1. The calculation of bathing load is partly determined by the _____.
   a. filter size
   b. pool capacity
   c. pump type
   d. turnover rate

2. The minimum distance that an emergency-pump shutoff switch should be installed from the pool is _____.
   a. 3'
   b. 5'
   c. 7'
   d. 9'

3. If a pool's gutter line drains into the sewer system, the line should be fitted with a _____.
   a. skimmer
   b. backwater valve
   c. backflow preventer
   d. filter

4. The proper pH range for a swimming pool is between _____.
   a. 7.4 and 7.8
   b. 6.4 and 6.8
   c. 5.4 and 5.8
   d. 4.4 and 4.8

2.0.0 Hot Tub and Spa Systems and Components

Performance Task

2. Identify the components of piping for a spa.

Objective

Identify hot tub and spa systems and their components.

a. Identify hot tub and spa components.
b. Explain how to install hot tubs and spas.

The terms hot tub and spa are often used interchangeably. However, there are some important technical differences between them. You should be aware of these differences because they can affect the successful installation, use, and servicing of the hot tub or spa. Hot tubs are filled with hot water and are typically made of molded fiberglass or acrylic, though wood is still sometimes used. The water in a hot tub circulates slowly through a heater. A spa uses jets, or air injectors, to provide additional massaging action.

Spas are usually made of fiberglass, steel, or concrete. This section reviews the components used in hot tubs and spas, including the components that are used in both and those that are used specifically in one or the other. It will then summarize the important installation steps required for hot tubs and spas. As always, refer to manufacturers' instructions and the local applicable code when installing hot tubs and spas. Proper installation is essential for ensuring that hot tubs and spas are both safe and healthy for the people who use them.

2.1.0 Identifying Hot Tub and Spa Components

CAUTION

Store chlorine in a well-ventilated area.

Except for the spa's air-injection ports, the components and installation requirements for hot tubs and spas are very similar. However, the differences could be important when ordering new or replacement parts. Before installing a spa or hot tub, ensure that you and the customer are referring to the same type of installation. The following sections review key hot tub and spa components that plumbers are responsible for installing.

2.1.1 Hot Tub Components

In most hot tubs, water is circulated slowly through a natural gas or electric heater. Self-contained hot tubs that are purchased as complete units are typically equipped with electric heaters. In addition to the heater, the plumbing components of a typical hot tub include the pipe and fittings that connect to the water supply, and a pump to circulate the water to the heater.

Approved piping materials for use in hot tubs as well as in spas include:

- Acrylonitrile butadiene styrene (ABS) pipe
- Chlorinated polyvinyl chloride (CPVC) pipe
- Copper or copper-alloy tubing
- Polyvinyl chloride (PVC) hose and pipe
- Stainless-steel pipe

Direct-contact water heaters are often used in hot tubs because of their greater efficiency, smaller footprint, and reduced complexity when compared with other types of water heaters. Unlike conventional water heaters that are used to provide potable hot water, direct-contact water heaters do not use a heat exchanger to heat water. Instead, the water is pumped directly over the heating element. Heating elements in direct-contact water heaters must be resistant to corrosion from exposure to water and any chemicals in the water, such as chlorine or ozone. Some hot tubs use heat pumps to heat water.

Heat pumps require the circulating pump to operate much more frequently than for a water heater, and they also take longer to heat the water. Furthermore, the temperature of heat-pump water heaters can be harder to regulate under certain conditions.

Some hot tubs are fitted with separate pumps for circulating and filtration, while others are fitted with a single pump that handles both functions. Unlike other types of pumps, which can dissipate the heat caused by their operation into the surrounding air or water, hot-tub pumps typically operate in temperatures that are at least as hot, if not hotter, than their typical operating temperature. In addition, the pump must be able to withstand the corrosive effects of both the heated water and any chemicals that are dissolved in the water. Therefore, it is important to ensure that a pump is specifically designed for use in high-temperature, corrosive environments, such as a hot tub.

Pumps and heaters are typically operated by electronic controls that are sealed against damage from humidity. These controls may include sophisticated digital features, such as password protection, to guard against unauthorized use. Optional features include insulating covers and saltwater filtration.

2.1.2 Spa Components

Spa systems come in a variety of sizes and shapes. Most systems have the same basic components (*Figure 14*). A pump circulates the water through a filter that traps particles and organisms. The water is then pumped to a heater, which warms it to a preset temperature. A thermostat in the tub controls the heater. From the heater, the water circulates into the tub by way of a **hydrojet**. A hydrojet is a nozzle that squirts a mixture of air and water at high pressure into the tub (*Figure 15*). Some spas use jets of water with no air mixed in. As the water circulates in the tub, the pump draws it out through drains in the sides and bottom of the tub, and the cycle begins again.

Hydrojet: A nozzle that sends a high-pressure stream of air and water into a spa.

An air blower provides the compressed air for the hydrojets. Install the air blower above the water level to prevent water from flooding the air-blower system. Refer to the manufacturer's instructions for the proper placement of the air blower. In most cases, 12" above the waterline is sufficient.

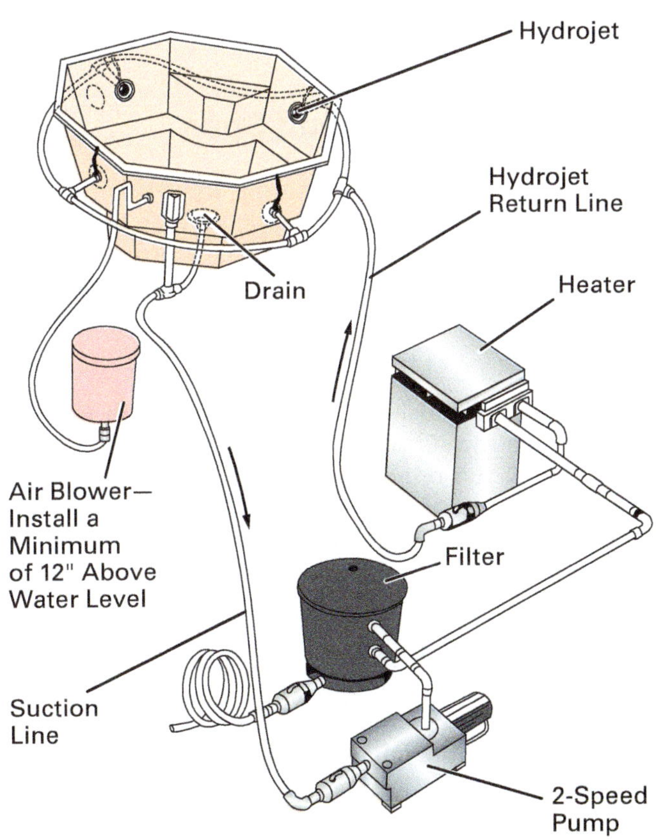

Figure 14 Typical spa installation.

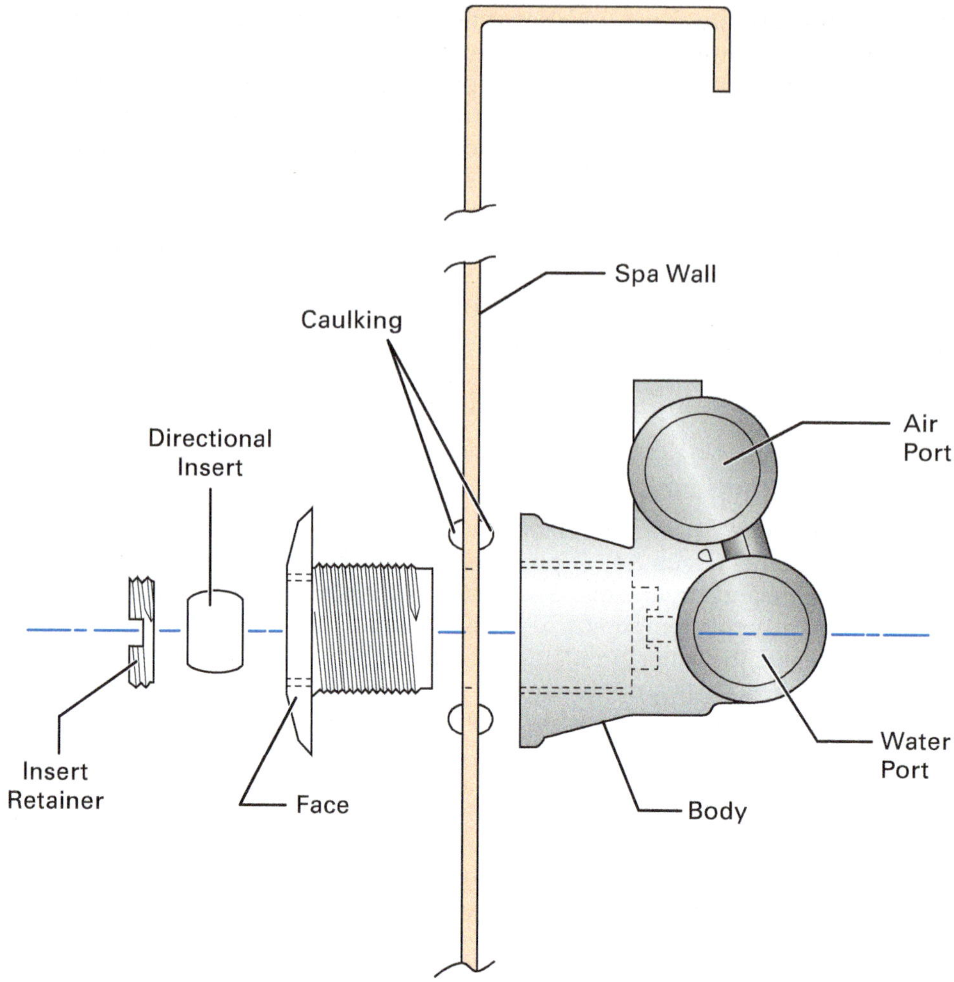

Figure 15 Exploded view of a hydrojet.

2.2.0 Installing Hot Tubs and Spas

The procedures for installing a hot tub or a spa are similar to those for a swimming pool. Most instructions come with detailed illustrations to help you install the components correctly. If you need additional information, contact the manufacturer or visit the manufacturer's website. Many companies have product information available to print or download.

2.2.1 Before Installing a Hot Tub or Spa

Many hot tubs and spas are installed indoors. They will have to be carefully located to fit in with existing supply and drainage systems. The wholesaler may deliver the complete hot tub or spa system directly to the jobsite, or you may have to pick it up yourself from a supplier or a contractor's warehouse. In either case, after you get the system, install the system according to the manufacturer's instructions, keeping the following guidelines in mind:

- Check the brand and model to ensure that the system meets the job specifications. Make sure that all components are included. Test-fit the components before final assembly.
- Check the condition of the tub and its components. Open all containers delivered to the jobsite while the delivery person is there. Doing this eliminates questions about who is at fault if damage is discovered later. If you pick up the system from a supplier or contractor's warehouse, inspect it before you leave. Check the contents of previously opened containers for missing items.

- Verify that the rough-in dimensions are correct. This is a critical step. Errors could have occurred in the initial rough-in. Changes in the building, such as wall placement and wall covering, could affect your work. You might have to install a hot tub or spa that is different than the one originally specified.
- Protect the tub and its components before installation. Ideally, you will be able to schedule delivery of the fixtures right before you are ready to install them. If that is not possible, store the fixtures in a clean, dry space that can be locked. After the components are installed, cut and fit cartons over them to protect surfaces from scratches and chips.
- Save all of the packing materials. You can use the package to protect the installed components from weather and other work. You can also use the package to return an unsuitable or damaged component.
- Save all the manufacturer's installation and care instructions, and warranty materials. Turn them over to the customer when the job is completed.
- Protect your customers. They expect to be the first to use their new hot tub or spa. Turn off the water supply and place a Do Not Use sign on the tub.

Follow this procedure on every installation. You will reduce (and even eliminate) mistakes, and save time and money.

2.2.2 Hot Tub and Spa Installation Procedures

Spa tubs may come with precut holes for the hydrojets. If not, follow manufacturers' directions for cutting them. Carefully locate and size the holes. If the holes are too big, the hydrojet fixture will fit loosely on the tub wall. Oversized holes could cause excess vibration and wear. Seal the fitting on both the inside and the outside of the tub wall.

Always follow the manufacturer's instructions when positioning the required blocking, grout, or other supporting material for the tub body. Trial-fit the unit to ensure that the loadbearing points are properly supported. Fiberglass tends to crack under stress, so you must correctly fit the framing around fiberglass units. Avoid bending the material during installation. Bending may cause cracks that will not become visible until later. After installation is complete, fill the tub to the overflow line and check for leaks. Turn off the water, drain the tub, and cover it with packing material to protect it.

NOTE

Turnover rates for hot tubs and spas vary with the manufacturer and the size of the hot tub or spa. Always refer to the manufacturer's instructions for guidance on calculating the appropriate turnover for the particular application.

Did You Know?

The Jacuzzi Brothers

The Jacuzzi™ name is synonymous with whirlpool baths. However, Valeriano and Francesco Jacuzzi also designed and built an airplane in 1920 to carry mail for the US Post Office. The plane was very advanced for its day. It was a monoplane (meaning it had only one set of wings) and had an enclosed cabin for the pilot. At the time, most planes were biplanes, and the pilot sat in an open cockpit, exposed to the wind.

The Jacuzzi brothers also invented and marketed deep-well pumps. In the late 1950s, they turned their attention to hydrotherapy (curing physical ailments with water), developing a portable whirlpool pump for bathtubs. In 1968 Roy Jacuzzi invented the first self-contained whirlpool bath, and two years later the company developed its first spa model. Today, the company has manufacturing plants around the world, and the name Jacuzzi is synonymous with whirlpool baths.

Hydrostatic pressure can also float empty hot tubs and spas out of the ground. Install a header pipe under the tub (*Figure 16*). Use perforated pipe (with $^1/_4$" holes every 90 degrees around the circumference of the pipe). Embed the pipe in gravel to allow water to flow into the pipe and to relieve the hydrostatic pressure. Plug the top of the header pipe when it is not in use, to keep the pipe clear of debris.

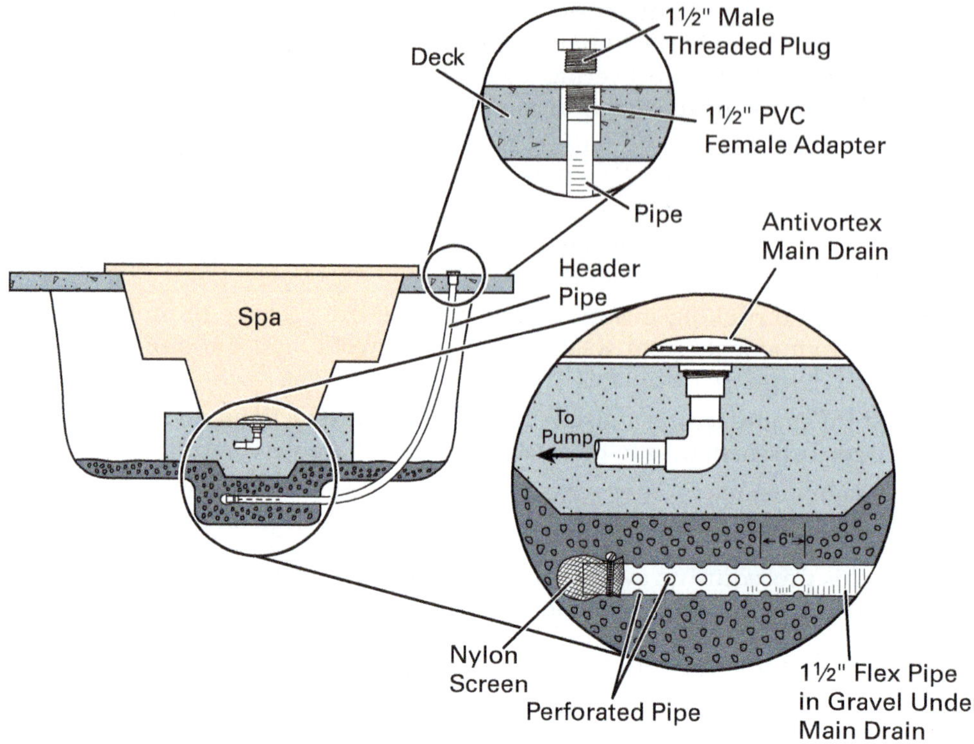

Figure 16 Perforated pipe installed underneath a spa.

If the main drain is directly connected to the sanitary drainage system, trap and vent the drain according to local code. If the drain empties into a floor drain, provide an air gap between the tub drain outlet and floor drain. Install the required number of skimmers per the local applicable code. The ISPSC, for example, requires the installation of one skimmer for every 150 ft^2 of water surface area in a hot tub or spa.

Refer to the installation instructions to ensure that the heaters and pumps are sized properly. If the heater will be turned off when the hot tub or spa is not in use, ensure that the heater is large enough to heat the water quickly. If the heater will be left on all the time, a smaller unit can be used. Spas with hydrojets require larger pumps than those with water jets that do not mix air into the stream.

Proper pH and chlorine levels are even more important in a spa or hot tub than in a swimming pool. The water is warmer, and most spas and hot tubs are not under direct sunlight. The combination of heat, humidity, and darkness is ideal for bacteria and fungus. Hot tubs and spas should be inspected regularly. Ensure that filters and chlorinators are well maintained. Provide the customer with a maintenance schedule. This information may also be included in the owner's manual.

For public spas, local codes require the installation of shutoff switches and alarms on pumps to indicate when the pump has stopped. The alarm should generate an audible sound of no less than 80 decibels and have a visible light. Both the audible and visible alarms must operate continuously until turned off using the shutoff switch. Codes also require public spa facilities to display a sign explaining the purpose of the alarm and warning people to not use the spa when the alarm is activated. The sign must be visible from the spa. Ensure that air jets are fitted with backflow preventers according to the local applicable code.

Like swimming pools, hot tubs and spas lose water because of splashing and evaporation. They require make-up water to maintain the water level. If the water level falls too low, the pump may lose its prime and become damaged. Connecting hot tubs and spas to a water supply line will prevent this. Install a float valve to control the water level in the tub, and install a backflow preventer

Did You Know?

Restrictions on Plastic Pipe

Refer to your local code before installing plastic water supply pipe. The code may restrict the use of plastics in commercial fire-rated buildings or in multi-story plumbing installations.

on the supply line to guard against back siphonage. If the tub is equipped with a fill pipe or faucet, install the fixture above the tub's flood-level rim.

Think About It

Energy Conservation

Pools and hot tubs can be expensive to operate. When installing a unit, it's common to offer the customer some tips on how to save energy and keep costs down. For example, run the filtration system during off-peak hours. When leaving for more than a week, turn off the heater and extinguish its pilot light, if applicable. If the hot tub is used only on weekends, lower the thermostat setting during the week.

Erect windbreaks such as fences, hedges, or cabanas around outdoor spas and hot tubs. This minimizes heat loss caused by wind. At the beginning of "hot tub season," make sure the pump is working before adding chemicals to the water. Backwash or clean the filter regularly to obtain maximum filtration. Consider replacing heaters that are more than five years old with more energy-efficient models.

2.0.0 Section Review

1. Unless the manufacturer's instructions specify otherwise, air blowers are typically installed in a spa at _____.
   a. 12" above the waterline
   b. the waterline
   c. 12" below the waterline
   d. 12" above the base of the spa

2. When installing a header pipe under a spa, the circumference of the pipe should be perforated with _____.
   a. $^{1}/_{8}$" holes every 90 degrees
   b. $^{1}/_{8}$" holes every 45 degrees
   c. $^{1}/_{4}$" holes every 90 degrees
   d. $^{1}/_{4}$" holes every 45 degrees

Module 02410 Review Questions

1. According to the 2024 International Swimming Pool and Spa Code, which are considered residential pools?
   a. Apartments, dormitories, and hotels
   b. Detached one- and two-family dwellings and townhouses 3 stories or less in height
   c. Pools of any size in school buildings
   d. Hotels and private clubs

2. A floater is a device that floats in a private swimming pool and releases _____.
   a. chlorine
   b. bacteria
   c. alga
   d. alkaline

3. The purpose of the recirculation system in a swimming pool is to _____.
   a. measure the water's capacity
   b. remove chlorine from the system
   c. increase the hydrostatic pressure
   d. sanitize the water

4. The capacity of a swimming pool is measured as _____.
   a. volume in gallons
   b. number of people it will hold
   c. square feet
   d. maximum depth

5. A residential swimming pool with a length of 20', a width of 12', and a height of 6' has a volume of _____.
   a. 1,440 gallons
   b. 10,771 gallons
   c. 748 ft^2
   d. 25,832 liters

6. When using a permanent fill spout, the outlet and water overflow line must have an air gap of at least _____.
   a. 3"
   b. 4"
   c. 5"
   d. 6"

7. According to the ISPSC, a residential pool must have one skimmer per _____.
   a. 200 ft^2
   b. 350 ft^2
   c. 550 ft^2
   d. 800 ft^2

8. Pool pumps are designed to run _____.
   a. every 2 hours
   b. every 24 hours
   c. continuously
   d. occasionally

9. Pool recirculation systems use filters to reduce _____.
   a. pressure
   b. turbidity
   c. friction
   d. capacity

10. High acid levels in a pool can cause _____.
    a. scale to form on plumbing components
    b. ultraviolet radiation
    c. iron and copper pipes to corrode
    d. contamination in the leach field

11. Most hot tubs are made of acrylic or molded fiberglass, although some are made of _____.
    a. steel
    b. concrete
    c. wood
    d. plastic

12. The massaging action found in spas is produced by _____.
    a. air injectors
    b. hydrostatic pumps
    c. recirculation systems
    d. vacuum fittings

13. When installing hot tubs and spas, refer to your local code and the _____.
    a. customer
    b. plumber
    c. public health code
    d. manufacturers' instructions

14. To relieve hydrostatic pressure in a hot tub or spa, install a _____.
    a. loadbearing point
    b. perforated pipe
    c. hydrojet
    d. saltwater filtration system

15. A spa pump's audible alarm should be no less than _____.
    a. 50 decibels
    b. 60 decibels
    c. 70 decibels
    d. 80 decibels

Answers to odd-numbered review questions are found in the Module Review Answer Key at the back of this book.

Answers to Section Review Questions

Answer	Section	Objective
Section One		
1. d	1.1.0	1a
2. b	1.2.2	1b
3. b	1.3.0	1c
4. a	1.4.0	1d
Section Two		
1. a	2.1.2	2a
2. c	2.2.2	2b

MODULE 02411

Plumbing for Mobile Homes and Travel Trailers

Objectives

Successful completion of this module prepares you to do the following:

1. Describe water supply and DWV systems for mobile home parks.
   a. Describe water supply systems for mobile home parks.
   b. Describe DWV systems for mobile home parks.
2. Describe water supply and DWV systems for travel trailer parks.
   a. Describe water supply systems for travel trailer parks.
   b. Describe DWV systems for travel trailer parks.

Performance Task

Under supervision, you should be able to do the following:

1. Size the water supply and DWV systems for a mobile home park using information provided by your instructor.

Overview

Plumbers may install water supply and drain, waste, and vent (DWV) systems to serve mobile homes and travel trailers. Mobile homes are considered independent vehicles that can be connected to public utilities for extended periods of time using flexible pipe and quick-disconnect fittings. Travel trailers that are classified as independent vehicles are fitted with storage tanks for fresh water and wastewater. Plumbers may be called on to install sanitary dump stations in the parks to receive wastes from travel trailers' holding tanks. Travel trailer parks must also provide bathroom and laundry facilities.

CODE NOTE

Codes vary among jurisdictions. Because of the variations in code, consult the applicable code whenever regulations are in question. Referring to an incorrect set of codes can cause as much trouble as failing to reference codes altogether. Obtain, review, and familiarize yourself with your local adopted code.

1.0.0 Designing Systems for Mobile Home Parks

Performance Task	Objective
1. Size the water supply and DWV systems for a mobile home park using information provided by your instructor.	Describe water supply and DWV systems for mobile home parks. a. Describe water supply systems for mobile home parks. b. Describe DWV systems for mobile home parks.

Mobile home: A manufactured home built on a wheeled chassis that has no fixed foundation and that is typically no smaller than 8' wide by 32' long.

Independent vehicle: A mobile home or recreational vehicle (RV) equipped with bathroom facilities. Independent vehicles can be connected to public utilities for long periods of time.

A **mobile home** is a portable dwelling that is built on a chassis without a fixed foundation. Most codes further define a mobile home as a trailer that is at least 8' wide and 32' long (*Figure 1*). Owners are responsible for maintaining the mobile home's supply and drainage systems. Mobile homes are considered **independent vehicles**. An independent vehicle has bathroom facilities and can be connected to public utilities for extended periods of time, using flexible pipe and quick-disconnect fittings (*Figure 2*).

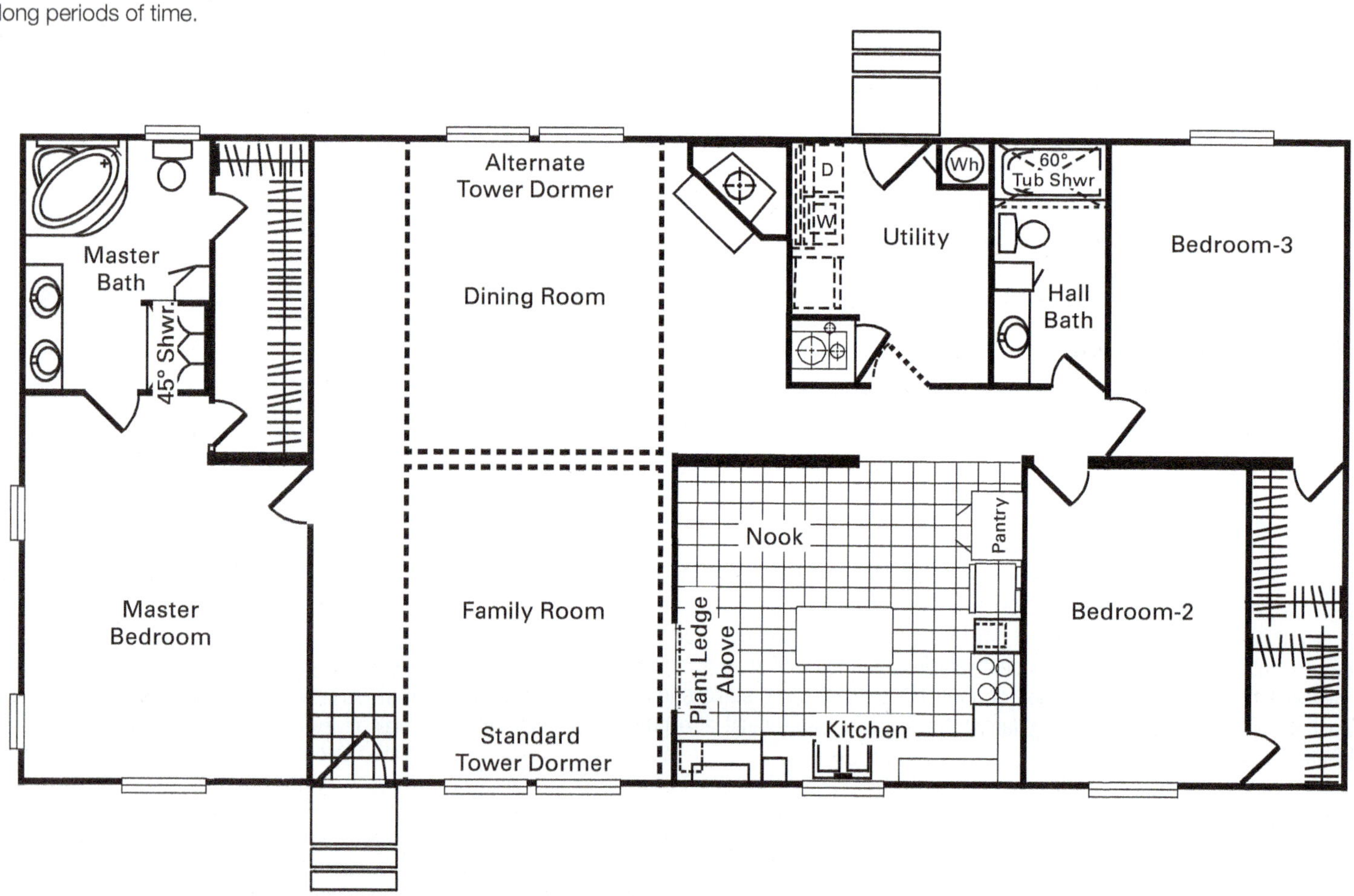

Figure 1 Floor plan of a typical mobile home.

Travel trailer: A portable dwelling that is towed by a motorized vehicle.

NOTE

Since 1977, mobile homes have been classified as "manufactured homes." Some codes classify such homes as permanent dwellings and **travel trailers** as temporary dwellings. Refer to the local applicable code, and follow the classifications that apply to your jurisdiction.

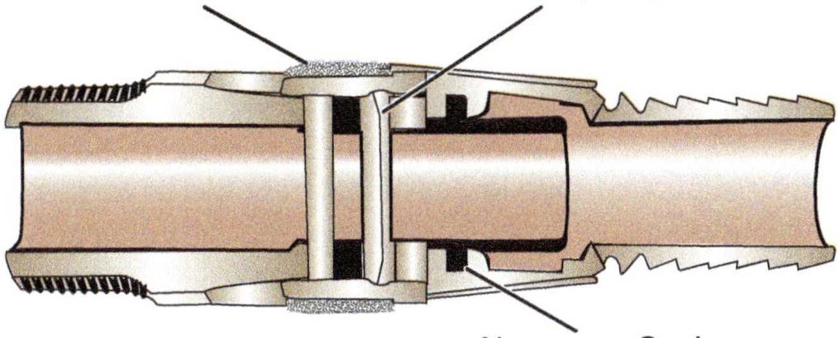

Figure 2 Quick-disconnect fitting for connecting to a supply line.

While manufacturers install the plumbing systems in mobile homes, plumbers design and install water supply and drain, waste, and vent (DWV) systems in the parks that serve mobile homes. The two sets of systems must be compatible. Otherwise, one or the other could suffer damage when it is connected. The materials and equipment used are the same as those used in other installations. However, the plumbing systems in mobile home parks differ from systems that plumbers are used to installing. Flow rates and pressures differ from those of other types of plumbing systems. Further, there are specialized methods for connecting the vehicles to the mains.

Construction standards for mobile homes vary widely. However, they must meet all the local plumbing codes enforced where they are installed. They must also use approved pipe and fittings. Their supply and discharge systems must be able to handle the pressures and flow rates of the main system. All connections to public utilities must be in good working order.

Piping for the water supply line must be at least $^1/_2$" in diameter. Some codes require dielectric fittings on water supply lines. The drainage line must be at least 3" in diameter. Ensure that the drainage-line connections are both gastight and watertight. The lines must be as short as possible. The waste system can be vented using air admittance vents (*Figure 3*).

Did You Know?

The HUD Code

In 1974 Congress passed a law providing for national construction and quality standards for all manufactured homes (a term that includes mobile homes, modular homes, and manufactured homes) in the United States. The National Home Construction and Safety Standards (commonly known as the HUD Code, after the US Department of Housing and Urban Development) is followed nationwide and preempts all local building codes for manufactured housing. Each manufactured home must have a red HUD label, certifying that the home has been factory constructed, tested, and inspected to comply with strict, uniform federal standards.

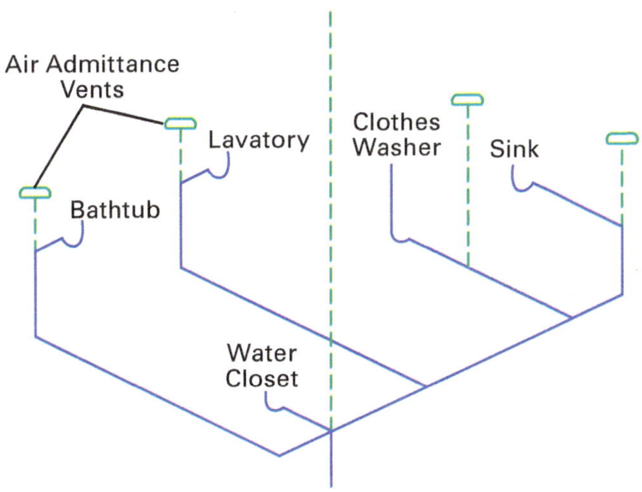

Figure 3 Typical waste-system installation in a mobile home.

Think About It

Temporary Housing and Natural Disasters

Starting in September 2005 the federal government ordered nearly 25,000 mobile homes and more than 114,000 travel trailers from both retailers and manufacturers to provide temporary housing for residents of the Gulf States who were left homeless by Hurricane Katrina. Federal, state, and local governments work hand in hand with builders of mobile and manufactured homes and travel trailers to respond to natural disasters—not only hurricanes, but tornadoes, floods, and wildfires—to provide immediate shelter solutions for those whose homes are damaged or destroyed.

If the mobile home uses natural gas, refer to the local applicable code to select the gas piping. Size the gas system according to the anticipated use; the local code will provide tables that will allow the system size to be calculated. Do not install gas lines under a mobile home unless they are enclosed in a gastight conduit. These conduits are vented at both ends, which allows gas to escape without posing a risk. Each home lot in the park should have a gas shutoff valve. Ensure that the shutoff valve is readily accessible, not under the mobile home. Plug or cap the lot's gas line when it is not in use. Most jurisdictions also require a flexible connection from the gas riser to the home's gas stub out.

Did You Know?

Manufactured Homes and Affordable Housing

As of June 2023, 11% of new single-family homes were manufactured homes, ranging from homes as small as 500 square feet to as large as 3,000 square feet.

Provide the owner with a maintenance schedule for the plumbing systems. Do not alter water service lines without referring to the local code. In some cases, prior approval from local code officials may be required. Before using a chemical drain cleaner, ensure that it is safe to use on the pipe material. Remember to flush sediment screens and water heaters regularly. These steps should help keep the mobile home's plumbing system trouble-free.

When installing plumbing systems in a mobile home park, ensure safe and reliable connections between the main system and each house. The first step is to develop plans and specifications for the installations. Before laying out the components of the plumbing system, complete the plot plan. The plot plan must show the following:

- Individual lots in the park
- Addresses and lot numbers
- Fittings at each lot (*Figure 4*)
- Nearby roads and utilities
- Park service buildings
- Specifications and layouts of supply and drainage lines
- Location of fire-protection systems and/or fire hydrants

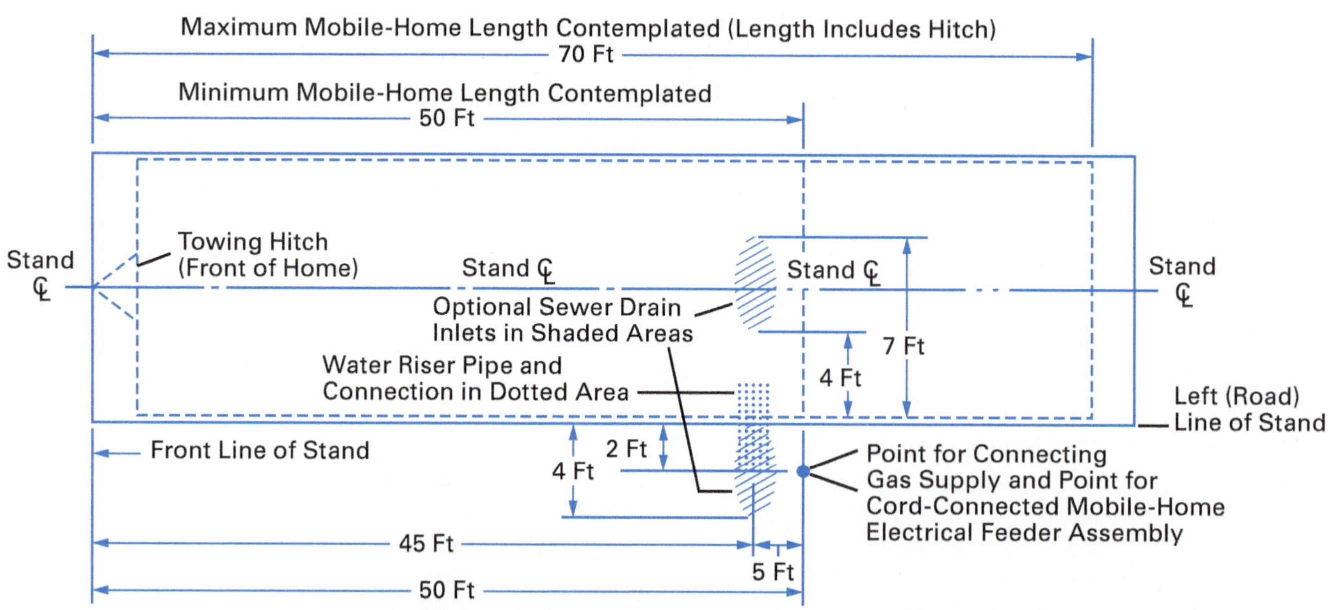

Figure 4 Typical mobile-home lot with locations of supply and drain connections.

- Site elevations and property lines
- Rights of way
- Utility easements

Once the plot plan has been drawn up, file copies with local building officials. Contact the local building office for the proper procedure. Once the plans have been approved, building officials will issue a permit to begin work. Be sure to follow all applicable local codes. Use approved piping materials and fittings. The following sections provide information on sizing and installing supply and drainage systems that are standard for most mobile home parks. Refer to the local applicable code for specific guidelines.

1.1.0 Water Supply Systems for Mobile Home Parks

Size each main supply line according to the number of mobile home lots to be serviced. Most codes require the system to provide at least 150 gallons of water per day to each home. The system must also provide water for the park's own needs, such as lawn care, fire protection, and laundry services. Connect the supply line to the public water supply or an approved private well.

Most codes require the supply system to provide water at a minimum pressure of 20 pounds per square inch (psi). In climates that are not subject to freezing, install a reduced pressure zone principle backflow preventer on the supply line; then, install a shutoff valve on the supply side of the backflow preventer and a pressure-relief valve on the discharge side.

Install shutoff valves, backflow preventers, and pressure-relief valves on all branch lines. For protection against cold weather, use leak-proof valves and locate them in a freeze-proof enclosure. Install a pressure-reducing valve on each line to an individual lot, after the water meter but before the shutoff valve (*Figure 5*).

Codes require that each mobile home lot be connected to the supply line. The minimum size for this line is usually $^3/_4$", but local requirements may vary. The local applicable code will specify the minimum rate of flow in gallons per minute (gpm) to be provided to each lot. If the water pressure in the system is low, resize the line to provide the required gpm.

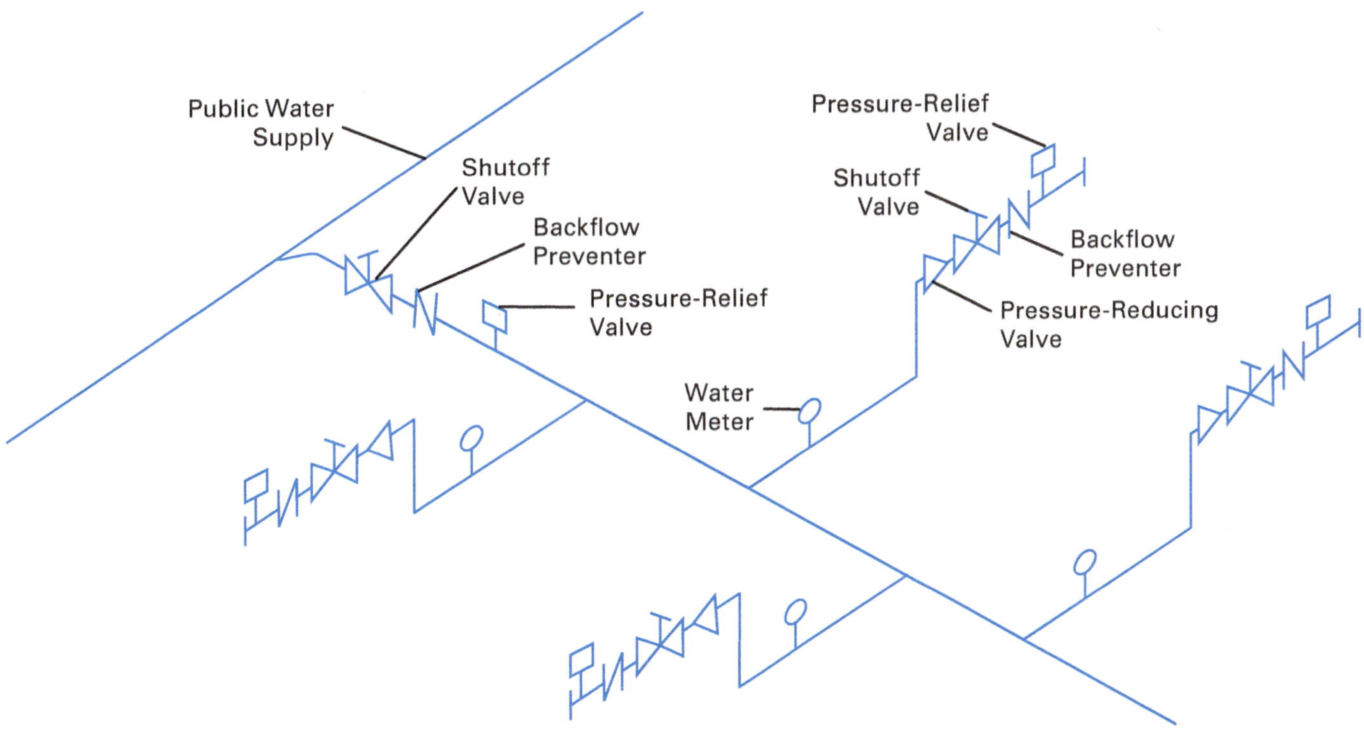

Figure 5 Water supply line installation with controls.

Install a shutoff valve on each line serving an individual lot. Most codes require a backflow preventer or check valve, and a relief valve on the discharge side of the shutoff valve (*Figure 5*). Locate the valves near the service connection so they can be accessed easily. Terminate the water supply line on the left (curb) side of the mobile home lot.

Install lawn and yard fixtures in accordance with local code. Provide hose-connection vacuum breakers (*Figure 6*) for all hose bibbs. Yard hydrants should be frost-proof in regions where freezing occurs (*Figure 7*). The hydrant has a small drain at its base, below grade. When the hydrant is shut off, water in the pipe empties out of the drain into a gravel bed. The drain prevents water from freezing in the pipe and shattering it. Set the hydrant in concrete to prevent accidental damage. Refer to the local applicable code for other requirements related to lawn and yard fixtures.

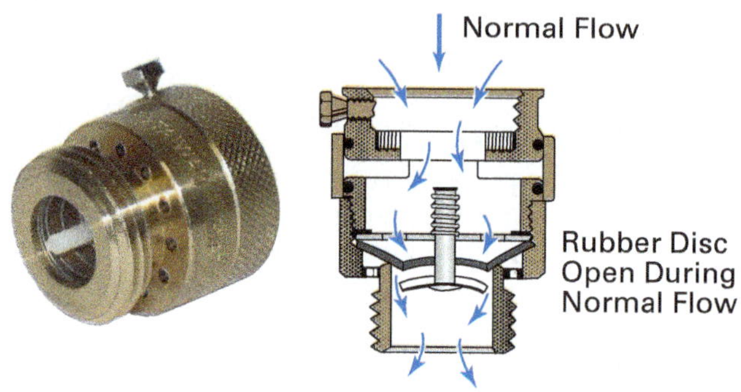

Figure 6 Hose-connection vacuum breaker.

Figure 7 Frost-proof yard hydrant.
Source: Courtesy of Simmons Manufacturing Company LLC

Make every effort to keep the water supply piping as clean as possible before installing it. Sand, gravel, or mud trapped in the pipes will flow through the piping system, lodge in the valves and faucets, and damage the washers and O-rings; the faucets will then leak. Store pipe in a clean, dry area, and cap all ends of the pipe at the end of each workday. Be sure to flush the line as thoroughly as possible to remove any sand, mud, or gravel from the system.

1.2.0 DWV Systems for Mobile Home Parks

The process for calculating fixture-unit loads in a mobile home park is different from the process used to calculate loads in a building. Normally, the load can be found by totaling up each fixture in the system. In a mobile home park, each home lot has a drain line already installed, even though the number of fixtures in a home can vary widely. For this reason, assign a fixed number of drainage fixture units (DFUs) per lot. Many codes require an estimate of 15 DFUs per lot. Refer to the local applicable code for the correct DFU rating in the local area.

Use pipe of at least 4" in diameter for the branch that connects the home to the main drain line. The pipe slope should allow for a minimum flow rate of 2' per second. Ensure that all pipes and fittings are gas- and watertight. The branch drain inlet should extend 3" to 6" above grade and should connect to the rear third of the mobile home. To protect the inlet pipe from damage, set it in concrete. Install a gastight plug on the inlet when the lot is unoccupied. If the branch is longer than 30', install a vent and a cleanout that extend above grade. Install a trap on the branch line (*Figure 8*).

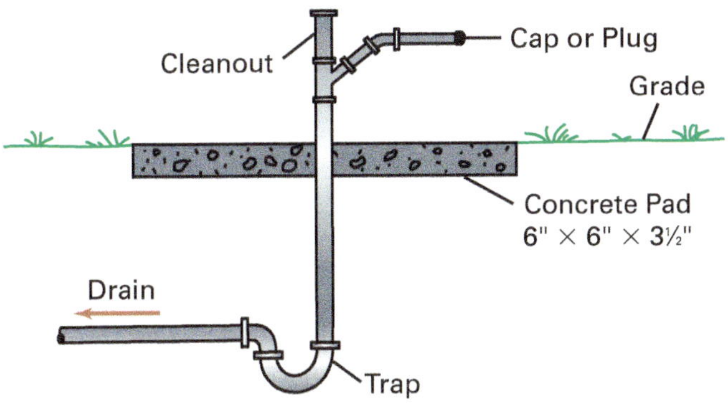

Figure 8 Typical branch drain connection.

WARNING!

Toxic and flammable vapors may be present when a DWV line connects to a sewer line. The public sewer is also a biohazard. Wear appropriate personal protective equipment when working with DWV and sewer lines.

Size each main drainage line to accommodate all of the mobile homes and buildings that will empty into it. Consider the potential for future expansion of the park when sizing the line. Note that the park's service building must have its own connection. Install vents on each main drainage line according to the local applicable code.

Each vent should have a minimum diameter of 4" and should be located no more than 5' downstream from the highest trapped branch on the line. Install relief vents every 200' along the main drainage line and cleanouts at regular intervals. Finally, connect the main drainage line to the public sewer according to local code.

WARNING!

Check the local code before modifying any building structural members. Improper modification can seriously weaken a building's support structure.

1.0.0 Section Review

1. Gas lines installed under a mobile home must be _____.
   a. enclosed in a gastight conduit
   b. made from PVC or CPVC
   c. buried at least 6' under grade
   d. fitted with check valves at both ends

2. When connecting a mobile home to a main drain line, the diameter of the pipe used in the branch should be at least _____.
   a. 1" in diameter
   b. 2" in diameter
   c. 3" in diameter
   d. 4" in diameter

2.0.0 Designing Systems for Travel Trailer Parks

Performance Tasks

There are no Performance Tasks in this section.

Objective

Describe water supply and DWV systems for travel trailer parks.

a. Describe water supply systems for travel trailer parks.
b. Describe DWV systems for travel trailer parks.

Tent trailer: A collapsible travel trailer with canvas walls.

Dependent vehicle: An RV that is not equipped with bathroom facilities. Dependent vehicles are not intended to be connected to public utilities for an extended time.

Travel trailers are among the most popular types of recreational vehicles (RVs). They are portable dwellings that are towed by a car or truck. Travel trailers come in a wide range of sizes and styles. Many travel trailers can be collapsed for easy travel and storage. Collapsible trailers with canvas walls are also called **tent trailers** or pop-up trailers. Most travel trailers can be classified as **dependent vehicles**. Dependent vehicles do not have bathrooms and are not intended to be connected to public utilities for long periods of time. Some trailers are independent vehicles. These trailers are fitted with storage tanks for fresh water and wastewater (*Figure 9*).

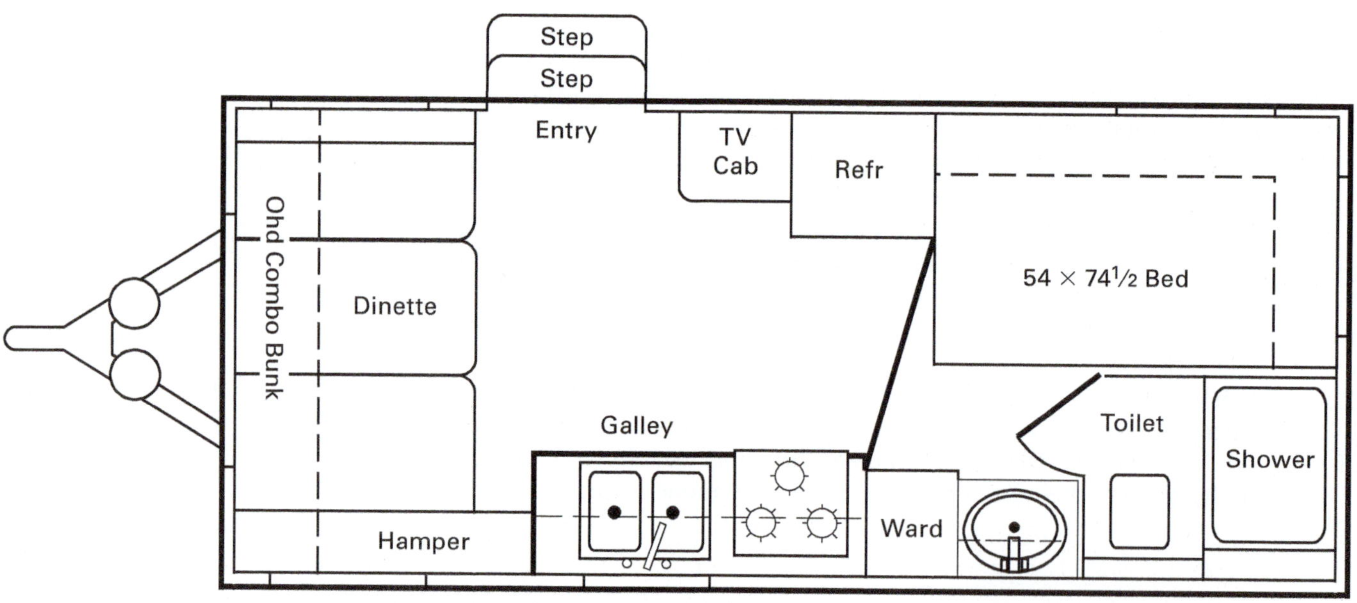

Figure 9 Floor plan of an independent travel trailer.

Job-Site Office Trailers

Source: Photo courtesy of GeneralContractor.com

Many construction projects require an on-site office. These offices are towed to the project site at the outset and towed away when the project has been completed. Some portable offices are small with very few amenities and resemble travel trailers. Others are larger with more amenities and are more similar to mobile homes. Regardless of the size of the office trailer, plumbers will be called on to connect their water supply and DWV systems once they have been situated on site. Be sure to correctly identify the trailer's inlet and outlet connections before making the connections.

As with the systems that serve mobile homes, plumbers are responsible for designing and installing water supply and DWV systems in the parks that serve travel trailers. The plumbing systems used in travel trailers are basic and simple. Use flexible pipe and quick-disconnect fittings when filling or emptying the storage tanks.

Did You Know?

Recreational Vehicles

An estimated 11.2 million households in the United States own an RV today, a number that has increased dramatically since 1980. RV rentals are also a popular choice; over 40 million Americans regularly go RVing. People are clearly attracted to the affordability and flexibility of RV travel (and have been for quite some time). Twenty years ago, the online auction house eBay reported that in 2004, "RV" was the most popular search term among its then-125 million users. Other studies have shown that even with high gas prices, RVs remain an economical mode of travel.

Codes only permit waste tanks to be emptied at approved dump stations in travel trailer parks. Like mobile homes, travel trailers are required to meet the local codes of the areas where they are parked. Discourage owners from modifying the trailer's supply and waste lines. If modifications are required, the owner should hire a trained plumber.

Think About It

RV Classification

For legal and insurance purposes, RVs are classified into two groups: motorized and towable. The motorized RV category is further broken down into three subcategories. These are large motorhomes (Type A), van campers (Type B), and small or mini motorhomes (Type C). Towable RVs, by contrast, are broken down into folding camping trailers, truck campers, fifth-wheel travel trailers, and conventional travel trailers.

The requirements for supply and drainage systems in travel trailer parks are very similar to those for mobile home parks. Travel trailer parks offer only basic facilities because trailers are there for only a short time. Most codes do not require these parks to have individual water and sewer connections for each lot. In most cases, trailers do not even have fixtures that need to be hooked up. Refer to the local applicable code for the standards for safe operation in the local area.

WARNING!

When installing a drainage system, keep in mind that the trench from the building to a sewer is usually the deepest trench on the jobsite. Cave-ins and falls are a danger. Shore up the trench properly before working in it. Observe all Occupational Safety and Health Administration (OSHA) trench safety guidelines.

2.1.0 Designing and Installing Water Supply and DWV Systems for Travel Trailer Parks

In most local codes, the regulations governing water supply to mobile home parks also cover travel trailer parks. However, travel trailer parks do not have to supply water at the same rate as mobile home parks. Size the supply system so it provides water to each branch at a rate of 8 gpm. Individual branches should be a minimum of $^1/_2$" in diameter. Refer to the local applicable code's guidelines for mobile home parks when installing the main supply line. Use only piping materials and fittings that have been approved by the local applicable code.

WARNING!

Always use water supply piping that is made from materials specified in the local code. All water supply piping must be resistant to corrosion. The amount of lead allowed in piping materials is strictly limited. Failure to use approved materials could contaminate the potable water supply, causing illness or death.

Sanitary dump station: A container designed to receive wastes from a travel trailer holding tank.

Graywater and Blackwater

Most travel trailers are fitted with separate holding tanks for wastewater, called graywater tanks and blackwater tanks. Graywater is water from sinks, showers, and similar fixtures, while blackwater is the wastewater from toilets. Both tanks are discharged into the same receptacle at the travel trailer park. Typically, blackwater will be emptied first.

Plumbers may be called on to install **sanitary dump stations** in parks that serve independent vehicles. Dump stations receive wastes from a travel trailer's holding tank. Ensure that the dump tank meets the local code's environmental standards. The hose connection on the dump tank must provide a tight fit with the trailer's waste line. Empty the dump station regularly. The station can be connected to the public sewer, or it can be cleaned by a pumper truck. Some cities provide municipal dump stations that are connected to the public sewer system. The city is responsible for maintaining these systems.

Travel trailer parks must provide bathroom and laundry facilities. The number and type of facilities depend on the type of trailers being served. If the park serves only independent trailers, minimal facilities are required. The trailers will have their own bathrooms. If the park serves either only dependent trailers or a mix of both types, then the park must offer services that meet its occupancy level.

Most codes require travel trailer parks to offer at least the following facilities for every 10 trailer lots:

- One laundry tray
- Separate bathroom facilities for men and women
- One shower per bathroom facility
- One lavatory per bathroom facility
- Two water closets for each women's bathroom facility
- One water closet and one urinal for each men's bathroom facility

Provide additional facilities for every set of 10 lots in the park. If the number of lots in the park is between multiples of 10, round up to the next multiple. For example, if the park has 32 lots, provide facilities for 40 lots. Refer to the local applicable code for specific requirements.

WARNING!

Incorrectly installed sanitary dump stations can cause and spread diseases such as typhoid, scarlet fever, infectious hepatitis, poliomyelitis, and dysentery. Remember, correct installations are essential to public health and safety.

2.0.0 Section Review

1. Size the water supply systems that serve travel trailers to provide water to each branch at a rate of _____.
   a. 6 gpm
   b. 8 gpm
   c. 10 gpm
   d. 12 gpm

2. In a travel trailer park, provide a complete set of facilities for every group of _____.
   a. 5 lots
   b. 10 lots
   c. 15 lots
   d. 20 lots

Module 02411 Review Questions

1. A mobile home is defined by most codes as a trailer that has a width of at least 8' and a length of at least _____.
   a. 18'
   b. 24'
   c. 32'
   d. 48'

2. A mobile home is typically connected to public utilities using flexible pipe and _____.
   a. quick-disconnect fittings
   b. flared fittings
   c. compression fittings
   d. threaded brass fittings

3. Some codes require the installation of dielectric fittings in mobile home park _____.
   a. gas supply lines
   b. drainage lines
   c. water supply lines
   d. water supply and drainage lines

4. Most codes specify that water supply mains in a mobile home park should be sized to provide each home with a daily minimum of _____.
   a. 100 gallons
   b. 150 gallons
   c. 200 gallons
   d. 250 gallons

5. The water supply line serving a mobile home should be provided with a reduced pressure zone principle _____.
   a. relief valve
   b. surge protector
   c. shutoff valve
   d. backflow preventer

6. Yard hydrants with a drain at the base that empties the pipe when the hydrant is shut off are described as _____.
   a. autodraining
   b. freeze free
   c. frost-proof
   d. discharge type

7. Many codes require drainage systems for mobile home parks to be sized using a per-unit allowance of _____.
   a. 11 DFUs
   b. 15 DFUs
   c. 18 DFUs
   d. 24 DFUs

8. A vent and a cleanout that extend above grade should be installed on any drainage line longer than _____.
   a. 30'
   b. 50'
   c. 80'
   d. 100'

9. A vent on a main drainage line should be located downstream from the highest trapped branch by a distance of no more than _____.
   a. 2'
   b. 5'
   c. 7'
   d. 10'

10. Water supply and waste lines of a trailer should only be modified by _____.
    a. the original manufacturer
    b. the trailer owner
    c. a factory-certified technician
    d. a trained plumber

Answers to odd-numbered review questions are found in the Module Review Answer Key at the back of this book.

Answers to Section Review Questions

Answer	Section	Objective
Section One		
1. a	1.0.0	1
2. d	1.2.0	1b
Section Two		
1. b	2.1.0	2a
2. b	2.1.0	2b

MODULE 02412

Introduction to Medical Gas and Vacuum Systems

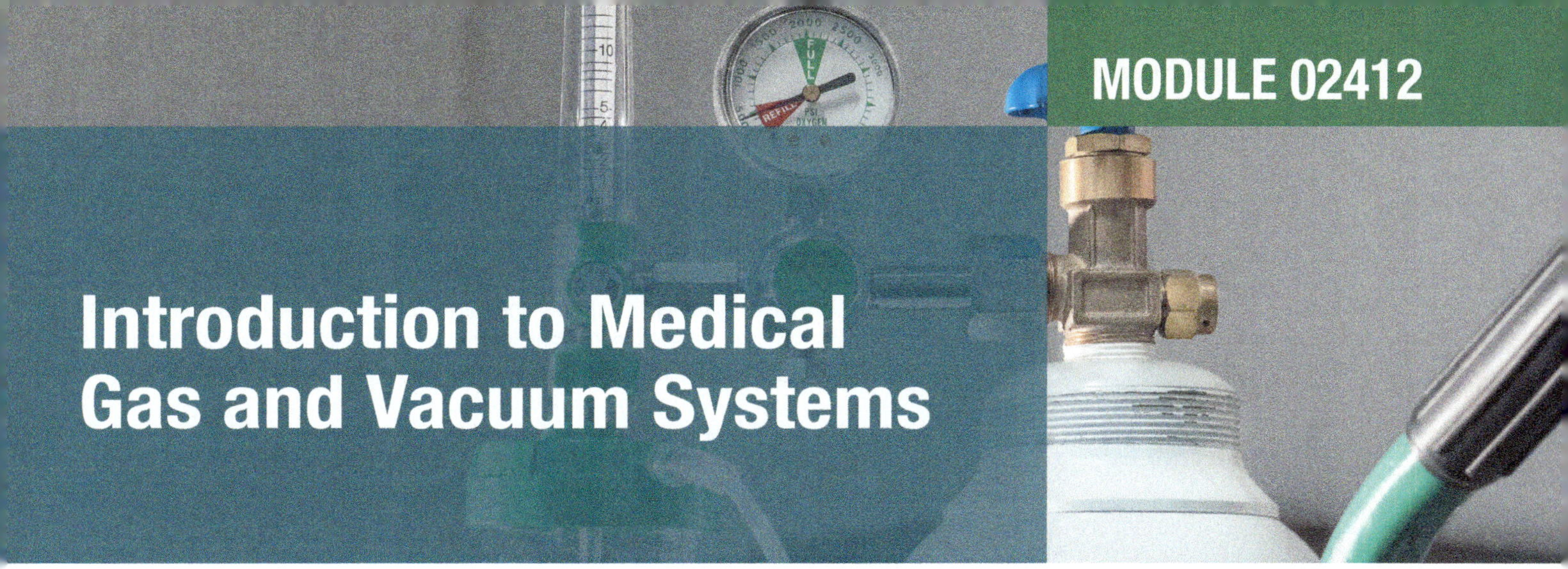

Source: Toa55/iStock/Getty Images

Objectives

Successful completion of this module prepares you to do the following:

1. Identify the building system requirements for medical gas and vacuum systems as outlined in the latest edition of NFPA 99, *Health Care Facilities Code,* and the professional qualifications as outlined in the latest edition of ASSE/IAPMO/ANSI Series 6000, Professional Qualifications Standard for Medical Gas Systems Personnel.
   a. Explain NFPA 99 building system categories.
   b. Explain ASSE/IAPMO/ANSI Series 6000 professional qualifications.
2. Describe the various medical gas and vacuum systems.
   a. Identify medical gas and vacuum systems.
   b. Identify control panels and alarm systems.
   c. Identify zone valves and station outlets/inlets.
3. Identify the safety issues related to medical gas and vacuum system installation.
   a. Identify personal protective equipment.
   b. Identify fire extinguishers, and establish a fire watch.
   c. Explain the use of shutdown and hot-work permits.
   d. Recognize life safety for the end user.
4. Identify the materials and tools for medical gas and vacuum systems and their storage and handling requirements.
   a. Explain how materials and tools are stored and used.
   b. Identify labels and lettering used on medical gas and vacuum systems.
5. Identify the equipment required for brazing copper tube with and without purging.
   a. Explain how to braze a joint with a purge.
   b. Explain how to braze a joint without a purge.
6. Describe the process for testing and verifying a medical gas and vacuum system.
   a. Explain how to conduct installation tests.
   b. Explain how to conduct system verification tests.

Performance Tasks

Under supervision, you should be able to do the following:

1. Braze copper tube with purging.
2. Braze copper tube without purging.

CODE NOTE

Codes vary among jurisdictions. Because of the variations in code, consult the applicable code whenever regulations are in question. Referring to an incorrect set of codes can cause as much trouble as failing to reference codes altogether. Obtain, review, and familiarize yourself with your local adopted code.

Digital Resources for Plumbing

Scan this code using the camera on your phone or mobile device to view the digital resources related to this craft.

SCAN ME

Overview

Medical gas and vacuum systems are installed, tested, and serviced by plumbers who have been specially trained and certified to work with them. This module covers the system requirements and professional qualifications required by code, describes common types of medical gas and vacuum systems, and introduces trainees to the safety requirements that must be observed when installing, testing, and servicing these systems. It explains special requirements for storing and preparing system components, explains labeling and identification requirements, reviews techniques for joining copper tubing and fittings, and explains the various safety tests required by code.

NCCER Industry-Recognized Credentials

If you are training through an NCCER-accredited sponsor, you may be eligible for credentials from NCCER. The ID number for this module is 02412. Note that this module may have been used in other NCCER curricula and may apply to other level completions. Contact NCCER at 1.888.622.3720 or go to **www.nccer.org** for more information.

You can also show off your industry-recognized credentials online with NCCER's digital credentials. Transform your knowledge, skills, and achievements into credentials that you can share across social media platforms, send to your network, and add to your resume. For more information, visit **www.nccer.org**.

1.0.0 Building System Requirements and Professional Qualifications

Performance Tasks

There are no Performance Tasks in this section.

Objective

Identify the building system requirements for medical gas and vacuum systems as outlined in the latest edition of NFPA 99, *Health Care Facilities Code*, and the professional qualifications as outlined in the latest edition of ASSE/IAPMO/ANSI Series 6000, Professional Qualifications Standard for Medical Gas Systems Personnel.

a. Explain NFPA 99 building system categories.
b. Explain ASSE/IAPMO/ANSI Series 6000 professional qualifications.

Respiration: The breathing cycle.

Anesthetic: A gas that is inhaled to inhibit sensitivity to pain.

Medical gas system: Equipment and piping that supply nonflammable medical gases under pressure, along with the associated pumps and compressors, filters and purifiers, valves, alarms, gauges, and controls.

Medical vacuum system: Equipment and piping that use negative pressure to remove waste gases, liquids, and solids, along with the associated pumps and compressors, filters and purifiers, valves, alarms, gauges, and controls.

Hospitals, doctors' and dentists' offices, clinics, and other medical facilities are equipped with special piping systems that supply oxygen, nitrogen, helium, carbon dioxide, nitrous oxide, and on-site compressed air under pressure (including medical air, which is contaminant-free compressed air used for patient **respiration**). These nonflammable gases are used for many things, including the following:

- Drying medical instruments
- Operating pneumatic tools such as booms and drills
- Assisting patients with breathing
- Delivering an **anesthetic**, which is a gas that helps patients temporarily lose their ability to feel pain
- Calibrating medical devices that are used for patient respiration and anesthetic delivery

The equipment and piping that supply these gases under pressure, along with their associated pumps and compressors, filters and purifiers, valves, alarms, gauges, and controls, are called **medical gas systems**. Medical gas systems are often paired with systems that use negative pressure to remove waste gases, liquids, and solids. These systems are called **medical vacuum systems**. Think of medical gas and vacuum systems as similar to the way water supply systems and drain, waste, and vent (DWV) systems work, but to supply gas instead of water.

Medical gas and vacuum systems are installed, tested, and serviced by plumbers who are specially trained and certified to work with these systems. In this section, you will learn about the three main categories of medical gas systems. You will also learn about the professional qualifications that are required to install, inspect, and verify, or test, medical gas and vacuum systems.

Always consult your local applicable code before installing, testing, or servicing medical gas and vacuum systems. Never work on a medical gas or medical vacuum system without proper training, certification, and supervision according to your local applicable code. Medical gas can injure and even kill people if the delivery and waste removal systems malfunction or are installed or used improperly. Safety, for yourself and others, must be your primary concern when working around medical gases and medical gas and vacuum systems.

WARNING!

Never use a medical gas system to supply flammable gases such as natural gas, hydrogen, or acetylene. Medical gas systems are designed for use only with nonflammable gases.

Supply Systems and the Law

NFPA 55, *Compressed Gases and Cryogenic Fluids Code* governs cryogenic fluid supply systems for oxygen, bulk nitrous oxide central systems, and carbon dioxide central supply systems that use permanently installed containers. There are Compressed Gas Association (CGA) codes governing such systems as nitrous oxide and carbon dioxide at customer sites, both above and below 1,000 lb of capacity. Other codes by these agencies govern supply systems for inert gases.

1.1.0 Understanding NFPA 99 Building System Categories

The requirements for medical gas and vacuum systems are outlined in the National Fire Protection Association's model code NFPA 99. The NFPA classifies medical gas and vacuum systems into four broad categories. The category to which a medical gas system belongs will depend on the health and safety risks that the system poses to patients and medical personnel. The facility owner is responsible for determining the specific category to which medical gas and vacuum systems belong. The owner does this by conducting a risk assessment for each system.

The four categories for medical gas and vacuum systems follow:

- **Category 1 systems** – Equipment or system failure is likely to cause serious injury or even death.
- **Category 2 systems** – Equipment failure is likely to cause minor injury.
- **Category 3 systems** – Equipment failure is unlikely to cause injury but may cause some discomfort to patients.
- **Category 4 systems** – Equipment failure would not affect patient care.

NFPA 99 provides detailed design requirements. The installation and testing requirements for each of the systems are similar, but each system has its own specific requirements as well. Plumbers who install and service medical gas and vacuum systems need to be familiar with all these requirements to ensure that the system is installed and tested correctly.

This module will cover the procedures for installing and testing Category 1, 2, and 3 medical gas systems.

Category 1 system: A medical gas system in which equipment or system failure is likely to cause serious injury or even death.

Category 2 system: A medical gas system in which equipment failure is likely to cause minor injury.

Category 3 system: A medical gas system in which equipment failure is unlikely to cause injury but may cause some discomfort to patients.

Category 4 system: A medical gas system in which equipment failure would not affect patient care at all.

NOTE

The categories for dental gas and vacuum systems differ from those for medical gas and vacuum systems. Category 1 dental piped gas and piped vacuum systems are those involving general anesthesia and deep sedation. Category 2 applies to moderate or minimal sedation procedures, while Category 3 applies to facilities where minimal or no sedation is used.

1.1.1 Category 1 Systems

A medical gas and vacuum system is designated a Category 1 system if the failure of the entire system or any part of it could lead to serious injury or death. Category 1 systems are considered essential for providing life support to patients. Therefore, codes require them to be capable of operating or being available to operate at all times. Examples of medical facilities that would require Category 1 systems include ambulatory surgical centers with operating rooms, critical care centers, and surgeons' offices equipped with general anesthesia.

1.1.2 Category 2 Systems

A medical gas and vacuum system is designated a Category 2 system if the failure of the entire system or any part of it could lead to minor injury. Category 2 systems do not provide critical life support. Therefore, codes require these systems to be highly reliable, but they also permit them to have brief periods of downtime. Examples of medical facilities that would require Category 2 systems include outpatient clinics that provide sedation and general care rooms.

1.1.3 Category 3 Systems

A medical gas and vacuum system is designated a Category 3 system if the failure of the entire system or any part of it could lead to discomfort, but not to injury. Category 3 systems support patients' needs, but their failure would not affect patient care. Therefore, codes permit Category 3 systems to have the same reliability as other building systems. Examples of medical facilities that would require Category 3 systems include dental offices without general anesthesia and basic care rooms.

1.2.0 Understanding ASSE/IAPMO/ANSI Series 6000 Professional Qualifications

Medical gas and vacuum systems are installed, inspected, and verified by people who have been specially trained and certified to perform these tasks. The minimum training standards that are used today were originally developed by the American Society of Sanitary Engineering (ASSE) in 1987 and adopted by the International Association of Plumbing and Mechanical Officials (IAPMO) and the American National Standards Institute (ANSI). The standard is called ASSE/IAPMO/ANSI Series 6000, Professional Qualifications Standard for Medical Gas Systems Personnel. The standard is called a series because it is a group of several related standards, one for each of the medical gas specialties. They include the following:

- Standard #6010: Medical Gas Systems Installers
- Standard #6020: Medical Gas Systems Inspectors
- Standard #6030: Medical Gas Systems Verifiers
- Standard #6040: Medical Gas Systems Maintenance Personnel
- Standard #6060: Medical Gas Systems Designer

This section provides an overview of the education, training, and general knowledge requirements for each of these categories. Note that the ASSE/IAPMO/ANSI Series 6000 standards provide the minimum requirements for people who work with medical gas and vacuum systems. Local codes may have additional requirements for training and certification above and beyond these standards. Be sure to review and understand the requirements that apply to where you work. Never attempt to install, inspect, test, or perform maintenance on a medical gas or medical vacuum system if you have not received the proper training and certification.

1.2.1 Medical Gas Systems Installer

Plumbers who install equipment and piping for medical gas and vacuum systems, including brazers, must follow the requirements established in ASSE/IAPMO/ANSI Standard #6010 when working in the field. Installers are required to know the codes, laws, and regulations that apply to installing medical gas and vacuum systems where they work. Medical gas systems installers are required to recertify when there are changes to the local codes that apply to medical gas and vacuum systems.

Brazers must recertify every six months by submitting an affidavit stating that they have brazed according to the requirements laid out in ASTM (American Society for Testing and Materials) B88, Standard Specification for Seamless Copper Water Tube, during the previous six months. The company's designated supervisor must sign the affidavit. Otherwise, the brazer must begin the certification process from the start.

Brazers may be required to report to the project architect or engineer of record, to job inspectors, the owner of the facility where the systems are being installed, and to the authority having jurisdiction (AHJ). They are also required to demonstrate that they understand the principles, components, and operation of the various types of medical gas and vacuum systems. This includes not only knowledge of the local code requirements, but also detailed knowledge of manufacturers' specifications and requirements for locating and ventilating medical gas and vacuum systems. It also includes making sure that system components are accessible for service and repair.

Medical gas systems installers should be able to describe basic test procedures and how test components are used, but they are only allowed to perform tests associated with the installation of medical gas and vacuum systems. Verification tests are performed by qualified **verifiers**. When installation tests are complete, the installer provides detailed written reports on each system test to the facility owner or the AHJ documenting the results of the test.

Verifier: A person who inspects medical gas and vacuum systems and verifies that they meet all applicable standards and codes, as well as manufacturers' specifications.

Brazers can be qualified under ASSE/IAPMO/ANSI #6010 if they demonstrate the ability to braze copper tube according to the requirement in NFPA 99. They must know how to clean the copper tube and fittings, ensure proper joint clearance and overlap, perform an internal purge, monitor the flow rate of purge gas, and use filler metal properly. The techniques for brazing with and without a purge are covered in detail in *Section 5.1.0.*

To be certified as a medical gas systems installer, a plumber must first be able to demonstrate at least four years' experience installing plumbing or other piping systems. Then, the plumber attends an approved training course of at least 32 hours that covers the ASSE/IAPMO/ANSI #6010 standard and the applicable sections of NFPA 99. Finally, the plumber must take and pass a written exam and a practical exam. To become recertified, a medical gas systems installer must attend a refresher course of at least four hours and take an exam on the contents of the latest edition of NFPA 99.

NOTE

Never allow noncertified personnel to work on the installation of a medical gas and vacuum system under your supervision. You will lose your certification.

1.2.2 Medical Gas Systems Inspector

Plumbers who inspect medical gas and vacuum system installations are required to follow the requirements established in ASSE/IAPMO/ANSI Standard #6020. As with installers, inspectors are required to know the codes, laws, and regulations that apply to installing medical gas and vacuum systems where they work. They are also required to demonstrate that they understand the principles, components, and operation of the various types of medical gas and vacuum systems. This includes not only knowledge of the local code requirements but also detailed knowledge of manufacturers' specifications and requirements for locating and ventilating medical gas and vacuum systems, as well as ensuring that system components are accessible for service and repair.

To qualify as a medical gas systems inspector, a plumber must be able to describe the test procedures required for system components, including (but not limited to) the following:

- Brazed joints
- Welded joints
- Alarm systems
- Medical air system components

- Medical gas manifolds and outlets
- Medical vacuum system components
- Valves used in medical gas and vacuum systems

To be certified as a medical gas systems inspector, a plumber should have at least two years' documented experience as a plumbing and/or mechanical inspector, or as a qualified medical gas systems installer. Candidates are required to take a training course of at least 24 hours on the ASSE/IAPMO/ANSI #6020 standard and the applicable sections of NFPA 99, as well as on applicable sections of NFPA 55. In addition, candidates must take and pass a written exam and a practical exam. As with medical gas systems installers, to become recertified, a medical gas systems inspector must attend a refresher course of at least four hours and take an exam on the contents of the latest edition of NFPA 99.

1.2.3 Medical Gas Systems Verifier

Plumbers who inspect medical gas and vacuum systems and verify that they meet all applicable standards and codes as well as manufacturers' specifications are called medical gas systems verifiers. Verifiers are required to follow the requirements established in ASSE/IAPMO/ANSI Standard #6030. Verifiers must be able to demonstrate the same knowledge of applicable codes, laws, and regulations, as well as the principles, components, operation, and manufacturers' specifications of the various types of medical gas and vacuum systems, as installers and inspectors.

Codes require verifiers to have no affiliation with either the installers or the health care facilities in which medical gas and vacuum systems are installed. This ensures that the verification tests are done without bias. A medical gas and vacuum system cannot be used until the verifier submits a detailed report approving the system for use. The report is submitted after a series of tests are performed satisfactorily. These tests will be discussed later in this module. The report includes information about any repairs or servicing that was performed, the function of alarm and equipment warning systems, and the circumstances surrounding any retests that were required. Copies of the report are provided to the facility's owner and to the inspector or AHJ.

Medical gas systems verifiers are required to have extensive knowledge about testing the components of medical gas and vacuum systems. This includes the items that installers and inspectors are familiar with, as well as the following:

- Carbon dioxide, oxygen, nitrous oxide, and dewpoint analyzers
- Flowmeters
- Gas sampling equipment
- Particulate filters
- Pressure and vacuum gauges

To be certified as a medical gas systems verifier, a plumber should have at least two years' documented experience as a verifier as well as appropriate certificates of general and professional liability insurance. Candidates are required to take a training course of at least 32 hours on the ASSE/IAPMO/ANSI #6030 standard and the applicable sections of NFPA 99, NFPA 55, and the Compressed Gas Association's (CGA) standard M-1, Guide for Medical Supply Systems at Consumer Sites. Like installers and inspectors, candidates for the medical gas systems verifier certification must take and pass a written exam and a practical exam. Recertification requires attending a refresher course of at least four hours and taking an exam on the contents of the latest edition of NFPA 99.

1.2.4 Medical Gas Systems Maintenance Personnel

Plumbers who maintain medical gas and vacuum systems that have already been installed, tested, and verified are called medical gas systems maintenance personnel. Maintenance personnel are required to follow the requirements established in ASSE/IAPMO/ANSI Standard #6040. They must also be able to demonstrate the same knowledge of applicable codes, laws, and regulations, as well as the principles, components, operation, and manufacturers' specifications of the various types of medical gas and vacuum systems, as the other categories of personnel discussed in this section.

Medical gas systems maintenance personnel are required to be able to identify system components and describe how they operate, as well as their performance characteristics and safety requirements. They are also required to demonstrate their familiarity with a variety of system components and must describe the various categories of medical gas and vacuum systems identified in NFPA 99. Their knowledge of system maintenance must include the following:

- Requirements of the local applicable code
- Location of system components
- System operating requirements
- Locations of system components, ventilation, and access points

Maintenance personnel are required to identify and describe the safety hazards to the system, to patients, and to themselves resulting from improper system maintenance. They are required to describe maintenance and test procedures for medical gas and vacuum systems, as well as the safety precautions that must be taken when working in confined spaces and shutting down electrical power. ASSE/IAPMO/ANSI #6040 also requires maintenance personnel to be able to document and record the status of system components such as valves, connections, alarms, and gas source equipment.

To be certified as medical gas systems maintenance personnel, plumbers should have at least one year's experience working for a health care facility on the maintenance of its medical gas and vacuum systems at the time of application for certification. Candidates are required to take a training course of at least 32 hours on the ASSE/IAPMO/ANSI #6040 standard and the applicable sections of NFPA 99 and take and pass both a written and a practical exam. As with the other specialties discussed in this section, recertification requires attendance in a refresher course of at least four hours and taking an exam on the contents of the latest edition of NFPA 99.

1.2.5 Medical Gas Systems Designer

To become a certified medical gas systems designer, a plumber must have two years of documented experience in the mechanical and plumbing systems industry, as well as practical knowledge of the design of medical gas and vacuum systems. The certification course is a 32-hour training course covering ASSE #6060, the essential requirements of NFPA 99, NFPA 55, and FGI Guidelines, as well as elements of design. There is also a 100-question written exam.

1.0.0 Section Review

1. Medical gas and vacuum systems that are not permitted to have brief periods of downtime are considered _____.
   a. Category 1 systems
   b. Category 2 systems
   c. Category 3 systems
   d. Category 4 systems

2. Knowledge of the requirements in CGA M-1, Guide for Medical Supply Systems at Consumer Sites, is required by _____.
   a. medical gas systems inspectors
   b. medical gas systems maintenance personnel
   c. medical gas systems installers
   d. medical gas systems verifiers

2.0.0 Types of Medical Gas Systems

Performance Tasks

There are no Performance Tasks in this section.

Objective

Describe the various medical gas and vacuum systems.

a. Identify medical gas and vacuum systems.
b. Identify control panels and alarm systems.
c. Identify zone valves and station outlets/inlets.

Medical gas systems are designed to distribute nonflammable gases from the source, or point of origin, to outlets at the point of use. Medical vacuum systems, on the other hand, are designed to remove waste gases, liquids, and solids from inlets at the point of use and carry them away for safe disposal. The equipment, valves, controls, and other components of medical gas and vacuum systems vary depending on the type of gas the system is designed to distribute. In this section, you will learn about the most common types of medical gas and vacuum systems used in health care facilities. This section also covers the control panels, alarm systems, zone valves, and system inlets and outlets that are commonly used to operate the various systems.

The shape and size of medical gas system outlets and vacuum system inlets vary depending on the manufacturer (*Figure 1*). Outlet shapes can also vary depending on the type of gas. This prevents the accidental connection of tools and equipment to the wrong type of gas. It is also why a plumber needs to carry a variety of adapters when using a **flowmeter** (*Figure 2*) to test the flow rate of gases within a system. Never use an adapter on a medical gas system outlet or vacuum system inlet for which the adapter is not designed. You can damage the system, which could result in a gas leak that can cause injury or even death.

Flowmeter: A meter that measures and indicates the volumetric flow rate of a gas or a liquid.

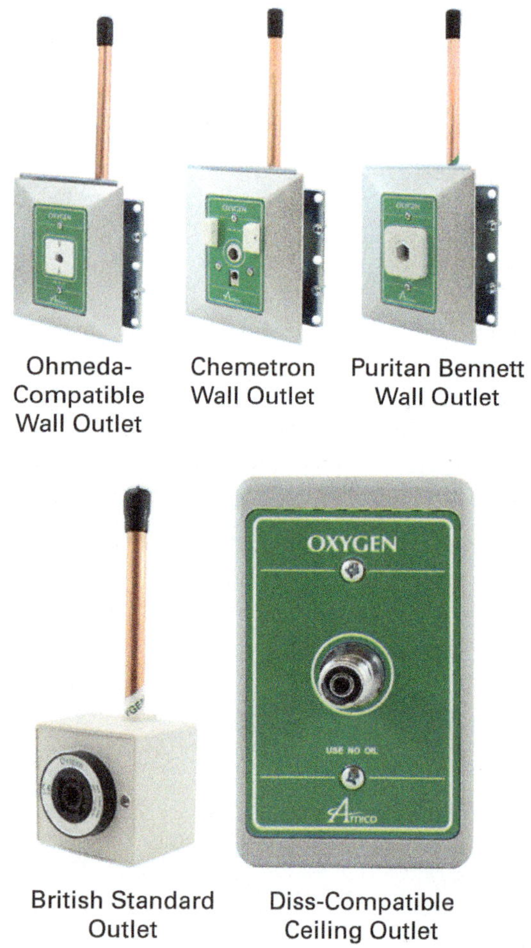

Figure 1 Common medical gas system outlets.
Source: Ramligallery/Shutterstock

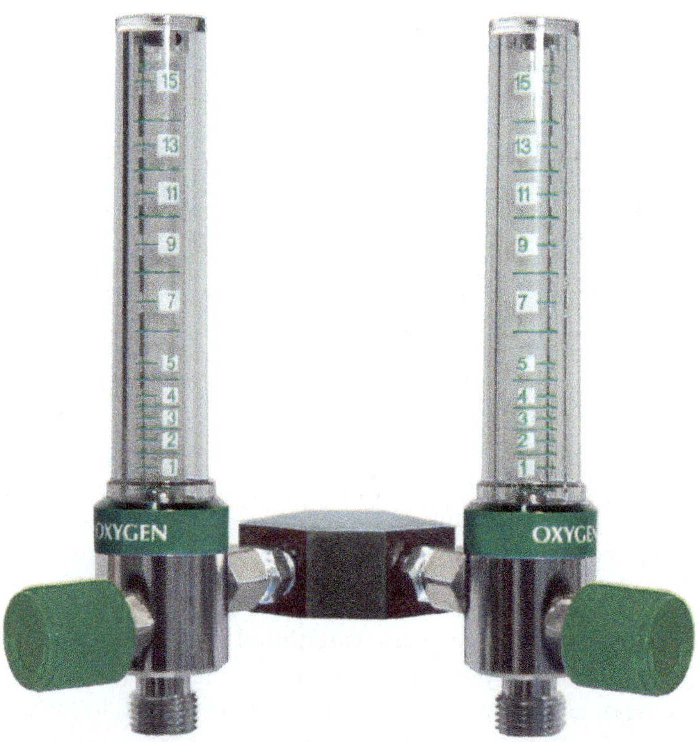

Figure 2 Oxygen flowmeters attached to a dual-port outlet.
Source: winnond/123RF

Equipment that delivers medical gas incorporates **Diameter Index Safety System (DISS) connectors**. These are vacuum and medical gas connectors that cannot be interchanged. They comply with CGA V-5, the Compressed Gas Association's Standard for Diameter Index Safety System (Noninterchangeable Low Pressure Connections for Medical Gas Applications).

DISS connector: A Diameter Index Safety System connector used for vacuum and medical gas. It is designed so it cannot be interchanged.

2.1.0 Identifying Medical Gas and Vacuum Systems

Medical gas and vacuum systems can be designed to deliver and carry away a wide variety of nonflammable gases that are used for patient respiration, clinical applications, power for pneumatic devices, and calibration. Medical gas and vacuum systems include such gases as the following:

- Oxygen
- Nitrogen
- Carbon dioxide
- Nitrous oxide
- Helium
- Medical air
- Vacuum
- Waste anesthetic gas disposal (WAGD)

When installing, testing, or servicing the components of a medical gas and vacuum system designed to deliver or dispose of a certain type of gas, never use fittings and components designed for a different type of gas. Some fittings are universal or designed for use with all types of gas. Unless universal fittings are specifically called for in the system specifications, use fittings and components that are designed for use with that type of gas. Codes specifically prohibit the use of adapters or conversion fittings (sometimes called "cheater" fittings) to adapt one gas-specific fitting to another.

> **Think About It**
>
> **Dental Air**
>
> Dental air was not always considered its own category, but it has since been added to distinguish it from other types of medical gas. Dental air refers to the compressed gases used in, and to drive, dental equipment. This includes pressurized carbon dioxide, nitrogen, and other compressed air systems.

2.1.1 Oxygen Systems

Oxygen is a colorless, odorless, nonflammable gas. Humans and many other living creatures require oxygen to sustain life. Around 21% of the air we breathe is made up of oxygen. Health care facilities use pure oxygen to help patient respiration before, during, and after medical procedures. The gas is delivered under pressure to hoses that are placed into a patient's nostrils, masks that are worn over the nose and mouth, and hoods that are placed over the head.

Although oxygen itself is not flammable, oxygen aids the combustion of other flammable gases, liquids, and solids. A material that catches fire in an **oxygen-enriched atmosphere**, which NFPA 99 defines as an atmosphere that contains more than 23.5% oxygen by volume, will burn faster and hotter than it would in regular air. To prevent fire, safety precautions against the accidental release of oxygen must be taken when installing, testing, using, and servicing oxygen systems.

Oxygen-enriched atmosphere: An atmosphere that contains more than 23.5% oxygen by volume.

Oxygen for medical gas systems can be provided as a gas or as a compressed liquid. Both types are stored in tanks. Pressure regulators attached to the oxygen tanks reduce the gas to the pressure required by the system. The equipment used to store oxygen and distribute it to the supply line at system pressure is called the **bulk oxygen system**. From there, the gas flows through manifolds (*Figure 3*) and piping to reach the various outlets where it is used. The standards for bulk oxygen systems are covered in NFPA 55.

Bulk oxygen system: The equipment used to store oxygen and distribute it to the supply line at system pressure.

2.1.2 Nitrous Oxide Systems

Nitrous oxide is a medical gas that is widely used as an anesthetic in health care facilities. It is a compound of oil-free, nitrogen NF and oxygen. Like pure oxygen, nitrous oxide is not flammable but will allow materials to burn hotter and faster when exposed to it. The gas has a sweet odor.

Figure 3 Manifold in medical gas system.
Source: Courtesy of Wendy Becovitz-McKinzie

WARNING!

Nitrous oxide supplies can be stored near oxygen supplies, but take care when placing these two gas sources in the same room as motor-driven equipment. A leak of either gas can result in a fire if ignited by a spark from the motor. The local applicable code will specify the proper placement of source gas containers and motor-driven equipment.

The typical operating pressure for a nitrous oxide system is 5 pounds per square inch gauge (psig) below the operating pressures of oxygen and medical air. The temperature of the central supply system for nitrous oxide should never be allowed to fall below –20°F or rise above 125°F. The specifications provided by the supply system's manufacturer may specify a narrower temperature range. A WAGD inlet is required to be installed where nitrous oxide is administered, to safely carry away the waste nitrous oxide. You will learn more about WAGD systems in *Section 2.1.5*.

WARNING!

When nitrous oxide is used in an anesthesia machine that blends the gas with other gases, never allow the pressure of the nitrous oxide to exceed the pressure of the other gases. In the event of backflow caused by reduced pressure in the other line, nitrous oxide could be allowed to flow through oxygen lines, potentially exposing patients to lethal doses of the gas.

2.1.3 Medical Air Systems

Medical air is purified atmospheric air (78% oil-free, nitrogen NF; 21% oxygen; and trace amounts of many other gases). It is used to provide support for patient respiration and can be supplied from cylinders, storage tanks, or compressors. Medical air can also be supplied by blending special types of compressed oxygen and oil-free, nitrogen NF. Medical air is sometimes confused with **instrument air**, which is compressed air that is used to power pneumatic tools and clean instruments. NFPA 99 classifies instrument air as a medical support gas distinct from medical air.

Instrument air: Compressed air that is used to power pneumatic tools and clean instruments. Not to be confused with medical air USP.

Medical air must be treated before it can be used. The treatment process removes water and oil from the air. Water in medical gas piping and fittings can freeze when exposed to cold gas, slowing and even blocking the gas flow. Oil can pose a fire hazard if it is exposed to an oxygen-enriched atmosphere. Unlike other medical gases, medical air systems are not required to be vented to the outside.

NFPA 99 requires medical air to be produced at **pharmaceutical grade**, which means that the gas meets the specifications for medical purity established by the United States Pharmacopeial Convention (USP). The USP publishes the standards for all medicines, health care products, food ingredients, and dietary supplements sold in the United States. The set of standards is called the United States Pharmacopeia (also abbreviated USP). Pharmaceutical-grade medical air is therefore called **medical air USP**.

The USP standards for medical air require the following:

- It must be supplied from cylinders, bulk containers, or medical air compressors, or be blended from USP-grade oxygen and oil-free, nitrogen NF.
- It must be free of liquid hydrocarbons and have no more than 25 parts per million (ppm) gaseous hydrocarbons.

It must have no more than 6.85×10^{-7} pounds per cubic yard particulate matter larger than one micron across at standard atmospheric pressure.

In most cases, ambient (ordinary) air meets or exceeds the USP standards for medical air, so very little needs to be done to prepare air to serve as medical air beyond drying and filtering it at the supply source. Your local applicable code will specify the requirements for medical air where you work. Medical air compressors (*Figure 4*) cannot be used to provide air for applications other than patient respiration and calibrating respiratory equipment. Misuse of medical air compressors can damage the compressors as well as the devices to which they are connected.

Pharmaceutical grade: Meeting the specifications for medical purity established by the United States Pharmacopeial Convention.

Medical air USP: Medical air that has been certified as pharmaceutical grade according to the standards of the United States Pharmacopeial Convention.

Did You Know?

Duplex Systems

Some medical gas systems are designed so the failure of a single device on the system does not require the entire system to be shut down. Instead, if a device fails, it can be isolated with valves and the gas routed around the device to continue serving other devices further down the line. This is called duplexing. The system designer determines whether to install a duplex arrangement on a particular system.

Figure 4 Medical air compressor.
Source: BeaconMedaes

Codes require a minimum of two compressors on a Category 1 medical air system. Locate medical air system compressors indoors in a dedicated area equipped with proper ventilation and all required utility connections. The temperature of the equipment space should be maintained according to manufacturers' specifications.

Backflow preventers should be fitted to prevent backflow due to pressure differentials between compressors that are operating and those that are not. Each compressor should be fitted with a manual shutoff valve to isolate it from the medical air system without loss of system pressure. Pressure-relief valves should be set at 50% above the line pressure. Always follow the manufacturer's specifications to ensure the materials used on components and pipes are compatible with the compressor.

Medical air compressors that use a reciprocating (piston) engine are required to have an inspection hole that is at least 1.5 times the size of the piston shaft. This is so a failure in the crankcase seal can be easily spotted. Otherwise, crankcase oil can contaminate the medical air and cause a health or fire risk.

WARNING!

Medical air can become contaminated by chlorine gas due to the deterioration of certain types of plastics over time, as well as by high levels of chlorine in public water supplies. Chlorine gas can be deadly in sufficient quantities. If chlorine levels are higher than those allowed by the local applicable code, install filters on medical air compressors to remove the chlorine from the air supply.

Codes require medical air system piping to be installed so continuous service can be maintained if a component fails. The piping should be sized to serve the peak calculated demand with one compressor offline. Codes also require a minimum of two air dryers, filters, and regulators on each medical air system. Dryers, filters, and regulators must be equipped with manual shutoff valves upstream and check valves downstream to allow for servicing. The valves can be installed to allow the component to be either isolated or bypassed, depending on the system design.

2.1.4 Vacuum Systems

Medical-surgical vacuum system: A special type of medical vacuum system designed to suction away bodily fluids during surgery.

Medical vacuum systems use negative pressure to provide suction to remove waste gases away from patient treatment areas and to a place where they can be vented safely. A special type of vacuum system is the **medical-surgical vacuum system**, which is designed to suction away bodily fluids during surgery.

Category 3 medical vacuum systems are further classified as either "wet" or "dry" systems. Wet systems remove liquids, air-gas mixtures, and solids from the work area. Dry systems trap liquids and solids before the service inlet and allow only air-gas mixtures to pass through the inlet. The inlets used in Category 3 vacuum systems are called service inlets, while the outlets used in Category 3 medical gas systems are called service outlets. The different types of inlets and outlets are discussed in *Section 2.3.2*.

The elements of a medical vacuum system include the vacuum-producing equipment, controls for pressure and operation, shutoff valves, alarms, gauges, and piping that extends to station inlets in patient care areas. Pumps are used to create low pressure in a medical vacuum system. Codes require pumps to be mounted on pads to absorb high-frequency vibrations that could damage the piping, fittings, and valves. Connect pumps to inlet and outlet piping using flexible connectors. As with medical gas systems, the piping in a Category 1 system should be installed to ensure continuous service in the event of a component failure.

Pump motors must be fitted with dedicated disconnect switches, motor starters, overload protection, backup transformers or voltage controls, and automatic restart. Control circuits for vacuum pumps should be wired so the failure of a pump will not interrupt the operation of the other pumps. Medical vacuum systems are required to have their exhaust lines run outdoors at least 10' from any doors, windows, or other openings or air intakes. They must be located at a different level than air intakes to avoid accidental re-ingestion of waste gases into the building. The end of the exhaust piping should be screened and turned face down, toward the ground. This will prevent entry by pests, dirt, and rain or snow. Refer to your local code for procedures applicable to your area.

Install exhaust piping to avoid dips or loops that would allow condensate oil to collect. Install a drip leg and valved drain at the lowest point of the exhaust line. Codes allow the exhaust piping from multiple pumps to be joined through a manifold into a single exhaust line, provided the line is sized properly, according to the local applicable code. This can save time and money during the installation process.

2.1.5 WAGD Systems

As previously mentioned, WAGD is the capture and removal of vented gases exhaled by patients when under anesthesia. WAGD is sometimes called scavenging or evacuation, though NFPA 99 discourages the use of those terms because medical professionals use them to refer to other things. WAGD systems are considered by code to be part of the medical-surgical vacuum system. Depending on the system design, WAGD may share the same piping as the medical-surgical vacuum system, or it may be a separate system. However, NFPA 99 encourages the use of separate WAGD systems for various reasons, including the following:

- The maximum safe negative pressure for patient respiratory systems is much lower than the minimum negative pressure of a standard medical-surgical vacuum system.
- The alarm systems required for medical-surgical vacuum systems cannot be used to monitor WAGD systems.
- High airflows between the two systems may cause problems for vacuum pumps.

WARNING!

Never label WAGD terminals "suction" or use suction terminals for WAGD.

WAGD systems may be located adjacent to the sources for medical air compressors, medical-surgical vacuums, and instrument air. Make sure pumps and motors that provide the source negative pressure for WAGD systems are adequately ventilated. NFPA 99 distinguishes between WAGD systems based on the horsepower of the engine or motor producing the low pressure in the system. Systems that use an engine or motor with a total power output of less than 1 horsepower may be located near the inlet or inlets they serve. Systems with engines or motors that produce 1 horsepower or more are considered central systems and must be installed according to the same local applicable code requirements used for locating central supply systems.

Vacuum pumps used to provide WAGD service are required to meet the same criteria as pumps used for medical-surgical vacuum systems. They must be made from materials that are inert (meaning they won't react chemically) in the presence of medical gases. If a WAGD system is to be connected to another vacuum system, the connection must be at least 5' from any vacuum inlet. Install WAGD exhaust systems using the same requirements as used for medical-surgical vacuum systems.

The location of WAGD inlets will vary depending on the needs of the facility. The growing use of injected, rather than gas, anesthetics means that patient treatment rooms may not require WAGD piping. However, NFPA 99 recommends that system designers designate at least one location in the facility where WAGD will be provided in case it is needed.

2.2.0 Identifying Control Panels and Alarm Systems

Medical gas and vacuum systems are fitted with controls and alarms that allow operators to monitor and control their operation. They are also equipped with alarm systems that indicate when line pressure rises above or falls below the specified range. Certified medical gas and vacuum systems installers must install control panels and alarm systems. The local applicable code will specify the minimum standards for control panel and alarm systems in medical gas and vacuum systems. Additional requirements may be noted in the system design specifications.

This section provides a general overview of control panels and alarm systems that are commonly used on medical gas and vacuum systems. The controls and layout of the control and alarm panels will vary depending on the manufacturer. Always follow the manufacturer's instructions when installing, testing, and servicing a control panel or alarm panel. Follow all required safety precautions, such as turning off the power and disconnecting alarms. Ensure that the medical gas or vacuum system can continue to operate safely and reliably during servicing.

2.2.1 Control Panels

Manufacturers provide control panels (*Figure 5*) for medical gas and vacuum systems as pre-assembled units that can be roughed in and joined to the system piping according to manufacturer instructions. Control panels may include the following:

- Gauges that indicate supply and outlet pressures
- Outlet pressure regulator
- Supply-valve open/close switch
- Gas outlet for testing and piping to additional outlets

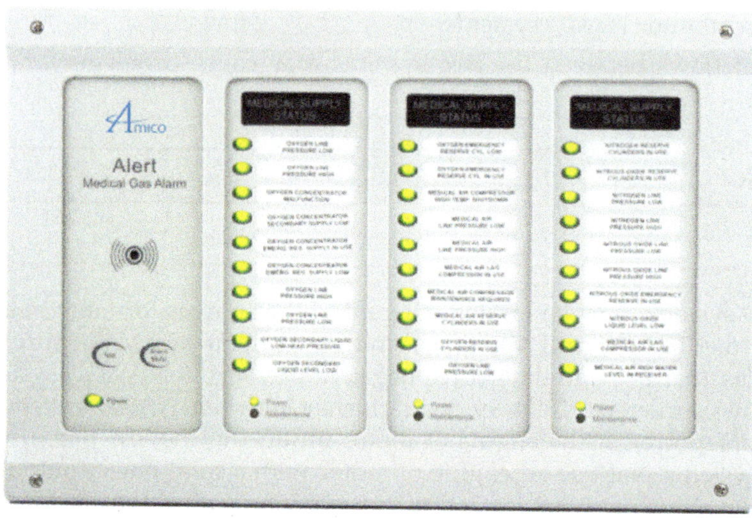

Figure 5 Typical medical gas system control panel.
Source: Adi Catur/Shutterstock

The location of a medical gas and vacuum system control panel will be specified in the design drawings. When installing the control panel box, ensure that it is supported on rigid mounting on the sides or the top and bottom. The outlet and inlet extensions on the box should be brazed to the system lines using the techniques discussed in *Section 5.1.0*.

Medical gas and vacuum system sources such as compressors, pumps, and air-drying equipment are also equipped with control panels that are specifically used to control these devices. These control panels are attached to the source equipment and are not installed by the plumber. All controls must be checked and tested to ensure they are working properly before the control panel can be put into service.

2.2.2 Alarm Systems

Alarm systems are used to warn operators when part of a medical gas and vacuum system is functioning incorrectly, or no longer functioning. Alarms provide both a visual warning (for example, a flashing light) and an audible warning (for example, a horn or buzzer) when there is a problem. Alarm systems also indicate when a system has switched to the reserve gas supply, which requires the primary supply to be recharged. NFPA 99 specifies three types of alarms for Category 1 and Category 2 medical gas and vacuum systems:

Master alarm: An alarm that continually monitors the gas supply source and mainline pressure for all systems and is directly connected to each of the devices it monitors.

Area alarm: An alarm that continually monitors medical gas and vacuum systems located in a specific area, such as a treatment room.

Local alarm: An alarm that continually monitors one system and is located near the system being monitored.

- **Master alarms**, which continually monitor the gas supply source and mainline pressure for all systems and are directly connected to each of the devices they monitor (*Figure 6*)
- **Area alarms**, which continually monitor medical gas and vacuum systems that are located in a specific area, such as a treatment room
- **Local alarms**, which continually monitor one system and are located near the system being monitored

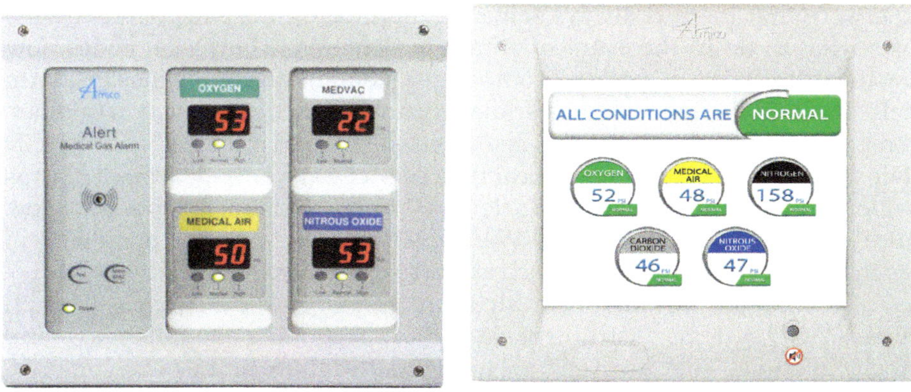

Analog Area Alarm Panel

Digital Area Alarm Panel

Figure 6 Medical gas master alarm panel.
Source: Nuaehnaja/Shutterstock

Alarm systems in Category 3 medical gas and vacuum systems are different from those used in Category 1 and Category 2 systems. These alarms are designed to indicate high and low pressures and changeovers to secondary systems on oxygen and nitrous oxide systems. Pressure switches and sensors for Category 3 alarm systems are fitted at the source equipment and downstream from shutoff valves.

Think About It

Where Do You Need an Alarm System?

NFPA requires the installation of alarm systems in every critical care or vital life-support area, but the definition of those areas can vary from facility to facility. Thanks to advances in medical technology, many activities that required doctors to administer medical gas can now be done without the use of any medical gas. That can make it hard for designers to know where to place alarm systems.

The NFPA's Medical Gas and Vacuum System Installation Handbook offers designers a new criterion for determining whether to install a medical gas or medical vacuum system. According to NFPA, the designer should ask: "If gases were to fail in the area or room during a procedure, would the procedure have to terminate prematurely? Is it possible that patient outcomes would be negatively impacted?" If the answer is yes, then the room can be used to provide critical care that includes the use of medical gas and vacuum systems, and thus the room warrants the installation of an alarm system.

NFPA 99 requires all master, area, and local alarms to have visual indicators for each component or system being monitored, and visual alarm indicators that must be manually switched off when the problem is resolved. Alarm systems must also have audible alarms that can be turned off at the discretion of the operator, and that produce a sound of at least 80 decibels at 3'. However, if a new alarm condition occurs after the audible alarm has been manually turned off, the audible alarm must be able to override the shutoff and provide an audible warning of the new problem.

There must be a way to identify whether a warning light has failed and when the alarm has been disconnected to a device being monitored. All indicators must be clearly labeled, as well as the area of coverage of the alarm panel itself. When room numbers or names change, the labels on the alarms must also be changed to ensure that the alarm coverage is properly identified.

Electrical power for alarms and sensors should be provided by the same electrical branch that powers the air compressor system. Always follow the local applicable electrical code when installing alarm wiring, including protection against fire. Alarm systems should also be equipped with an automatic restart capability after a power loss of 10 seconds or longer.

Codes require the installation of two separate master-alarm systems, one in the nurse's station and the other in another location that is constantly staffed, such as a security office, maintenance office, or telephone switchboard. This is to ensure that there is always someone monitoring the alarm system who knows what to do in the event of an alarm situation. Many new codes now allow the installation of a central computer system for one of the master-alarm panels. The signals from one master-alarm panel cannot be relayed to another.

Area alarms monitor medical gas, medical-surgical vacuum, and WAGD systems where anesthesia and critical care services are provided. Area alarms provide visual and audible warning signals for high and low line pressure, low medical-surgical vacuum, and low WAGD vacuum. Area alarms are placed in areas that are constantly staffed, such as nurses' stations.

Local alarms are installed near air compressors, medical-surgical vacuum pumps, WAGD systems, instrument air, and proportioning equipment to monitor their functions. They are designed to signal the failure of the main compressor, vacuum pump, or proportioning system; high carbon monoxide levels; high dew point in the mechanical air system; and abnormally high temperature in the discharge air. Master-alarm panels are required to have at least one warning signal to indicate an alarm condition with source equipment.

Transducer: An electrical device that converts an analog signal, such as line pressure, into an electrical signal that can be read by an electrical device, such as an alarm circuit board.

Alarm panels may include a power supply, line pressure gauges, an alarm circuit board, and **transducers** that convert line pressure into electrical signals that can be read by the alarm circuit board. Displays indicate the pressures on the various lines and the status of the line. An audible alarm speaker will sound if the status becomes abnormal. The alarm control panel will include buttons to test and silence the alarm, and keys to program the various functions. Sensors for area alarms should be located in critical care areas alongside zone-valve boxes.

2.3.0 Identifying Zone Valves and Station Outlets/Inlets

Zone valves are quarter-turn ball valves designed to isolate station outlets and inlets in a medical gas and vacuum system (*Figure 7*). Station outlets and inlets are connected to terminal points along the medical gas and medical vacuum lines to which instruments and devices can be connected. In this section, you will learn how zone valves and station outlets and inlets work, and you will be introduced to the code requirements that affect their installation and use. As with all other aspects of medical gas and vacuum systems, strict adherence to the local applicable code and the manufacturer's instructions are required to ensure that the system can be operated safely.

2.3.1 Zone Valves

NFPA 99 does not define the term *zone*, but in practice a zone is considered one branch of outlets or inlets serving a designated area. A zone valve, therefore, is designed to quickly isolate a set of outlets or inlets serving a zone from the rest of the system in the event of a malfunction or an emergency. Codes require the installation of zone valves on all station outlets and inlets. Zone valves are accessible through a zone-valve box (*Figure 8*), which may house a single valve or multiple valves, depending on the design of the system.

NFPA 99 requires zone valves to be placed in easily accessible areas where they are always visible. They must also be installed so there is a wall between the valve and the outlets or inlets that it controls. This is to allow a person to operate the valve from behind a safe barrier that protects the person from fire, smoke, and escaped gas. If a zone has more than one operating room, a separate shutoff valve is required for each room. Zone valves must be installed on the same floor as the zone they control. Closing a zone valve should not affect the supply of medical gas or vacuum to other zones in the system. Never place zone valves in series.

Figure 7 Ball valve used for gas shutoff.
Source: Heather Smith - The Alloy Valve Stockist's photo gallery, CC BY 3.0, https://commons.wikimedia.org/w/index.php?curid=9631221

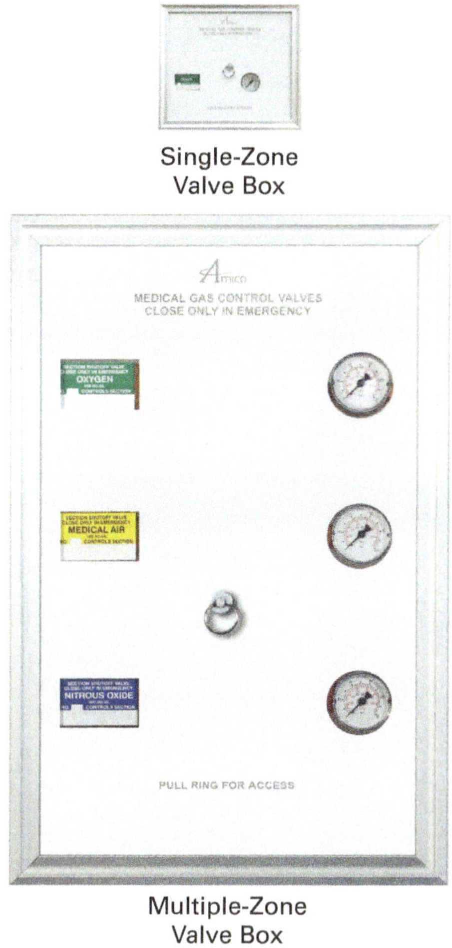

Figure 8 Zone-valve box.
Source: pixmeeup/Shutterstock

WARNING!

Misconnected gas lines can cause serious injury or even death. After shutoff valves have been installed, check to make sure that all outlets and inlets are connected to the correct service lines.

Install a pressure or vacuum indicator gauge on the outlet or inlet side of the zone valve. The zone valve should be located upstream of all gas delivery columns, hoses, booms, medical instrument pendants, and other equipment served by the outlets and inlets in the zone. Additional in-line shutoff valves may be installed to disconnect gas and vacuum service to individual rooms or areas within a zone.

In certain types of facilities where security may be an issue, such as in pediatric hospitals or psychiatric wards, zone-valve boxes may be modified to protect against vandalism or damage. Acceptable forms of modification include lockable valve-box doors or location of the valve box inside a locked room.

2.3.2 Station Outlets/Inlets

Medical gas outlets and vacuum system inlets may be located on walls or ceilings, depending on the design of the system and the equipment that the outlet is designed to serve. To guard against accidentally using the wrong gas, the shape of the outlets and inlets for each gas type are mechanically different to prevent anyone from accidentally connecting the wrong hose to the outlet or inlet.

Codes require outlets and inlets to be placed high enough so they will not damage equipment connected to them. They must also be placed high enough to prevent damage by furniture or equipment in the room (such as the headboard of a patient's bed). Each outlet or inlet must be clearly labeled according to the labeling standards discussed in *Section 4.2.0*. The local applicable code will specify the dimensions of the copper tubes used on system inlets. Never allow a station outlet to be used for anything other than the delivery of medical gas.

When locating station outlets and inlets, be sure to account for the equipment that will be connected to them. Allow sufficient space between the outlet or inlet and the space above and below the outlet or inlet. WAGD inlets should be provided wherever anesthetic gas is administered. They should be clearly noninterchangeable with other medical gas and vacuum system outlets or inlets and clearly marked for WAGD service.

2.0.0 Section Review

1. Category 3 medical vacuum systems are classified as either _____.
   a. medical air or instrument air systems
   b. outlet or inlet systems
   c. high-pressure or low-pressure systems
   d. wet or dry systems

2. Alarms that continually monitor medical gas and vacuum systems that are located in a specific area are called _____.
   a. local alarms
   b. area alarms
   c. master alarms
   d. Category 3 alarms

3. NFPA 99 requires that zone valves be installed so the valve and the outlets or inlets that they control are separated by _____.
   a. at least one floor
   b. a wall
   c. fireproofing
   d. at least 3'

3.0.0 Medical Gas System Installation Safety

Performance Tasks

There are no Performance Tasks in this section.

Objective

Identify the safety issues related to medical gas and vacuum system installation.

a. Identify personal protective equipment.
b. Identify fire extinguishers, and establish a fire watch.
c. Explain the use of shutdown and hot-work permits.
d. Recognize life safety for the end user.

The safety of patients and medical personnel is a primary concern when working with medical gas and vacuum systems. Exposure to gas in sufficient quantities can cause injury or even death. The pneumatic tools that use instrument air can injure patients and medical personnel if the pressure is too high or too low. The failure of a pump, compressor, or other mechanical device at the system source can cause a fire. Improper exhaust can cause suffocation. As a plumber you are already used to working safely, but medical gas and vacuum systems require additional safety precautions that you must observe to ensure the system is not damaged and that patients and medical personnel remain safe. Whether installing, testing, or servicing components in a gas or vacuum system, therefore, it is important to follow the safety requirements as spelled out in the local applicable code.

In this section, you will learn about the personal protective equipment that should be used when installing, testing, and servicing medical gas and vacuum systems. You will also learn about proper fire prevention and fire safety techniques that should be followed when working with these systems. This section also covers permits that are required for shutting down an existing medical gas or vacuum system, and for performing **hot work**, which, as its name suggests, is work performed on metal while it is being heated. Finally, this section addresses the important issue of **life safety** for users of medical gas and vacuum systems.

Hot work: Work performed on heated metal.

Life safety: Essential for ensuring the safety of patients.

Always refer to your local code for the standards that apply to medical gas and vacuum system installations in your area. Remember that the standards in NFPA 99 are minimum standards established for safe operation and that local codes may be more stringent.

3.1.0 Using Personal Protective Equipment

As you have already learned in the Plumbing curriculum, personal protective equipment (PPE) is designed to protect you from injury. Many plumbers are injured on the job because they do not use PPE. Working around medical gas and vacuum systems requires the use of PPE for brazing and working with metals. PPE will greatly reduce your chances of getting hurt. The best PPE is of no use unless you do the following four things:

- Inspect it regularly and replace any PPE that is damaged or worn.
- Care for it properly.
- Use it properly when it is needed.
- Avoid altering or modifying it in any way.

Wear eye protection (*Figure 9*) when installing, testing, and servicing medical gas and vacuum systems. There are different types of eye protection. Safety goggles give your eyes the best protection from all directions. Regular safety glasses will protect you from objects flying at you from the front, such as large chips, particles, sand, or dust. Face shields offer better protection than safety glasses and protect your entire face. You can add side shields for further protection.

Figure 9 Tinted safety glasses.

Prescription safety glasses may be substituted if they provide the same protection as regular safety glasses. If welding is part of your job, you must use safety goggles with tinted lenses or a welding hood. Tinted lenses protect your eyes from the bright flame when brazing copper pipe. Tinted eye protection should have a tint value of 3 or 4.

Wear heavy-duty gloves to protect your hands from cuts and burns when brazing pipe and fittings (*Figure 10*). Work gloves are usually made of cloth, canvas, or leather, but they may also be made of metal mesh or Kevlar®, which protects against metal cuts. Never wear cloth gloves around rotating or moving equipment. Inspect your gloves every day to ensure that they are in good condition. Immediately replace gloves that are worn, torn, or no longer fit. Wear long sleeves to protect your arms.

Figure 10 Work gloves.
Source: Courtesy of Ridge Tool Company

In addition to a visual test, an effective way to check rubber gloves is to conduct an air test. The following steps are outlined in *OSHA Standard 29 CFR 1910.137.*

Step 1 Hold the glove by its cuff, flip it several times to make a seal, and roll the glove toward its fingers. An air pocket will form inside the glove.

Step 2 Hold the rolled portion of the glove tightly. Inspect the inflated exterior of the glove for cracks or degradation of the insulating material surface. Forcing air into the glove will expose damage to the insulation that cannot be seen during visual inspections.

Step 3 Inspect the glove for holes in the insulating material. Hold the glove close to your ear. If you hear air escaping from the glove or if the glove will not hold pressure, the glove is damaged. Damaged gloves must be removed from service.

CAUTION

Protective equipment that is good one day may be damaged the next. Always do a daily visual inspection of PPE to prevent injuries from defective or worn equipment.

WARNING!

Always thoroughly wash and disinfect your hands if you have been exposed to sewage, even if you were wearing gloves. Doing so can prevent disease or illness. In general, it is best to use an alcohol gel instead of an antibacterial cleaner. Recent research indicates that overuse of antibacterial cleaners may make bacteria more resistant to antibacterial agents.

3.2.0 Using Fire Extinguishers and Establishing a Fire Watch

Always have an approved fire extinguisher available before starting your work. You are not expected to be an expert firefighter. But you may have to deal with a fire to protect your safety and the safety of others. You need to know the locations of firefighting equipment when performing hot work. You also need to know which equipment to use on different types of fires. However, only qualified personnel are authorized to fight fires.

Most companies tell new employees where fire extinguishers are kept. If you have not been told, be sure to ask. Also, ask how to report fires. The telephone number of the nearest fire department should be clearly posted in your work area. If your company has a company fire brigade, learn how to contact it. Learn your company's fire safety procedures. Know what type of extinguisher to use for different kinds of fires and how to use each type. Make sure all extinguishers are fully charged. Never remove the tag from an extinguisher; it shows the date the extinguisher was last serviced and inspected (*Figure 11*).

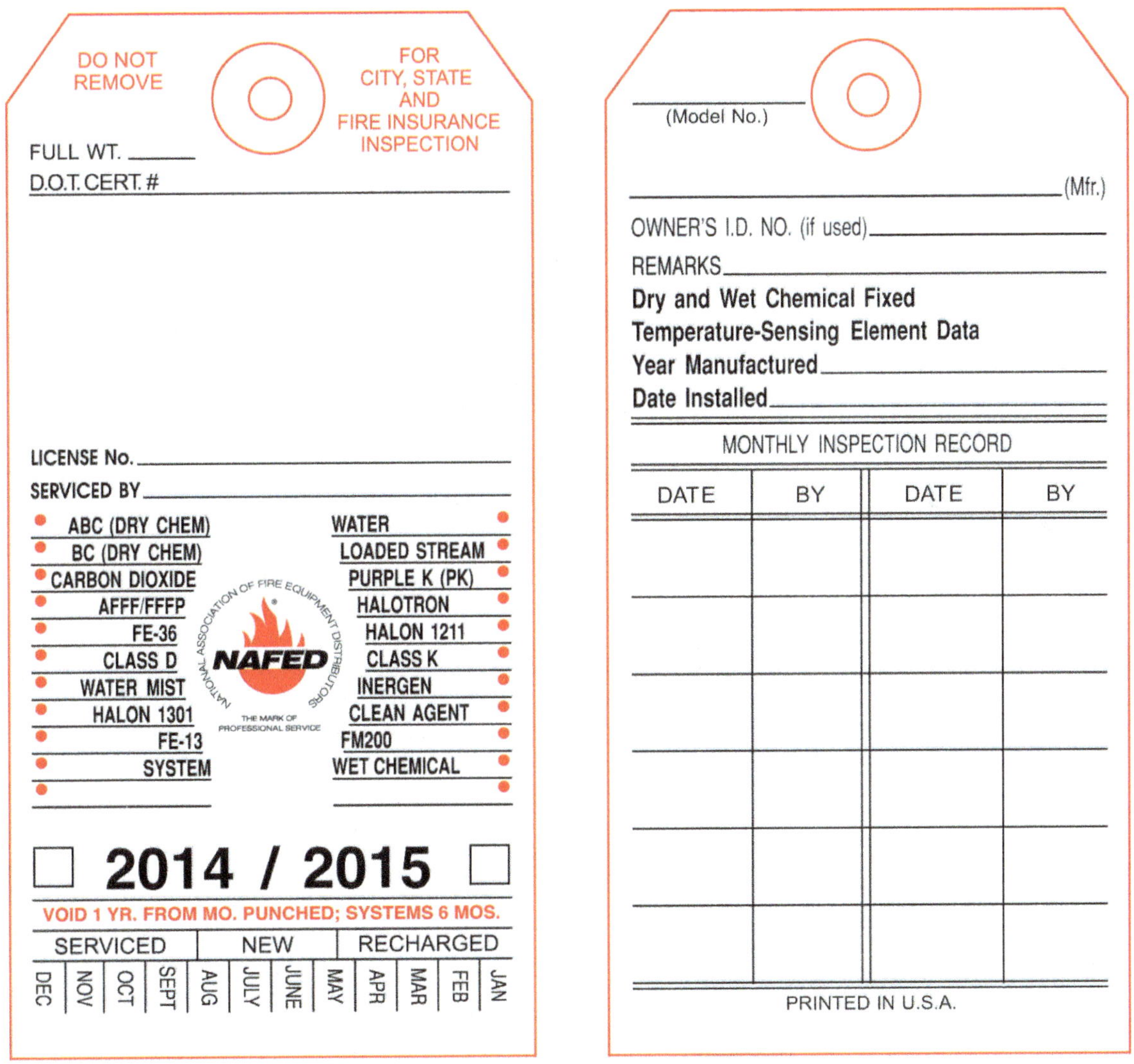

Figure 11 Fire-extinguisher tag.
Source: National Association of Fire Equipment Distributors (NAFED)

Several classes of fuels can be involved in fires. These include liquids (code A), gases (code B), ordinary combustibles (code C), and combustible metals (code D). (Flammable cooking liquids are their own class.) Each class of fuel requires a different method of firefighting and a different type of extinguisher. The label on a fire extinguisher (*Figure 12*) clearly shows the class of fire on which it can be used.

Figure 12 Fire-extinguisher labels.
Source: Badger Fire Protection

> **Did You Know?**
>
> **Type K Extinguishers**
>
> Type K fire extinguishers are intended for use on fires involving flammable cooking liquids, such as cooking oil and animal- or vegetable-based greases. You will likely not encounter them unless you are passing through a cafeteria area or the food preparation area of a medical facility, but it's useful to know they exist.

> **Fire-Watch Duration**
>
> Many companies mandate a minimum amount of time a fire watch must remain in place. A flame that touches flammable material, or a small ember that lands on insulation in a wall, might take an hour or more to start a fire. Even if you think the danger of fire has passed, check with your company or the general contractor to verify how long a fire watch must remain in place.

When you check the extinguishers in your work area, you will see that some are rated for more than one class of fire. You can use an extinguisher coded ABC to fight a class A, B, or C fire. But remember, if the extinguisher has only one code letter, it cannot be used on any other class of fire, even in an emergency. You could make the fire worse and put yourself in great danger.

Post a fire watch when you are brazing or cutting. One person other than the welding or cutting operator must constantly scan the work area for fires. Fire-watch personnel should have ready access to fire extinguishers and alarms and know how to use them. Welding and cutting operations should never be performed without a fire watch. The area where welding is done must be monitored afterward until there is no longer a risk of fire.

3.3.0 Using Shutdown and Hot-Work Permits

When working on medical gas and vacuum systems, it is important to complete and submit an appropriate permit before shutting the system down or performing hot work. The permit form should be completed and submitted to the authorized facility representative, typically a director of facility services, for approval. The permit request must be approved before the work can take place.

Refer to the local code for the standards that apply in your area. In some cases, you may also be required to post notices to inform facility staff that the permitted work will be performed. These notices will allow people to stay safely away from the work area and to avoid using the system being worked on.

Shutdown permits (*Figure 13*) include a checklist of the systems that will be shut down during the work. The health care facility will specify the date and time of the shutdown on the permit. The contractor will be asked to identify the location and expected duration of the shutdown.

Spring Hills Health Care Facility
Utility System Shutdown Permit

Personnel performing installation, maintenance, or repair work on any utility system supplying a SHHC building will be required to apply for a written shutdown permit and receive approval for said shutdown. This permit must be completed and submitted to the Director of Facilities Services or his/her representative in a timely manner as indicated by the attached notification matrix. Utility systems will not be shut down under any circumstance without a properly completed and approved permit. All departments effected must receive prior notification. A separate shutdown permit must be submitted for each unrelated utility system manipulation.

Note: All valves for any Infrastructure system are to be opened and/or closed by Facility Services personnel only unless specific written consent is given.

Project Name:

Check the box of the utility system the shutdown is being requested for:

	Electric - Normal Power		Fire Alarm System
	Electric - Emergency Power		Mechanical Air
	Chilled Water		Sanitary Sewer
	Medical Gas: Specify________________		Exhaust System
	Medical Vacuum		Air Handlers
	Steam and Condensate		Communications - Nurse Call
	Heating Hot Water		Communications - Telephone
	Domestic Hot Water		Communications - Public Address
	Domestic Cold Water		Communications - Data
	Fire Sprinkler System		Air Distribution
	Fire Suppression		Other:
	Tube System		Other:

1. Date of Shutdown: ____________________ 2. Time of Shutdown: ________________

3. Duration of Shutdown: ________________ 4. Specific Location of Shutdown: __________

5. Provide a detailed description of the work: (note - attach map & POA if appropriate)

____

____

____

Page 1 of 3

Figure 13 Example of a shutdown permit.

(*continued*)

Figure 13 (*continued*)

Spring Hills Health Care Facility
Utility System Shutdown Permit

6. List all building(s), floor(s), room(s), and department(s) that will be impacted by the shutdown.

7. Will mitigation measures be required to accommodate utility system loss during the shutdown? If yes, attach mitigation plan.

8. Are any special inspections or testing required? If yes, list inspector, company, and contact phone number.

9. Are there any post disruption/connection procedures required (such as chemicals, flushing, air elimination, etc.)? Please describe.

10. Are any other permits or notifications required for this work?

a. Hot Permit ___________ d. ILSM ___________

b. Over Ceiling Permit ___________ e. Fire Department ___________

c. ICRA Permit ___________ f. Other ___________

11. Has a contingency plan for a prolonged outage or other factors been developed? If yes, please attach plan or describe. Please include backup supplies.

Page 2 of 3

Figure 13 Example of a shutdown permit. (*continued*)

Figure 13 (*continued*)

Spring Hills Health Care Facility
Utility System Shutdown Permit

I have received, read, and fully filled out the **Spring Hills Health Care Facility Utility System Shutdown Permit.** I will ensure all personnel are properly trained, have appropriate equipment, and shall be responsible for the successful shutdown and reactivation of the said utility systems.

General Contractor / Company Name	Date Request Submitted
General Contractor / Personnel Name	Contact Telephone Number
General Contractor / Company Name	Date Request Submitted
General Contractor / Personnel Name	Contact Telephone Number
Subcontractor / Company Name	Date Request Submitted
Subcontractor / Personnel Name	Contact Telephone Number
Subcontractor / Company Name	Date Request Submitted
Subcontractor / Personnel Name	Contact Telephone Number

REQUIRED APPROVALS:

Facility Services Approval	Date Approved
Safety Approval	Date Approved
Infection Control Approval	Date Approved
Project Management Approval	Date Approved

Reason for Disapproval:

Page 3 of 3

Figure 13 Example of a shutdown permit.

If a detailed description of the work to be performed is requested, write out the steps. Describe any special inspections or testing that will be required, contingency plans, and mitigation methods, if required. The form must be signed and dated by the general contractor and the personnel performing the work. The facility will designate the appropriate people to countersign and approve the shutdown permit. All affected organizational units, such as departments that are affected by the shutdown, must be notified prior to the shutdown. This can be done by posting notices of service interruption (*Figure 14*).

SPRING HILLS HEALTH CARE FACILITY

IMPORTANT NOTICE
SERVICE INTERRUPTION

Service Affected: ______________________

Date ______________________

Time: ______________________

Area(s) Affected: ______________________

Remarks: ______________________

____

____

For Information, Call : ______________________

____

Date Issued: ______________________

By: ______________________

Title: ______________________

Distribution: ______________________

____

Figure 14 Example of a service-interruption notice.

Shutdowns in medical gas and vacuum systems are required when tying in new zones to an existing system or modifying existing piping. Before the shutdown, the health care facility is responsible for identifying the systems to be shut down and making all preparations to ensure continued facility operation during the shutdown. This may include arranging for alternative gas-supply cylinders, suction equipment, and other necessary tools and equipment. The health care facility is responsible for posting notices of the shutdown throughout the facility.

Prior to the shutdown, the contractor should work with facility staff to identify areas and systems that will be affected by the shutdown, including alarm systems. The installer is responsible for scheduling the necessary personnel, tools, and equipment for the job. Immediately prior to the shutdown, the installer notifies the facility representative. Then, the installer completes the work in a timely fashion and notifies the facility representative when the work has been completed. The local applicable code will specify the necessary test and repair equipment that will be used when shutting down systems.

Hot-work permits (*Figure 15*) are required when brazing of copper tube is to be performed. The requirements for securing a hot-work permit are like those used for shutdown permits. Follow all safety requirements for hot work as specified in the local applicable code. Occupational Safety and Health Administration (OSHA) regulations restrict hot work in confined or enclosed spaces, adjacent to spaces that have combustible materials, near fuel tanks, or on pipelines that may contain fuel residue. For example, although you are permitted to take a torch into a confined or enclosed space, the fuel bottle cannot also be brought inside. A torch cannot be left unattended in a confined or enclosed space; when you leave, you must take the torch with you.

Always follow the appropriate safety guidelines when performing hot work. Be sure to wear appropriate personal protective equipment, as discussed in *Section 3.1.0*. Always have a standby person present during hot work. The standby person should know whom to contact in case of emergency and how to disconnect the power.

3.4.0 Understanding Life Safety for the End User

Life safety systems are medical gas and vacuum systems that are essential for ensuring the safety of patients and caregivers. These systems typically include the medical gases themselves, patient alarms, ventilation systems, and vacuum pumps. The National Fire Protection Association describes life safety requirements in NFPA 101, *Life Safety Code®*. As with NFPA 99, NFPA 101® classifies life safety according to the level of hazard from the health and safety risks posed by the system to patients and medical personnel. All systems required for patient life safety must meet the requirements of NFPA 101® and the local applicable code for medical gas and vacuum systems.

Life safety branches of a medical gas and vacuum system are specially designated to supply power for lighting and other equipment that is essential for ensuring the safety of patients. Life safety systems are designed to automatically switch to backup power sources if electrical power to the facility is interrupted. Master, zone, and local alarms are all required to be connected to life safety branches. NFPA 101® also provides the requirements for penetrations through interior fire-rated or smoke barrier walls.

WARNING!

NFPA 101® prohibits the installation of medical gas and vacuum piping in exit enclosures. IBC and IFC codes contain similar prohibitions.

Spring Hills Health Care Facility
Brazing/Welding/Hot-Work Permit

Valid from _____ to _____. __________ Master Card No. ______________
(am/pm) (am/pm) Date

1. Work Description
Equipment Location or Area ______________________________
Work to be done:

2. Gas Test	☐ None Required			
	☐ Instrument Check	Test Results	Other Tests	Test Results
	☐ Oxygen 20.8% Min			
	☐ Combustible % LFL			

Gas Tester Signature Date Time

3. Special Instructions ☐ None ☐ Check with issuer before beginning work

4. Hazardous Materials ☐ None What did the line/equipment last contain?

5. Personal Protection ☐ Standard Equipment: welder's hood with long sleeves; cutting goggles

☐ Goggles or Face Shield ☐ Respirator ☐ Forced Air Ventilation

☐ Standby Man ☐ Other, specify: ______________

6. Fire Protection ☐ None Required ☐ Portable Fire Extinguisher

☐ Fire Watch ☐ Fire Blanket ☐ Other, specify: ______________

7. Condition of Area and Equipment

Required Yes	No	These Key Points Must Be Checked
		a. Lines disconnected & blanked or if disconnecting is not possible, blinds installed?
		b. Lines steamed, purged, or otherwise properly cleaned of combustibles
		c. Area and equipment satisfactorily clean of oil or combustibles?
		d. Trenches, catch basins, & sewer connections properly covered or sealed?
		e. Immediate area and /or area under the work barricaded or roped off?
		f. Adjoining equip. & operations checked to have any effect on the job?
		g. Area fire suppression (fire water and sprinkler system) in service?

Comments

Figure 15 Example of a hot-work permit.

(continued)

Figure 15 (*continued*)

Spring Hills Health Care Facility
Brazing/Welding/Hot-Work Permit

8. Approval	Permit Authorization			Permit Acceptance		
	Area Supv.	Date	Time	Maint. Supv./Engineer Contractor Supv.	Date	Time
Issued by						
Endorsed by						
Endorsed by						

9. Individual Review

I have been instructed in the proper Hot Work Procedures

	Signed	Signed
Persons Authorized to Perform Hot Work	__________	__________
	__________	__________
	__________	__________
	__________	__________
	__________	__________
Fire Watch	__________	__________

10. Job Completion

☐ Yes ☐ No Is the work on the equipment completed?

☐ Yes ☐ No Has the worksite been cleaned and made safe?

Worker answering above questions ____________________

Issuer's Acceptance ____________________

Forward to Production Superintendent within 7 days of job completion

Figure 15 Example of a hot-work permit.

3.0.0 Section Review

1. The tint value of eye protection used for brazing should be at least _____.
   a. 3
   b. 4
   c. 5
   d. 6

2. The classes of fuels that can be involved in fires are liquids, gases, combustible metals, cooking liquids, and _____.
   a. untreated organic materials
   b. fuels
   c. plastics
   d. ordinary combustibles

3. On behalf of the contractor, shutdown permits must be signed and dated by _____.
   a. the on-site supervisor
   b. the personnel performing the work
   c. the general contractor and the personnel performing the work
   d. the head of the contractor's legal department

4. The National Fire Protection Association specifies the requirements for medical gases, patient alarms, ventilation systems, and vacuum pumps that affect the safety of patients in _____.
   a. NFPA 150, *Fire and Life Safety in Animal Housing Facilities*
   b. NFPA 72®, *National Fire Alarm and Signaling Code*
   c. NFPA 99, *Health Care Facilities Code*
   d. NFPA 101®, *Life Safety Code*

4.0.0 Medical Gas System Materials, Tools, Storage, and Handling

Performance Tasks

There are no Performance Tasks in this section.

Objective

Identify the materials and tools for medical gas and vacuum systems and their storage and handling requirements.

a. Explain how materials and tools are stored and used.
b. Identify labels and lettering used on medical gas and vacuum systems.

Medical gas and vacuum systems use copper tube and fittings. Plumbers use brazing to connect tube and fittings in these systems. This section builds on that knowledge with information that is specific to the proper assembly of medical gas and vacuum systems. Although the techniques required for brazing copper tube and fittings may be familiar, medical gas and vacuum systems have specific safety requirements and special fittings that differ from other types of distribution systems that use copper.

For medical gas systems, NFPA 99 specifies the use of Type L hard-drawn seamless copper medical gas tube that has been manufactured according to ASTM B819, Standard Specification for Seamless Copper Tube for Medical Gas Systems. If the system operating pressure is above 185 pounds per square inch (psi) and the pipe size is above 3" OD (outside diameter), Type K tube should be

NOTE

When dealing with medical gas systems, pipe and fittings are referred to by their outer diameter (OD), not their inner diameter (ID), unless otherwise specified.

used. For medical-surgical vacuum systems, use hard-drawn seamless copper that has been manufactured to any of the following standards:

- ASTM B 819, Standard Specification for Seamless Copper Tube for Medical Gas Systems
- Type K, L, or M water tube manufactured according to ASTM B88, Standard Specification for Seamless Copper Water Tube
- ACR (air conditioning and refrigeration) tube manufactured according to ASTM B280, Standard Specification for Seamless Copper Tube for Air Conditioning and Refrigeration Field Service

NFPA 99 also allows the use of stainless-steel tube or pipe in medical-surgical vacuum systems. Corrugated medical tubing (CMT) is also allowed in positive-pressure systems provided it conforms to certain specifications. Codes specify hard-drawn copper tube because soft-drawn tube can be easily kinked, which can restrict gas flow in the system. Codes require medical gas and vacuum system tubing to be at least $^1/_2$" NPS ($^5/_8$" OD) in diameter. (Runouts to alarm panels and connecting tubing for gauges and alarm devices can be $^1/_4$" NPS, $^3/_8$" OD.) Codes require that medical gas and vacuum system piping be labeled properly by the manufacturer. The installer is required to provide documentation that shows the piping materials comply with the local applicable code.

This section covers the proper storage and use of materials and tools for installing medical gas and vacuum systems, and the proper use of labels and identification. Be sure to follow your local applicable code when using materials and tools to install medical gas and vacuum systems. Remember, NFPA 99 establishes the minimum safe standard for this work. The requirements in your local applicable code may be stricter. Always follow manufacturers' instructions when using tools and equipment. Wear appropriate personal protective equipment when installing medical gas and vacuum systems.

4.1.0 Storing and Using Materials and Tools

The tools required for installing medical gas and vacuum systems are the same basic tools that plumbers carry in their tool kits. However, the specific safety requirements of medical gas and vacuum systems require extra care in preparing and using tools. All materials and tools should be clean and free of oil or grease. Oil and grease can pose a fire hazard when exposed to oxygen in a medical gas system.

In addition to materials and tools for brazing copper tube, you will also need the following tools when working on medical gas and vacuum systems:

- Tubing cutter
- Pipe reamer
- Chamfer tool
- Purge regulator
- Abrasive pads
- Lint-free white cloth wipes
- Oxygen analyzer
- Adapters for fitting pressure gauges on each type of station outlet and inlet used throughout the system
- Leak-detection solution
- Nitrogen purge alarm

4.1.1 Storing Materials and Tools

Because of the strict safety requirements for medical gas and vacuum systems, plumbers need to take extra care to prevent contamination of tools, equipment, and materials that are used in medical gas and vacuum system installation, test, and servicing. Codes require tools and equipment to be stored separately from those used on other types of plumbing installations.

4.1.2 Preparing Pipe and Fittings

Prior to the publication of ASTM standard B819, plumbers had to clean medical gas and vacuum system tubing and fittings on the jobsite. Since then, however, manufacturers prepare the copper tube used in medical gas and vacuum systems to meet the standard's requirements for cleanliness. Once the tube has been formed, the inside of the tube is **dehydrated**. This means that any trace of water is removed from the inner surface of the tube. The tube is then **charged** (filled) with oil-free, nitrogen NF gas to a low pressure. The reason copper tube is charged with oil-free, nitrogen NF is to prevent copper atoms on the inner surface of the tube from reacting with oxygen in the atmosphere to form a layer of copper oxide on the inside of the tube.

Dehydrated: A condition in which all water has been removed.

Charged: Filled with an inert gas such as oil-free, nitrogen NF.

CAUTION

Remember to remove internal plugs from the copper tube when installing them. Leaving plugs in installed pipe will cause damage to the system. The plugs must be cut out and fixed before the system can be tested.

All pipe and fittings that are unbagged, cleaned, or prepared for installation must be installed within 8 hours of being cleaned for brazing. Once the tube has been charged, the ends are capped or plugged to keep the oil-free, nitrogen NF inside. Caps are fitted on the outside of the tube and are easily visible. Plugs fit flush with the end of the pipe and are harder to see. While codes may not specify the use of caps instead of plugs, it is good practice to use capped pipe to aid in remembering to remove them during the installation process. After the tube has been cut, the pieces are recapped.

Components such as valves and fittings are sealed in bags at the manufacturer's plant and must remain sealed until they are prepared for installation. Be sure to keep the documentation that comes with these sealed bags. They are the only proof that they were delivered sealed. On-site cleaning of medical gas tube, fittings, valves, and other system components is permitted only under certain special conditions.

These conditions are specified in the local applicable code. Codes prohibit any other types of cleaning of medical gas and vacuum system tube, fittings, valves, or components on the jobsite. This is because their cleanliness cannot be guaranteed outside of the controlled conditions of the manufacturer's facility, and because the chemicals used to clean them represent a potential environmental hazard.

4.1.3 Cutting and Reaming

Cut copper tubing with a handheld tube cutter or a mini cutter. The handheld tube cutter (*Figure 16*) is preferred because it makes a cleaner joint and leaves no metal particles (such as those left by a hacksaw, which also does not guarantee a straight cut). Use a tube cutter that is the right size for the copper being cut, and make sure that the proper cutting wheel is in place.

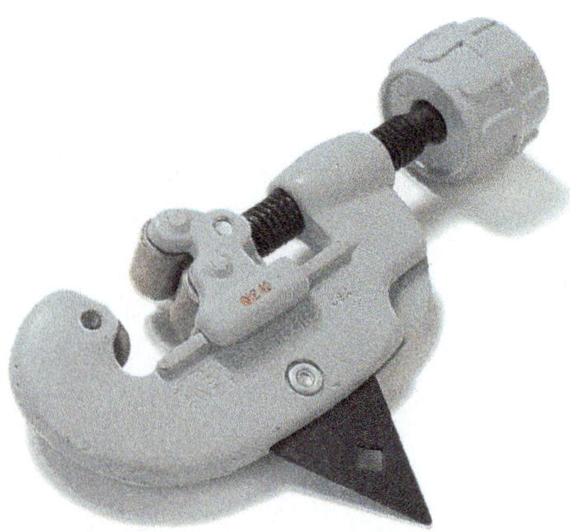

Figure 16 Handheld tube cutter.
Source: Courtesy of Ridge Tool Company

After cutting, ream all cut tube ends to the full inside diameter of the tube. Reaming removes the small burr created when the tube is cut. Burrs left on tubing can cause the tube to corrode. A tube that is reamed correctly provides a smooth inner surface for better flow. Remove burrs on the outside of the tube to ensure a good fit.

Tools used to ream tube ends include the reaming blade on the tube cutter, files (round or half round), a pocketknife, and a deburring tool. If the tube becomes deformed, use a sizing tool to bring the tube back to roundness. Refer to *Figure 16* for an example of a deburring blade on a tube cutter. A variety of models are available, and the cutting sizes range from $^1/_8$" to $4^1/_8$" OD.

To use the handheld tube cutter, follow these steps (*Figure 17*):

Step 1 Place the tube cutter on the tube at the point where you want to cut. Tighten the knob, forcing the cutting wheel against the tube.

Step 2 Make the cut by rotating the cutter around the tube under constant pressure.

Step 3 Use a deburring tool to remove any burrs from inside the tube. Some cutters have built-in deburring blades.

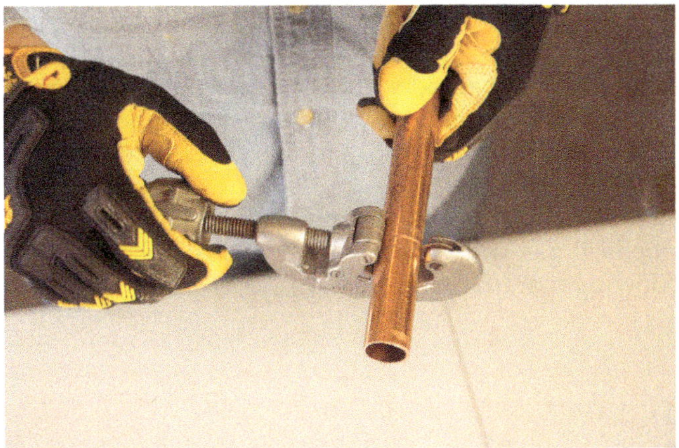

Cutting

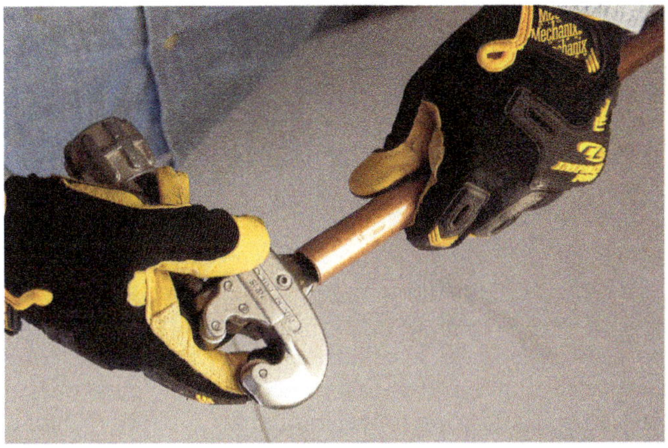

Reaming

Figure 17 Using a handheld tube cutter.

4.1.4 Cleaning the Joints

The exterior surfaces of tube ends must be cleaned with nonabrasive pads. Never use steel wool, sanding paper or cloth, or wire brushes to clean copper tube, fittings, valves, or components used in medical gas and vacuum systems.

When the surfaces have been cleaned, wipe them with a clean, lint-free white cloth. Then inspect the interiors of tube, fittings, valves, and components to ensure they are free of contamination, obstructions, and debris. Never use contaminated items in a system designed for oxygen service. Codes require joints to be brazed within eight hours of cleaning.

4.2.0 Using Labels and Identification

Codes require all piping, valves, station outlets and inlets, and alarm panels to be clearly labeled. Codes further specify the locations for these labels, so make sure to refer to the applicable regulations and guidelines. This is done to ensure that medical gas and vacuum systems are not accidentally cross-connected during installation, testing, and servicing. Labeling is also used to help identify problems in the event of a system failure or an emergency.

This section discusses the requirements for labeling medical gas and vacuum system piping, valves, and station outlets and inlets. The labeling of master-, zone-, and local-alarm panels is discussed in *Section 2.2.0*. As with all medical gas and vacuum system requirements, refer to your local applicable code for the labeling and identification standards enforced in your area.

Labels may be stencils or adhesive markings. The colors for the lettering and background indicate the type of gas or vacuum line. NFPA 99 lays out the standard colors for use on labels to help identify the different systems. See Table 5.1.11 in NFPA 99 for a list of medical gas and vacuum system standard designation colors.

4.2.1 Labeling of Copper Tube

Manufacturers use special markings and labels to indicate that copper tube is suitable for medical gas and vacuum applications. Tube must be labeled with one of the following manufacturers' marks:

- OXY
- MED
- OXY/MED
- OXY/ACR
- ACR/MED

Tube labels must follow the patterns described in Table 5.1.11 in NFPA 99. Labels must also include the following information:

- Name of the gas or vacuum system, or the appropriate chemical symbol
- Correct color code
- Operating pressure, if different from the standard gauge pressure for that type of gas

Labels on the tube cannot be more than 20' apart. Each system must have at least one label per room, on each side of every wall or partition that the pipe penetrates. Vertical risers must be labeled at least once on every floor. Tube that is labeled OXY is suitable for use with other nonflammable medical gas. This is because OXY tubing has been manufactured according to the standards of ASTM B88 for medical use.

4.2.2 Labeling of Shutoff Valves

Codes require shutoff valves to be labeled with the name or chemical symbol for the gas or vacuum system on which it is used, and the rooms or areas served. Valves must also be labeled with a warning notice that says the valve shall not be operated except in an emergency. Source, main line, riser, and service valves must all be labeled with the name of the line they are serving, and the area or areas served by that line. If the operating pressure of the line is different from the standard gauge pressure, the pressure on the line must also be labeled on the valve.

4.2.3 Labeling of Station Outlets and Inlets

As with valves, station outlets and inlets should be labeled with the name or chemical symbol for the gas or vacuum symbol that the line serves. If the operating pressure of the line is different from the standard gauge pressure, the line pressure must also be labeled on the station outlet or inlet.

4.0.0 Section Review

1. Codes require that the diameter of medical gas and vacuum system tubing should be at least _____.
   a. 1"
   b. $^3/_4$"
   c. $^1/_2$"
   d. $^1/_4$"

2. The minimum number of labels that each system must have per room is _____.
   a. one
   b. two
   c. three
   d. four

5.0.0 Brazing and Line Purging of Medical Gas Systems

Objective

Identify the equipment required for brazing copper tube with and without purging.

a. Explain how to braze a joint with a purge.

b. Explain how to braze a joint without a purge.

Performance Tasks

1. Braze copper tube with purging.
2. Braze copper tube without purging.

As you have already learned, brazing is a heating process like soldering that uses the combustion of oxygen and a fuel gas, often acetylene, to heat filler metals that melt between 1,100°F and 1,500°F. Plumbers use brazing to create joints that are very strong. Plumbers use brazing torches (*Figure 18*) to braze joints. A brazing torch kit consists of a burner wand connected by a hose to pressurized oxygen and fuel gas tanks via a regulator (to the oxygen tank) and valve (to the fuel gas tank). Codes require brazers to be qualified according to the standards outlined in the local applicable code.

Figure 18 Typical brazing torch being used on copper fitting.

5.1.0 Brazing a Joint

To ensure that a brazed joint is strong enough to withstand the positive and negative pressures in a medical gas and vacuum system, socket-type brazed joints are required. A socket joint involves coupling one end of a length of copper tube to another (*Figure 19*).

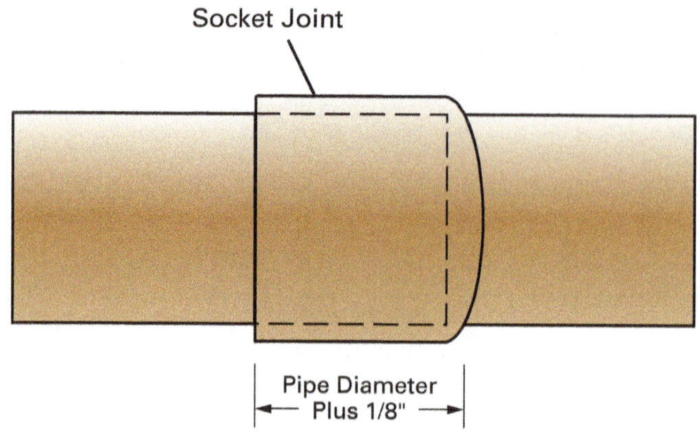

Figure 19 A coupled joint.

5.1.1 Brazing a Joint with a Purge

To ensure that the piping system will maintain its integrity in the event of a fire, use a brazing alloy that has a melting temperature greater than 1,000°F. Be sure to use filler metals that are compatible with the type of tube being brazed. When brazing copper-to-copper joints, use a copper-phosphorus or copper-phosphorus-silver filler metal without flux. Use flux sparingly to avoid contaminating the inside of the tube. Never field-braze cast metals. Joints that are brazed in place must be accessible for inspection.

Joints in medical gas systems must be continuously purged with oil-free, nitrogen NF during the brazing process. This is to prevent the formation of copper oxide on the insides of the joints. To further protect against contamination, always cap or plug all pipe openings to maintain the oil-free, nitrogen NF atmosphere inside the pipe prior to purging.

When purging, monitor the source gas supply. An audible alert is required to indicate low supply. Keep the flow rate of the oil-free, nitrogen NF purge gas low enough that it won't produce positive pressures in the system. Use a pressure regulator and a flowmeter to control the flow of the gas. Never use just a pressure regulator by itself.

To braze copper tube in medical gas and vacuum systems using a nitrogen purge, follow these steps:

Step 1 Insert the tube ends fully into the depth of the fitting socket or to a depth specified in the local applicable code.

Step 2 Make sure a discharge opening is present on the side of the joint that is opposite the point where purge gas will be introduced. Introduce the purge gas to the joint.

Step 3 Insert an oxygen analyzer into the end of the tube to confirm that the pipe is free of oxygen.

Step 4 Heat the joint to the temperature required for brazing, being careful to keep the joint from overheating.

Step 5 Maintain the flow of the oil-free, nitrogen NF purge gas until the brazing is complete and the joint is cool to the touch. Then cap or plug the purge discharge opening to prevent contamination and to maintain the internal nitrogen atmosphere.

5.1.2 Brazing a Joint without a Purge

When completing the installation of medical gas and vacuum systems, the last connecting joints on both systems will need to be brazed without a purge. This is commonly called a dirty joint. The steps for brazing copper tube in a medical gas and vacuum system without using a nitrogen purge are like the steps in *Section 5.1.0*. The differences are that brazing takes place at a much higher temperature (between 1,100°F and 1,500°F).

5.0.0 Section Review

1. Copper tube used in medical gas and vacuum systems is filled with oil-free, nitrogen NF to prevent the insides of joints from being contaminated with _____.
   a. carbon dioxide
   b. copper oxide
   c. sulfur dioxide
   d. ferrous oxide

2. The last connecting joint on a medical gas or vacuum system is commonly called a(n) _____.
   a. clean joint
   b. open joint
   c. dirty joint
   d. close joint

6.0.0 Testing and Verifying Medical Gas Systems

Objective

Describe the process for testing and verifying a medical gas and vacuum system.

a. Explain how to conduct installation tests.
b. Explain how to conduct system verification tests.

Performance Tasks

There are no Performance Tasks in this section.

To minimize the risk of injury or death resulting from the failure of a component, all medical gas and vacuum systems and their components are tested thoroughly. Two separate sets of tests are conducted on each system. The installer conducts the first set of tests while the system is being installed. The second set of tests is conducted by a verifier following the completion of the installer's tests. This ensures that the tests are conducted impartially.

The various installer and verifier tests are detailed in the following sections. Never attempt to perform a test for which you are not qualified. Always follow manufacturers' instructions and your local applicable code when testing medical gas and vacuum systems and their components. Always wear appropriate personal protective equipment when testing. Make sure all safety guidelines are observed. These steps will prevent damage to the system while protecting your health and safety (as well as the health and safety of others).

6.1.0 Conducting Installation Tests

During the installation process for Category 1, 2, and 3 medical gas and vacuum systems, the installer is required to perform and document a series of tests of the systems and their components. Tests are conducted using oil-free, nitrogen NF gas. Tests of manufactured assemblies should be performed after the installation of the distribution piping but before the system verification tests.

The tests must also be performed before installing any manufactured assemblies that are supplied through flexible hoses or tubing, and at each station outlet/inlet on the installed manufactured assemblies that are supplied through copper tubing. The following sections explain the purpose of each test and provide the general steps for performing the test. Always follow manufacturers' instructions and your local applicable code when conducting tests.

6.1.1 Initial Blowdown Test

Initial blowdown test: An installation test in which oil-free, nitrogen NF gas is blown through the distribution piping to remove dirt and other particulate matter.

The first installation test is called the **initial blowdown test**. In this test, oil-free, nitrogen NF gas is simply blown through the distribution piping to remove dirt and other particulate matter. Conduct the initial blowdown test after the distribution piping has been installed but before installing other system components such as station outlet/inlet rough-in assemblies, pressure and vacuum alarms and indicators, pressure-relief valves, and manifolds.

6.1.2 Initial Pressure Test

Initial pressure test: An installation test performed to ensure that there is no pressure loss in the distribution piping.

To ensure that there is no pressure loss in the distribution piping, the installer performs an **initial pressure test** of each section of piping. Initial pressure tests are conducted following the initial blowdown test and after station outlet/inlet rough-in assemblies have been installed, but before installing system components that could be damaged by the test pressure. These components include pressure and vacuum alarms and indicators, and pressure-relief valves.

Follow these steps when conducting the initial pressure test:

Step 1 Close the source shutoff valve, and ensure that it remains closed throughout the test.

Step 2 Pressurize the system to 1.5 times the system working pressure but not less than 150 psig (pounds per square inch gauge).

Step 3 Maintain test pressure while examining each joint for evidence of leakage. Use an ammonia-free leak detectant that is safe for use with oxygen systems.

Step 4 Repair or replace any leaks according to your local applicable code, and then retest the system.

WARNING!

Leaks represent fire hazards, but also potential health hazards. Some studies have found that excessive exposure to nitrous oxide and other medical gases can cause illness.

6.1.3 Initial Cross-Connection Test

Initial cross-connection test: An installation test performed to ensure that medical gas and vacuum systems are not cross-connected.

Medical gas and vacuum systems must never be cross-connected. To ensure this, installers perform an **initial cross-connection test** on each system after the initial blowdown and pressure tests.

Follow these steps to perform an initial cross-connection test:

Step 1 Reduce the pressure in all piping systems to the standard atmospheric pressure of 14.7 psi (pounds per square inch). Ensure that the test gas source is disconnected from all piping systems except for the system being tested.

Step 2 Pressurize the system with oil-free, nitrogen NF to 50 psig.

Step 3 Inspect each station outlet/inlet in all installed systems to ensure that only those on the system being tested are discharging gas.

Step 4 Repeat the steps for each medical gas and vacuum system in turn. While conducting the tests, confirm that the station outlets/inlets are labeled and identified properly.

Step 5 If you detect any cross-connection, or if any of the outlets/inlets are incorrectly labeled or identified, correct the problem before proceeding to test the next system.

6.1.4 Initial Piping Purge Test

Initial piping purge test: An installation test performed to purge the system of any remaining dirt and other particulate matter.

Following the completion of cross-connection tests, the system is purged to remove any remaining dirt and other particulate matter from the lines. This test is called the **initial piping purge test**. It provides a more thorough cleaning than the initial blowdown test.

Follow these steps to conduct the initial piping purge test:

Step 1 Using manufacturer-approved adapters that are designed for use with the system being tested, purge each outlet with intermittent high-volume bursts of oil-free, nitrogen NF. Begin the purge at the system outlet/inlet closest to the zone valve and continue outward until reaching the furthest system outlet/inlet in the zone.

Step 2 Continue the blasts until the lines are free of debris. You can determine this by holding a clean white cloth up to the outlet.

Step 3 When the nitrogen blasts cause no discoloration of the cloth, the line is clean.

6.1.5 Standing Pressure Test for Positive-Pressure Medical Gas Piping

To ensure that the distribution piping in a medical gas system will not experience excessive pressure loss when under service pressure, installers conduct a 24-hour **standing pressure test** following the initial pressure test. Standing pressure tests are conducted after the station outlet valve bodies, faceplates, and other distribution system components have been installed.

Standing pressure test: An installation test performed to ensure that the distribution piping in a medical gas system will not experience excessive pressure loss when under service pressure.

The source valve must remain closed during the standing pressure test. Follow these steps:

Step 1 Pressurize the positive-pressure portion of the system with oil-free nitrogen NF to 20% above the normal operating line pressure for the system, and allow the system to stand for 24 hours.

Step 2 At the end of that time, inspect the system pressure gauge to ensure that the test pressure has not changed because of leaks. However, changes in pressure caused by fluctuations in the ambient temperature are permitted.

Step 3 Locate any leaks, and conduct repairs or replacements as required by the local applicable code.

Codes require the AHJ to witness the standing pressure test. Following the successful completion of the test, the AHJ will complete a form confirming the test has been performed and completed according to the local applicable code and will provide the form to the verifier prior to the start of the system verification tests.

6.1.6 Standing Vacuum Test for Vacuum Piping

As with positive-pressure medical gas systems, negative-pressure vacuum distribution piping must also undergo a 24-hour test to ensure it will not experience excessive vacuum loss when in use. The test for vacuum systems, as you might expect, is called the **standing vacuum test**. This test is conducted following the successful completion of the standing pressure test and the installation of all vacuum system components.

Standing vacuum test: An installation test performed to ensure that vacuum system piping in a medical vacuum system will not experience excessive vacuum loss when in use.

Follow these steps to conduct a standing vacuum test:

Step 1 Ensure that the source of the test vacuum is disconnected from the piping system.

Step 2 Reduce the pressure in the vacuum piping until it reads between 12" of mercury and a full vacuum. Allow the system to stand for 24 hours.

Step 3 At the end of that time, inspect the system pressure gauge to ensure that the test vacuum has not changed because of leaks. Changes in vacuum caused by fluctuations in the ambient temperature are permitted.

Step 4 Locate any leaks, and conduct repairs or replacements as required by the local applicable code.

As with the standing pressure test for positive-pressure medical gas piping, codes require the AHJ to witness the standing vacuum test. Following the successful completion of the test, the AHJ will complete a form confirming the test has been performed and completed according to the local applicable code. The AHJ will then provide the form to the verifier prior to the start of the system verification tests.

6.2.0 Conducting System Verification Tests

When the installer tests have been completed to the satisfaction of the AHJ, a second set of tests is performed on the medical gas and vacuum systems. These tests are performed by a verifier, a technically competent and experienced professional who has been specially certified by ASSE to perform these tests.

The verifier must not be affiliated with the contractor that installed the systems. The verifier can be a member of the staff of the facility in which the systems have been installed, if properly trained and certified. Verification tests can be conducted using either nitrogen gas or the gas for which the system is designed to use in operation. Verifiers test each terminal connection point individually.

The following sections explain the purpose of each verification test and provide the general steps for performing the test. Always follow manufacturers' instructions and your local applicable code when conducting tests.

6.2.1 Standing Pressure Test

To ensure that the distribution piping is free from leaks, the verifier first conducts a standing pressure test. This test is like the pressure test conducted by the installer, but instead of lasting 24 hours, this test lasts 10 minutes. Follow these steps:

Step 1 Pressurize the system with oil-free, nitrogen NF or source gas until operating pressure is reached.

Step 2 Close the source valve and all zone valves. Let the system stand for 10 minutes.

Step 3 At the end of that time, inspect the system pressure gauge to ensure that the test pressure has not changed because of leaks. In pressure caused by fluctuations in the ambient temperature are permitted.

Step 4 Locate any leaks, and conduct repairs or replacements as required by the local applicable code.

6.2.2 Cross-Connection Test

Next, the verifier conducts a cross-connection test to verify that the distribution piping remains free of any cross-connections. This test is performed following the closure of the walls and the installation of all system components. There are two types of cross-connection tests: the individual pressurization method and the pressure-differential method. The choice of test depends on the type of system as well as on manufacturer and code requirements.

To conduct a cross-connection test using the individual pressurization method, follow these steps:

Step 1 Reduce the pressure in all piping systems to the standard atmospheric pressure of 14.7 psi. Ensure that the test gas source is disconnected from all piping systems except for the system being tested.

Step 2 Pressurize the system with oil-free, nitrogen NF to 50 psig.

Step 3 Inspect each station outlet/inlet in all installed systems to ensure that only those on the system being tested are discharging gas.

Step 4 Disconnect the test gas source, and reduce the system pressure to atmospheric pressure.

Step 5 Repeat the steps for each medical gas and vacuum system in turn until all systems have been tested for cross-connection.

To conduct a cross-connection test using the pressure-differential method, follow these steps:

Step 1 Reduce the pressure in all medical gas piping systems to the standard atmospheric pressure of 14.7 psi.

Step 2 Increase the test gas pressures in all systems according to NFPA 99, *Table 5.1.12.4.3.2(B)*, and maintain these pressures for the duration of the test. If a system operates at a nonstandard operating pressure, test the system at a pressure that is at least 10 psig higher or lower than that of other systems being tested.

Step 3 Ensure that vacuum systems are in operation during this test so they can also be tested at the same time.

Step 4 Use a test gauge to ensure that the test pressure at each station outlet is the correct pressure specified in NFPA 99, *Table 5.1.12.4.3.2(B)*. Ensure that the test gauges have been properly calibrated using the pressure indicator for the source pressure regulator.

Step 5 Identify each station outlet by label and, if required, by color marking. Ensure that the pressure indicated at each outlet matches the pressure specified in NFPA 99, *Table 5.1.12.4.3.2(B)*.

6.2.3 Valve Test

Verifiers test the system shutoff valves to ensure that they are functioning and that they are labeled properly. Refer to manufacturers' instructions for testing the valves, as the test procedure can vary. Record the rooms or areas that are controlled by each valve. This information is used to help verify the labeling of the valves.

6.2.4 Alarm Tests

To ensure that the master and area alarms installed in medical gas and vacuum systems are functioning properly and labeled correctly, the verifier tests them after completing the cross-connection and valve tests. An alarm test is also required when connecting a new piping system to an existing one, or when extending an existing system. Codes require the facility to maintain permanent records of all alarm tests on file. Use oil-free, nitrogen NF gas, system gas, or operating vacuum for these tests, as required. If the alarm system is computer-controlled, include the computer system in the alarm tests.

Test the master alarm to ensure that the audible and **noncancelable** (meaning that it cannot be turned off manually) visual signals are activated when the pressure in the main line fluctuates at least 20% above or below the normal operating pressure. Verify the operation of all master-alarm signals during this test.

Noncancelable: A signal that cannot be turned off manually.

Next, test the area alarms for individual positive- and negative-pressure systems to ensure that they are operating as designed. The area alarms in a positive-pressure piping system should activate if the pressure fluctuates at least 20% above or below the normal operating pressure. In a vacuum system, the alarm should activate if the vacuum falls below a gauge pressure of 12" of mercury.

6.2.5 Piping Purge Tests

Verifiers perform a piping purge test on medical gas and vacuum systems to remove any remaining dirt and particulate matter from systems' lines. This may be the first test conducted by a verifier. Follow these steps:

Step 1 Using an appropriate adapter supplied by the facility or the system's manufacturer, subject each outlet to a nitrogen gas purge of at least 8 scfm (standard cubic feet per minute).

Step 2 Interrupt the purge several times during the test to create intermittent spurts of gas. This helps dislodge particulates still in the lines.

Step 3 Continue the test until the gas output leaves no discoloration on a white cloth that is placed loosely over the adapter. Always use an appropriate adapter for a purge test.

6.2.6 Piping Particulate Test

To verify that distribution piping is completely dirt-free, the purge test is followed by a particulate test. Follow these steps:

Step 1 Filter no less than 35 ft^3 of oil-free, nitrogen NF through a clean, white 0.45-micron filter at a flow rate of no less than 3.5 scfm.

Step 2 Test 25% of the zones at the outlets that are the farthest away from the source.

Step 3 If the filter collects more than 0.001 gram of matter from any outlet, test the most remote outlet in every zone.

6.2.7 Piping Purity Test

In addition to being free of particulate matter, medical gas distribution piping must also be free of excessive water vapor and hydrocarbons such as petroleum and natural gas. Follow these steps to perform a piping purity test:

Step 1 Test the outlet that is farthest from the gas source for total nonmethane hydrocarbons. Compare these results to the quality of the source gas. If using system gas as the source gas, conduct the test at the source equipment.

Step 2 The difference in purity between the source gas and outlet gas should not exceed 5 ppm (parts per million) of nonmethane and halogenated hydrocarbons (hydrocarbons that contain halogen).

Step 3 Measure the moisture concentration at the outlet to ensure that it does not exceed 500 ppm or an equivalent pressure dew point of 10°F at a pressure of 50 psig.

6.2.8 Final Tie-In Test

The final tie-in test is designed to ensure that the distribution system is free of any leaks at the connection point, and that no new contamination has been added. This test must be performed before connecting any new, extended, or additional piping. Follow these steps:

Step 1 Before joining new work to an existing system, conduct a leak test on each joint in the final connection. Use system gas at the designated operating pressure, and test for leaks using an approved leak detectant that is safe for use with oxygen and that contains no ammonia.

Step 2 Test vacuum joints using an ultrasonic leak detector or other approved method, as specified by the manufacturer or the local applicable code.

Step 3 Test pressure joints following the brazing and leak testing of the final connection.

6.2.9 Operational Pressure Test

Positive- and negative-pressure distribution piping must be free from excessive loss of pressure or vacuum when operating. The operational pressure test is performed to ensure this. Perform pressure tests at each station outlet/inlet or terminal where connections are made, using either system gas or operating vacuum, depending on the system. NFPA 99 specifies the following delivery and draw requirements:

- Gas outlets with a pressure of 50 psig (including oxygen, nitrous oxide, medical air, and carbon dioxide) are required to deliver 3.5 scfm with a maximum pressure drop of 5 psi and a static pressure of between 50 to 55 psi.
- Support gas outlets are required to deliver 5 scfm with a maximum pressure drop of 5 psig and a static pressure of 160 to 185 psig.
- Medical-surgical vacuum inlets are required to draw 3 scfm without reducing the vacuum pressure below 12" of mercury at any adjacent station inlet.
- Oxygen and medical air outlets that serve critical care areas are required to permit transient flow rates of 6 scfm for 3 seconds.

6.2.10 Medical Gas Concentration Test

Verifiers conduct a medical gas concentration test to ensure that the correct concentration of system gas is available at each outlet. Once the system has been purged using system gas, analyze each gas source and outlet using appropriate measuring instruments operated according to manufacturers' instructions. Refer to NFPA 99, *Table 5.1.12.4.11* for the allowable gas concentrations for each type of gas.

6.2.11 Medical Air Purity Test

Test the quality and purity of medical air by taking a sample at the system's sample port. This test must be conducted before the source valve is opened. Refer to NFPA 99, *Table 5.1.12.4.2.3* for the maximum allowable impurities in medical air.

6.2.12 Labeling

Verifiers ensure that proper labeling has been affixed to all distribution piping, system outlets/inlets, shutoff valves, master and area alarms, and source equipment. As has already been described, the style and placement of labels is discussed in *Section 4.2.0.*

6.2.13 Source Equipment Verification

Finally, the verifier tests the complete gas supply, medical air compressor, and proportioning systems to ensure they are functioning correctly. These tests are performed once the interconnecting lines, accessories, and source equipment have all been installed.

To test a gas supply source, follow these steps:

Step 1 Ensure that the changeover from primary to secondary supply and the reserve are operating according to the manufacturer's instructions.

Step 2 Test the reserve unit's actuation and pressure-monitoring switches and signals, if installed.

Step 3 Test the bulk-supply and master signal-panel installations to ensure they are functioning correctly. Test the signal panels when changing or replacing storage units as well.

The medical air compressor system is tested to ensure air quality and to ensure that the alarm sensors are calibrated and installed according to the manufacturer's instructions. Test the medical air compressor system at the sample port. To test a medical air compressor system, follow these steps:

Step 1 Check the operation of system control sensors, and controls including dew point, air temperature, and air quality, before allowing the system to go into service.

Step 2 Verify the quality of medical air being delivered by the compressor air supply when installing new components and before patient use. Conduct air-quality tests following normal operation of the medical air source system and with the source valve closed, under a simulated load for at least 12 hours elapsed (not aggregate) time.

Step 3 Simulate loading by venting air continuously at a rate that is approximately 25% of the system's capacity.

Step 4 Create a demand of approximately 25% of the rated compressor capacity to cause the compressors to cycle and the dryers to operate.

Like the medical air compressor system, the proportioning system and its reserve are tested to ensure that they are functioning properly and that the alarms are calibrated and installed correctly. Conduct the test at the proportioning system's sample port. Lastly, test the medical-surgical vacuum systems according to manufacturers' instructions prior to putting them into service.

6.0.0 Section Review

1. The tests that the AHJ is required by code to witness are the installer's _____.
   a. initial cross-connection and initial piping purge tests
   b. initial blowdown and initial pressure tests
   c. standing pressure and standing vacuum tests
   d. medical gas concentration and medical air purity tests

2. In an operational pressure test, the maximum allowable pressure drop of a gas outlet with a pressure of 50 psig is _____.
   a. 20 psi
   b. 15 psi
   c. 10 psi
   d. 5 psi

Module 02412 Review Questions

1. The special piping system used in hospitals, clinics, and doctors' offices to supply nonflammable gases is called a(n) _____.
   a. anesthetic supply system
   b. medical gas system
   c. clinical gas system
   d. respiratory-assist supply network

2. The requirements for installing, testing, and maintaining medical gas and vacuum systems are outlined in the model code designated as _____.
   a. NFPA 101®
   b. NFPA 563, Part B
   c. NFPA 157.3
   d. NFPA 99

3. A surgeon's office equipped to administer general anesthesia would require a medical gas and vacuum system classified as _____.
   a. Category 1
   b. Category 2
   c. Category 3
   d. Category 4

4. A Category 2 medical gas and vacuum system is classified as one that if it failed, _____.
   a. could result in serious injury or death
   b. could result in discomfort, but not injury
   c. would not affect patient care
   d. could result in minor injury

5. To be certified for work with medical gas and vacuum systems, maintenance personnel must pass written and practical exams after completing a training course lasting _____.
   a. 8 hours
   b. 16 hours
   c. 32 hours
   d. 48 hours

6. Fittings designed for use with all types of gas in a medical gas system are called _____.
   a. standard fittings
   b. universal fittings
   c. cheater fittings
   d. adapter fittings

7. To meet the requirements of NFPA 99, medical air must be produced at a grade called _____.
   a. ultrarefined
   b. medical-surgical
   c. pharmaceutical
   d. premium

8. The abbreviation WAGD stands for _____.
   a. waste anesthetic gas disposal
   b. weighted average gas dispenser
   c. wide-area gas delivery
   d. water-aided gas dispersion

9. The location of a medical gas and vacuum system control panel will be specified in the _____.
   a. manufacturer's product manual
   b. local applicable code
   c. design drawings
   d. local building code

10. NFPA 99 specifies that audible alarms must produce a sound of 80 decibels at _____.
    a. 3'
    b. 12'
    c. 25'
    d. 50'

11. Valves designed to isolate a set of outlets or inlets from the rest of the system are called _____.
    a. isolation valves
    b. area valves
    c. sectional cutoff valves
    d. zone valves

12. The condition of rubber gloves can be checked visually, or with a(n) _____.
    a. tension test
    b. air test
    c. abrasion test
    d. conductivity test

13. When doing hot work, you should know the location of fire extinguishers and _____.
    a. the first-aid kit
    b. what type of extinguisher to use on different fires
    c. the date the extinguisher was last inspected
    d. the distance to the nearest telephone

14. Before brazing can be done on copper tubes, a contractor must obtain a(n) _____.
    a. site license
    b. brazing certificate
    c. hot-work permit
    d. OSHA Form 1012

15. NFPA 101® classifies life safety systems according to the _____.
    a. level of hazard posed
    b. number of zones involved
    c. flammability of the gas used
    d. ventilation required

16. For use in medical gas systems, NFPA specifies the use of hard-drawn seamless copper tubing designated as _____.
    a. Type A
    b. Type HD
    c. Type K
    d. Type L

17. In addition to basic plumbing tools, work on medical gas and vacuum systems requires _____.
   a. oxyacetylene cutting tools
   b. adapters for connecting a pressure gauge
   c. a set of metric wrenches
   d. a soldering gun

18. Before medical gas tubing is charged with oil-free, nitrogen NF gas and sealed at the factory, the inside must be _____.
   a. dehydrated
   b. coated with epoxy
   c. deoxidized
   d. irradiated

19. Forgetting to remove internal plugs from medical gas piping can cause damage when the system is _____.
   a. purged
   b. connected
   c. tested
   d. completed

20. Onsite cleaning of medical gas and vacuum tubing is generally prohibited because the chemicals used _____.
   a. are toxic
   b. can damage the copper tubes
   c. are very costly
   d. are a potential environmental hazard

21. After tubing is cut, burrs must be removed with a procedure called _____.
   a. surfacing
   b. reaming
   c. grinding
   d. debriding

22. When cutting and reaming copper tube, if the tube becomes deformed, bring the tube back to roundness using a _____.
   a. rounding tool
   b. round or half-round file
   c. sizing tool
   d. deburring tool

23. A medical gas line labeled with white type on a blue background is used for _____.
   a. nitrous oxide
   b. oxygen
   c. nitrogen
   d. carbon dioxide

24. A shutoff valve must be labeled with the line's operating pressure if _____.
   a. the line carries oxygen
   b. the pressure is different from the standard gauge pressure
   c. the pressure is above 14.7 psig
   d. the system has a pressure-relief valve

25. Medical gas and vacuum tubing must be joined with brazed _____.
    a. tapered joints
    b. butt joints
    c. socket joints
    d. compression joints

26. To protect against joint failure in a fire, the brazing alloy must have a melting temperature above _____.
    a. 1,000°F
    b. 1,200°F
    c. 1,600°F
    d. 1,800°F

27. During brazing operations, the joint must be continuously purged with oil free, _____.
    a. carbon dioxide
    b. nitrogen NF
    c. helium
    d. nitrous oxide

28. An initial pressure test is conducted by pressurizing the system to at least _____.
    a. 14.7 psig
    b. 75 psig
    c. 150 psig
    d. 195 psig

29. A standing pressure test conducted by a verifier has a duration of _____.
    a. 10 minutes
    b. 30 minutes
    c. 4 hours
    d. 24 hours

30. In a piping purity test, the moisture concentration at the outlet farthest from the source must not exceed _____.
    a. 100 ppm
    b. 500 ppm
    c. 1,250 ppm
    d. 1,500 ppm

Answers to odd-numbered review questions are found in the Module Review Answer Key at the back of this book.

Answers to Section Review Questions

Answer	Section	Objective
Section One		
1. a	1.1.1	1a
2. d	1.2.3	1b
Section Two		
1. d	2.1.4	2a
2. b	2.2.2	2b
3. b	2.3.1	2c
Section Three		
1. a	3.1.0	3a
2. d	3.2.0	3b
3. c	3.3.0	3c
4. d	3.4.0	3d
Section Four		
1. c	4.0.0	4a
2. a	4.2.1	4b
Section Five		
1. b	5.1.1	5a
2. c	5.1.2	5b
Section Six		
1. c	6.1.5, 6.1.6	6a
2. d	6.2.9	6b

APPENDIX 02406A

Local Code Citations for Performance Task

Use a copy of your local applicable code to answer the following questions about requirements in your area for various types of plumbing installations. Be sure to cite the reference following your answer.

1. What are the duties and powers entrusted to your local code officials? ____________________
2. What work is specifically exempt from the requirement for a permit? ____________________
3. What are the requirements for connecting the drainage piping to offsets and bases of stacks? ____________________
4. How are vacuum system station receptacles installed? ____________________
5. What standard is referenced for the installation of nonflammable medical gas systems? ____________________
6. What are the minimum required air gaps for lavatories and fixtures with similarly sized openings? ____________________
7. How is the maximum water-consumption flow rate determined for a public lavatory? ____________________
8. What is the maximum allowable amount of lead content in water supply pipe and fittings? ____________________
9. What are the requirements for joints between copper and copper-alloy tubing? ____________________
10. What are the size requirements for a shower compartment? ____________________

APPENDIX 02406B

APPENDIX 02410A

Exercises for Performance Task 1

Use the drawing of a hypothetical swimming-pool installation in *Figure A1* and the information in this module to answer the following questions. Round your answers up or down to the nearest whole number. Remember to show all your work.

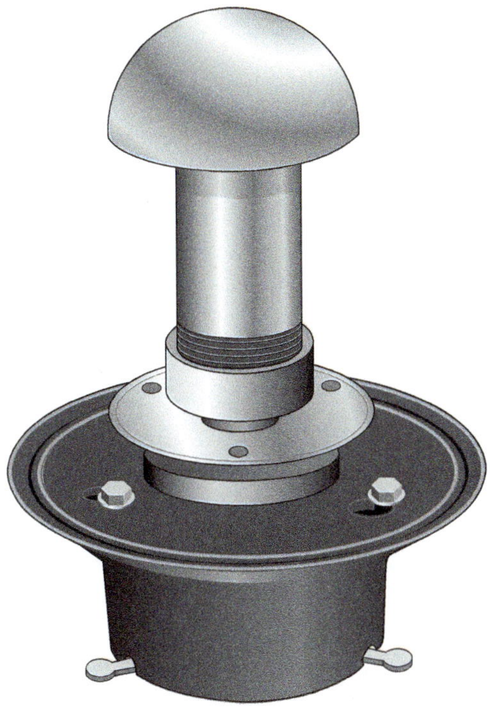

Figure A1 Swimming pool installation.

1. Calculate the total pool volume in cubic feet: ______________________
2. Calculate the total gallon capacity of the pool: ______________________
3. Given that turnover is said to occur once every 10 hours, the pool's bathing load would be: ______________________
4. Given that turnover is said to occur once every six hours, the pool's bathing load would be: ______________________
5. Enter the turnover in gpm for Questions 3 and 4: ______________________ and ______________________

Module Review Answer Key

MODULE 01 (02401)

Answer	Section
1. b	1.0.0
3. d	1.1.1
5. c	1.1.2
7. a	1.2.2
9. a	1.3.0
11. d	1.3.0
13. b	2.0.0
15. c	2.2.0

MODULE 02 (46100)

Answer	Section
1. b	1.1.1
3. d	2.1.3
5. c	2.3.1
7. b	3.1.2
9. c	3.2.5
11. a	4.1.0
13. c	4.3.1
15. d	4.5.2
17. c	5.1.2
19. b	5.3.1

MODULE 03 (02403)

Answer	Section
1.b	1.1.2
3.c	1.1.4
5.c	1.2.2
7.a	1.2.3
9.c	1.3.3
11.a	1.4.2
13.b	2.0.0
15.c	2.2.4

MODULE 04 (02404)

Answer	Section
1. b	1.0.0
3. a	1.1.0
5. a	1.1.1
7. a	2.1.0
9. b	2.2.0

MODULE 05 (02405)

Answer	Section
1. b	1.0.0
3. c	1.0.0
5. b	1.3.0
7. c	2.2.0
9. d	2.2.1
11. c	2.2.4
13. a	3.1.0
15. d	3.5.0

MODULE 06 (02406)

Answer	Section
1. a	1.0.0
3. b	1.1.2
5. d	1.2.2
7. a	2.1.0
9. a	2.2.2

MODULE 07 (02408)

Answer	Section
1. d	1.0.0
3. a	1.3.0
5. a	1.4.0
7. c	2.2.0
9. b	2.3.0
11. b	3.1.0
13. a	3.2.0
15. c	3.2.0

MODULE 08 (02409)

Answer	Section
1. d	1.0.0
3. c	1.1.1
5. c	1.3.0
7. a	2.2.2
9. c	2.2.1
11. d	3.0.0
13. d	3.1.0
15. c	4.0.0

MODULE 09 (02410)

Answer	Section
1. b	1.0.0
3. d	1.0.0
5. b	1.1.0
7. d	1.2.1
9. b	1.2.3
11. c	2.0.0
13. d	2.2.1
15. d	2.2.2

MODULE 10 (02411)

Answer	Section
1. c	1.0.0
3. c	1.0.0
5. d	1.1.0
7. b	1.2.0
9. b	1.2.0

MODULE 11 (02412)

Answer	Section
1. b	1.0.0
3. a	1.1.1
5. c	1.2.4
7. c	2.1.3
9. c	2.2.1
11. d	2.3.0
13. b	3.2.0
15. a	3.4.0
17. b	4.1.0
19. c	4.1.2
21. b	4.1.3
23. a	4.2.0
25. c	5.1.0
27. b	5.1.1
29. d	6.1.5

GLOSSARY

Accomplishment: An individual's need to set challenging goals and achieve them.

Accountability: In relation to job completion, accepting responsibility and being personally answerable for the results.

Accounting: The process of compiling and analyzing financial records.

Accounts receivable: Money that is owed to a business but has not been paid.

Active system: A solar heating system that uses a pump and controls to circulate hot water through the system.

AED (automated external defibrillator): A device used to help those experiencing sudden cardiac arrest.

Aeration system: A type of soil absorption system that uses an aeration tank to clarify effluent.

Aeration tank: A waste storage and separation tank used in an aeration system. In an aeration tank, anaerobic bacteria decompose sludge and then aerobic bacteria clarify the effluent.

Aerobic bacteria: Bacteria that need oxygen to live. They are used in aeration tanks to clarify effluent.

Air break: A piping arrangement (in a drainage system) in which a drain from a fixture, appliance, or device discharges indirectly into another fixture, receptacle, or interceptor at a point below the flood level rim and above the trap seal.

Air gap: The unobstructed vertical distance through the free atmosphere between the outlet of the waste pipe and the flood level rim of the receptacle into which the waste pipe is discharging.

Airtrol valve: In a hydronic system, a valve that purges air from the system into the expansion tank.

Anaerobic bacteria: Bacteria that do not need oxygen to live. They are used in septic tanks and aeration tanks to decompose solid wastes.

Anesthetic: A gas that is inhaled to inhibit sensitivity to pain.

Anodic inhibitor: A chemical applied to a pipe to prevent galvanic corrosion.

Aquifer: An underground reservoir containing water-saturated soil, sand, and rock.

Area alarm: An alarm that continually monitors medical gas and vacuum systems located in a specific area, such as a treatment room.

Artesian aquifer: A deep aquifer that does not lose water through runoff. Also called a confined aquifer.

As-built drawings: A revised set of drawings that reflect all changes made during the construction process and show the exact specifications of the completed work.

Asset: Any type of property owned by a business.

Authority: The right to act or make decisions in carrying out an assignment.

Automatic trap primer: A device that feeds water into a low-use trap, allowing the trap to maintain a constant seal.

Average flow: The flow rate in a water pressure booster system under average operating conditions, used to calculate off-peak loads.

Backwater valve: A valve that prevents the backflow of sewage into a pool drain.

Baffle: Partition inside the body of an interceptor. Baffles are designed to reduce turbulence caused by incoming wastewater and to block waste from escaping through the outlet.

Balance sheet: A form used to track assets and liabilities.

Balance: The condition of an account when assets and liabilities are equal.

Balancing valves: Valves in a recirculation system that ensure steady water flow.

Batch system: A solar heating system in which water is both heated and stored in the collector. The system uses the principle of convection to circulate water.

Bathing load: The maximum number of people allowed in a pool per hour, based on capacity and turnover rate.

Behavior: The way a person responds to a specific situation.

Bid: A formal offer to do work according to an estimate.

Biochemical oxygen demand (BOD): A measure of life-sustaining oxygen content in water.

Boiler: In a hydronic system, a device used to heat or boil water. Boilers differ from typical water heaters in that they have a greater capacity, operate at higher temperatures and pressures, and have a greater heat output.

Bond: A financial guarantee that work will be completed according to the terms of a contract.

Bored well: A type of well drilled by an auger and lined with a well casing.

Bulk oxygen system: The equipment used to store oxygen and distribute it to the supply line at system pressure.

Capacity: The volume of a pool in gallons.

Catch basin: A reservoir installed before the sanitary piping. It allows sediment to settle out of wastewater.

Category 1 system: A medical gas system in which equipment or system failure is likely to cause serious injury or even death.

Category 2 system: A medical gas system in which equipment failure is likely to cause minor injury.

Category 3 system: A medical gas system in which equipment failure is unlikely to cause injury but may cause some discomfort to patients.

Category 4 system: A medical gas system in which equipment failure would not affect patient care at all.

Cesspool: A lined, covered pit for storing solid and liquid wastes. Cesspools allow effluent to seep into the soil.

Change order: An agreement to perform work in addition to what was originally agreed to in a contract.

Charged: Filled with an inert gas, such as oil-free, nitrogen NF.

Chlorinate: To disinfect a well with a chlorine compound such as calcium hypochlorite or liquid bleach.

Circuit Setter™: A type of balancing valve used to regulate the flow on the return of an entire circuit, with a dial permitting fine adjustments.

Cistern: A water storage tank used to supply some private water supply well systems with potable water.

Closed loop system: Another name for an indirect system.

Closed system: A private waste-disposal system that stores wastes until they can be disposed of.

Code: A legal document enacted to protect the public and property, establishing the minimum standards for materials, practices, and installations.

Collector: A device used in a solar heating system to receive heat from sunlight and transfer it to water or another liquid.

Combined upfeed and downfeed system: A recirculation system in which hot water is supplied in the upward and downward flow through the system.

Communication: The process of exchanging facts, information, and ideas. Communication comes in many forms, including verbal, written, visual, and nonverbal.

Constant-speed system: A water pressure booster system in which water circulates continuously through the system and pumps replacement water from the main.

Construction safety plan: A document that outlines the procedures, rules, and regulations that are or will be put in place to protect workers during a construction project.

Continuous demand: Demand for water caused by outlets, pumps, and other devices with a relatively constant flow.

Contract: A legal agreement between a contractor and a client for work to be done.

Convection: Circulation caused by the sinking of dense cold water and the rising of hot water.

Cost control: The process that helps keep expenses under control by managing labor, material, and other costs to ensure that the project finishes on budget.

CPR (cardiopulmonary resuscitation): A lifesaving technique that is used when someone's heartbeat or breathing has stopped.

Cradle-to-grave responsibility: A contractor's legal responsibility for the management of hazardous wastes from creation to disposal.

Crew leader: A person appointed leader of a team or crew, usually an experienced craft professional who has demonstrated leadership qualities. Also known as *frontline leader, frontline supervisor, lead person*, or *foreman*.

Critical activities: Tasks in the critical path management process that could cause significant delay if not completed on time.

Critical path management (CPM): A task-planning process in which diagrams show the relationships between tasks and estimated completion times.

Daily work report: A report written by a job superintendent describing activities at a jobsite.

Debt: Money owed by a business for payment of direct costs and indirect costs.

Decision-making: The action or process of making a choice. Decision-making involves starting an action, stopping one, or choosing a different plan or action.

Deep well: A well that is drilled deeper than 25'.

Deep-well jet pump: A pump used in a deep well that uses a nozzle and venturi, usually located separately from the pump about 5' below the water line within the well, to accelerate the water flow.

Dehydrated: A condition in which all water has been removed.

Dependent vehicle: An RV that is not equipped with bathroom facilities. Dependent vehicles are not intended to be connected to public utilities for an extended time.

Differential thermostat: In a solar heating system, a control that activates a pump when water in the collector reaches a specified temperature above that of the water in the storage tank.

Direct cost: An expense for an activity or an item related directly to a project.

Direct return system: A two-pipe system in which the return water flows in the opposite direction from the flow in the supply pipe.

Direct system: A solar heating system that uses a pump and temperature controls to control the flow of hot water. It is also called an open loop system.

DISS connector: A Diameter Index Safety System connector used for vacuum and medical gas. It is designed so it cannot be interchanged.

Distractions: Anything that disrupts the communication flow between sender and receiver. Also called *noise*.

Distribution box: A watertight container that directs effluent to different parts of a leach field.

Diversity: The fact of individuals having a variety of abilities, skills, backgrounds, and experiences.

Diverter tee: In a one-pipe system, a tee that diverts the water flow by using an internal baffle to create a pressure drop.

Double-suction impeller: A pump impeller with two cavities, one each on opposite sides of the impeller. It is used to channel water in high-flow conditions in water pressure booster systems.

Downfeed system: A recirculation system that supplies hot water as it travels down the system.

Drain-down system: A direct system in which water drains out of the collector when the pump shuts off.

Draw-off hose: A hose used to siphon grease and oil from interceptors.

Drilled well: A type of well created by chisels or rotary drills.

Driven well: A type of well drilled by a pointed well head driven by a hammer.

Drop pipe: Another name for the discharge pipe that connects the well to the water storage tank.

Drywell: A covered pit filled with loose sand and gravel that collects roof and basement drainage and allows it to percolate into the surrounding soil.

Dug well: A type of well excavated by shovel and backhoe and lined with brick, stone, or concrete.

Duration: The length of time of an activity.

Duty: A work task an employee is expected to perform.

Effluent: Liquid waste that separates from solid waste. In soil absorption systems and cesspools, effluent is allowed to percolate into the soil.

Elevated-tank system: The oldest and simplest type of water pressure booster system, featuring a water tank on the roof operated by the weight of the water column.

Equity: A dedication to providing opportunities for all.

Equity: The value of the business if all debts are paid.

Estimating: Calculating the costs to complete a specific project, which are submitted as part of a bid.

Estimating: The process of calculating the cost of a project.

Ethics: The moral principles that guides an individual's or organization's actions when dealing with others.

Ethylene glycol: An alcohol-based liquid used in an indirect system to transfer heat from a collector to a heat exchanger.

Expansion joint: A mechanical device installed on a pipe that allows the pipe to expand and flex as hot water flows through it.

Expansion tank: A tank attached to the supply pipes in a recirculation system that allows water to cool and slow down as it flows.

Feasibility study: An analysis that considers all of a project's relevant factors—including budget, technical elements, legal aspects, and scheduling—to determine the likelihood that the project can be completed successfully.

First aid: Help given to a sick or injured worker on the worksite immediately after an injury occurs.

Floater: A device that releases chlorine through vents while floating in a pool. These types of chemical treatments are not permitted in public pools.

Flow control fitting: A device that regulates the flow of wastewater into grease and oil interceptors.

Flowmeter: A meter that measures and indicates the volumetric flow rate of a gas or a liquid.

Foot valve: A valve used to prime water-lubricated pumps in wells deeper than 32'.

Forced circulation system: A recirculation system that uses a pump to circulate hot water through the system.

Forced heat system: A type of hydronic system that uses fans and ducts to circulate air heated by passing it over coils containing hot water.

Formal organization: A group formed to serve a specific purpose or meet set goals. Each member has a role designed to help achieve that goal.

Gravity draw-off system: A system that automatically drains oil from an interceptor into a storage tank.

Gravity return system: A recirculation system in which the force of gravity circulates hot water throughout the system. These systems are no longer permitted by code.

Gross income: All money earned before expenses are subtracted.

Harassment: Unwelcome conduct that is based on race, color, religion, sex, national origin, age, disability, or genetic information. Harassment includes verbal or physical conduct that impacts a person's work performance or creates an intimidating, hostile, or offensive work environment.

Hazards: Conditions or activities that, if left uncontrolled, have the potential for harm.

Header pipe: A pipe used to relieve hydrostatic pressure. The pipe runs underneath the pool to a pump connection on the pool deck.

Heat exchanger: A device used in an indirect system to transfer heat from one liquid to another in such a way that the two liquids do not come into contact with each other.

High-limit aquastat: A safety device that deactivates a hot water boiler when the water reaches the maximum temperature.

Hot work: Work performed on heated metal.

Humus: In an organic system, the dried solid waste that results from decomposition. Humus can eventually be added to soil to fertilize it.

Hydrojet: A nozzle that sends a high-pressure stream of air and water into a spa.

Hydrologic cycle: The process through which water cycles through the environment as both a liquid and a vapor.

Hydronic system: A plumbing system that uses water or steam to heat a building. It is also called a radiant system.

Hydropneumatic tank system: A water pressure booster system that uses air pressure to circulate the water.

Hydrostatic pressure: The force applied to the bottom of a pool, spa, or hot tub by high groundwater levels beneath it.

Impairment: The point where someone's intake of substances affects her or his ability to perform appropriately.

Incidents: Unplanned events that result in property damage, fatalities, injuries, illnesses, or near misses.

Inclusion: A feeling of belonging. Inclusion means everyone on the team feels welcomed, valued, and included in the group.

Independent vehicle: A mobile home or recreational vehicle (RV) equipped with bathroom facilities. Independent vehicles can be connected to public utilities for long periods of time.

Indirect cost: An expense for an activity or an item that supports the administrative and overhead activities of a business.

Indirect system: An active system that uses a liquid other than water to distribute heat from the collector. It is also called a closed loop system.

Indirect waste system: A drainpipe attached to a fixture or appliance that is separated from the regular drainage system by a backflow preventer.

Indirect waste: Wastewater that flows through an indirect waste pipe.

Informal leader: An individual who influences the decisions of others, although they hold no formal position of authority.

Informal organization: A group of people that interact casually and develop connections through ongoing interactions.

Initial blowdown test: An installation test in which oil-free, nitrogen NF gas is blown through the distribution piping to remove dirt and other particulate matter.

Initial cross-connection test: An installation test performed to ensure that medical gas and vacuum systems are not cross-connected.

Initial piping purge test: An installation test performed to purge the system of any remaining dirt and other particulate matter.

Initial pressure test: An installation test performed to ensure that there is no pressure loss in the distribution piping.

Inspection: The process of examining the worksite and identifying any workplace hazards, so these hazards can be addressed before work proceeds.

Instrument air: Compressed air that is used to power pneumatic tools and clean instruments. Not to be confused with medical air USP.

Interceptor: A device designed and installed to separate and retain for removal, by automatic or manual means, deleterious, hazardous, or undesirable matter from normal wastes, while permitting normal sewage or wastes to discharge into the drainage system by gravity.

Intermittent demand: Demand for water caused by fixtures that are used no more than about five minutes at a time.

International Plumbing Code® (IPC): The model plumbing code of the International Code Council (ICC). It was first issued in 1995.

Invoice: An itemized list of materials for purchase or rent.

Job description: A description of the scope and responsibilities of a worker's job so that the individual and others understand what the job entails.

Job diary: A written record that a supervisor maintains periodically (usually daily) of the events, communications, observations, and decisions made during the course of a project.

Job safety analysis (JSA): A technique that focuses on job tasks as a way to identify hazards before they occur. It focuses on the relationship between the worker, the task, the tools, and the work environment. Also known as *job hazard analysis (JHA)* or *activity hazard analysis (AHA)*.

Leach field: In a soil absorption system, an area of soil that is designed to accept effluent from distribution pipes. Leach fields are also called absorption fields or drain fields.

Leach: To percolate into soil and be absorbed.

Level 4 schedules: Schedules that offer additional detail to provide working schedules to the crews doing the work, usually within three to six weeks of completing specific work.

Liability: In accounting, money owed by a business. In insurance, a legal responsibility.

Life safety: Essential for ensuring the safety of patients.

Limit switch: In a solar heating system, a switch that activates when the water in the collector reaches a preset temperature.

Local alarm: An alarm that continually monitors one system and is located near the system being monitored.

Lookahead schedule: For planning and scheduling work that is happening in the short-term, usually within one to three weeks; used for anticipating material, labor, tool, and other resource requirements, as well as identifying potential schedule conflicts or other problems. Also called a *short-term schedule*.

Low-water cutoff device: A device that deactivates a pump when the water level in a well falls to a preset low point. See also tail pipe assembly.

Manual draw-off system: A system that drains oil out of an interceptor through a valve-operated draw-off hose.

Master alarm: An alarm that continually monitors the gas supply source and mainline pressure for all systems and is directly connected to each of the devices it monitors.

Maximum flow: The theoretical total flow in a water pressure booster system if all outlets are simultaneously opened.

Maximum probable flow: The flow in a water pressure booster system during periods of peak demand.

Mechanical grease interceptor: A plumbing appurtenance, fitted with a timer-activated self-cleaning capability, installed in the sanitary drainage system to separate free-floating fats, oils, and grease from wastewater.

Medical air USP: Medical air that has been certified as pharmaceutical grade according to the standards of the United States Pharmacopeial Convention.

Medical gas system: Equipment and piping that supply nonflammable medical gases under pressure, along with the associated pumps and compressors, filters and purifiers, valves, alarms, gauges, and controls.

Medical vacuum system: Equipment and piping that use negative pressure to remove waste gases, liquids, and solids, along with the associated pumps and compressors, filters and purifiers, valves, alarms, gauges, and controls.

Medical-surgical vacuum system: A special type of medical vacuum system designed to suction away bodily fluids during surgery.

Mentor: An experienced and trusted adviser who can help an employee with guidance, motivation, and support.

Mobile home: A manufactured home built on a wheeled chassis that has no fixed foundation and that is typically no smaller than 8' wide by 32' long.

Mobilization: The final stage of preplanning, during which all items required to start construction are moved to the jobsite, arrangements are made with the other trades to coordinate jobsite activities, and all jobsite details are organized in preparation for construction.

Model code: A set of comprehensive, general guidelines that establishes and defines acceptable plumbing practices and materials, and lists prohibited installations.

Modified waste line: A vertical pipe at the terminal of a waste line that allows indirect waste to mix with air.

Morale: An individual's emotional outlook toward work and the level of satisfaction gained while performing the jobs assigned.

Motivate: To make someone feel determined or enthusiastic or to make someone behave in a particular way.

Net income: Another term for profit.

Noncancelable: A signal that cannot be turned off manually.

Nonverbal communication: Things the speaker and the receiver see or hear while communicating. Examples include facial expressions, body movements, hand gestures, tone of voice, and eye contact.

Objectives: Goals or desired outcomes stated at the beginning of a project.

Occupational Safety and Health Administration (OSHA): A national public health agency that is part of the US Department of Labor. To promote a safe and healthy work environment, OSHA issues standards and rules for working conditions, facilities, equipment, tools, and work processes.

One-pipe system: A hydronic system in which water or steam is fed to the heating units and returned to the boiler through a single run of pipe.

Open loop system: Another name for a direct system.

Open-hub waste receptor: A waste drainpipe or pipe hub that extends at least 1" above the floor line.

Organic system: A private waste-disposal system that allows waste to decompose through exposure to aerobic bacteria. The resulting dried waste is called humus.

Organization: A group of people working together.

Organizational chart: A diagram that shows how the various management and operational responsibilities relate to each other within an organization. Named positions are ranked top to bottom, from those functional units with the most and broadest authority to those with the least. Lines connect positions and indicate chains of authority and other relationships.

Overhead: The costs of maintaining a business that are not applicable to a specific job. Office and administrative expenses are common examples of overhead.

Oxygen-enriched atmosphere: An atmosphere that contains more than 23.5% oxygen by volume.

Packer-type ejector: For jet pumps, an alternative discharge method to the two-pipe system in which the space between the water supply pipe and the well casing acts as a pressure pipe.

Parallel: A one-pipe system in which each heating unit draws only the amount of water it requires from the supply line.

Passive system: A solar heating system that relies on convection to circulate water.

Peer: Someone who is in a position or level in an organization equal or similar to yours.

Perched water table: A small water table that is cut off from a larger reservoir.

Personal growth: Improving one's talents and potential, both in and out of the workplace.

pH value: A measure of the hydrogen ions in water. The term is an abbreviation for "potential hydrogen."

Pharmaceutical grade: Meeting the specifications for medical purity established by the United States Pharmacopeial Convention.

Pitless adapter: A device that permits a discharge pipe to exit a well and connect to a water storage tank. Pitless adapters protect well water from contamination and freezing.

Planning: The process of laying out all details of a construction project before construction begins.

Policy: A general statement establishing guidelines for a specific activity.

Premium: What the contractor pays the insurance company to ensure that the project is covered.

Preplanning: The phase immediately following the awarding of a contract but before the start of construction, during which subcontractors are selected, purchase orders are issued, construction schedules are developed, and estimates are reviewed in depth.

Pressure control valve: A valve that diverts a portion of a pump's discharge through the ejector, thereby maintaining adequate pressure in the ejector.

Pressure distribution system: A soil absorption system in which effluent is pumped through the distribution pipes under pressure.

Pressurized air-bladder storage tank: A tank used in a water pressure booster system that uses compressed air to maintain water pressure.

Primary-secondary system: A one-pipe system with a parallel arrangement.

Problem-solving: A process used to identify possible solutions to a situation.

Procedures: Formal instructions to carry out and meet policies.

Production figure: An estimate of the time required for a crew to perform a task.

Production: The amount of construction materials put in place. It is the quantity of materials installed on a job.

Productivity rate: A measure of how much work gets done over a period of time, also called *output per labor hour*.

Professional growth: Gaining new skills and work experience that can help one to reach a goal in one's career.

Profit and loss statement: A list of all income and expenses over a given period.

Profit: The amount of money that the contractor will make after paying all the direct and indirect costs.

Project delivery methods: Different ways to organize the services of development, planning, and construction in order to execute a project.

Project manager: Person responsible for managing one or more projects.

Purchase order: An order form submitted to a supplier.

Radiant loop: A heating unit that consists of a network of hot water pipes in the floor or ceiling.

Radiant system: Another name for a hydronic system.

Radiation: The transfer of heat between bodies through space.

Radiator: A heating unit that emits heat from water via a series of metal fins.

Recapitulation sheet: A summary of material, labor, and equipment costs for each task in a project. Also called a recap sheet.

Receptor: A floor, sink, standpipe, hub drain, or floor drain that receives the discharge of one or more indirect waste pipes.

Recirculation system: A plumbing installation that circulates hot water within a building, providing customers with hot water on demand.

Recognition: Acknowledgement of an employee's accomplishments through verbal and nonverbal communications.

Redline drawings: Marked-up versions of construction drawings documenting any changes made during construction or installation.

Respiration: The breathing cycle.

Responsibility: The act of accepting a task and completing it to the best of one's ability.

Reverse return system: A two-pipe system in which the return water flows in the same direction as the flow in the supply pipe.

Reverse thermosiphoning: In a thermosiphon system, a condition in which water flows backward through the system.

Rewards: Tangible items given to an employee as recognition for good work.

Rework: To redo a previously completed task.

Root cause analysis (RCA): A method of problem-solving used for identifying the root causes of faults or problem.

Root cause: A core issue or fundamental reason for the occurrence of a problem.

Safety Data Sheet (SDS): Document listing information about a material or substance that includes common and proper names, chemical composition, physical forms and properties, hazards, flammability, handling, and emergency response in accordance with national and international hazard communication standards.

Sanitary dump station: A container designed to receive wastes from a travel trailer holding tank.

Scheduling: Involves defining which activities happen in what order and assigning dates and times to those tasks.

Scope: The list of the commitments and activities that all contractors, subcontractors, and suppliers are expected to do during a project. Also called *scope of work*.

Sediment bucket: A removable basin in a drain that prevents solid wastes from entering a sanitary system.

Septic system: A soil absorption system that uses a septic tank to separate out sludge.

Septic tank: A tank used in a soil absorption system to settle out sludge and scum and to decompose them using anaerobic bacteria.

Series: A one-pipe system in which all the water in the supply loop goes through each heating unit.

Seven-minute peak demand period: An estimate of the greatest possible water demand in a private water supply well system.

Shallow well: A well that is no more than 25' deep.

Shallow-well jet pump: A pump used in a shallow well that uses a nozzle and venturi attached to the pump to accelerate the water flow.

Sidewall drain: A drain installed on a swimming pool wall below the waterline.

Single-suction impeller: A type of impeller used in centrifugal pumps that have a single-suction cavity. Single-suction impellers are used in low-flow water pressure booster systems.

Site-specific safety plan (SSSP): A safety plan that includes a focused set of information that accounts for a jobsite's unique hazards.

Situational leadership: A leadership style that involves adjusting to the specific situation or task and the leadership needs of the group.

Skimmer: A device that traps pool debris in a removable basket.

Soil absorption system: A private waste-disposal system that uses tanks to separate out effluent, then allows it to drain into a leach field.

Soil mottle: Spots and streaks in soil that indicate water saturation.

Soil profile description: A report that specifies the depths and thicknesses of soil layers in a proposed leach field.

Solar thermal heating system: A hydronic system that uses the sun to heat water. Also referred to simply as a solar heating system.

Sole source: A type of bid in which a project is offered to a contractor of the client's choice.

Special waste: Waste that must be removed, diluted, or neutralized before it can be discharged into a sanitary system.

Specific heat: A measure of how much heat a liquid can hold per pound, measured in British thermal units (Btus).

Stage: In a vertical turbine pump, the arrangement of a single impeller and its water passage.

Standing pressure test: An installation test performed to ensure that the distribution piping in a medical gas system will not experience excessive pressure loss when under service pressure.

Standing vacuum test: An installation test performed to ensure that vacuum system piping in a medical vacuum system will not experience excessive vacuum loss when in use.

Steam dome: In a steam hydronic system, the place in the top of the boiler where steam collects.

Stop Work Authority (SWA): A program that outlines when and how employees and contractors can stop work when an identified hazard could result in an incident.

Submersible pump: A pump consisting of a multistage centrifugal pump and a submersible electric motor. Submersible pumps are usually installed in the bottom of deep wells.

Substance abuse: The inappropriate overuse of drugs and chemicals, whether they are legal or illegal.

Suction outlet fitting assembly: A pool drain that serves as a connection for cleaning hoses.

Superintendent: An on-site supervisor who is responsible for one or more crew leaders or frontline supervisors.

Tail pipe assembly: A device that matches a pump's discharge rate to a well's recovery rate. See also low-water cutoff device.

Task planning: The process of organizing activities and materials so that tasks are completed in the right order and on time.

Tempering mixing valve: A valve that adds cold water to the hot water flow to control water temperature.

Tent trailer: A collapsible travel trailer with canvas walls.

Thermal purge valve: A valve that allows heated water to drain from the recirculation system when no-flow conditions exist.

Thermistor: In a solar heating system, a control that adjusts a pump's speed according to changes in the water temperature.

Thermosiphon system: A passive system in which hot water is stored in a tank located higher than the solar collector.

Thermosiphoning: In a thermosiphon system, the process whereby cold water flows back to the collector to be reheated.

Three-way communication: A form of communication with an initial message from a sender, which is then repeated by the receiver to confirm accuracy, and then validated by the sender. Also called *repeat-back communication*.

Toolbox talks: Short, informal safety meetings held at the start of each day or shift. Sometimes referred to as *tailgate meetings* or *safety briefings*.

Trait: A distinguishing quality of a person.

Transducer: An electrical device that converts an analog signal, such as line pressure, into an electrical signal that can be read by an electrical device, such as an alarm circuit board.

Travel trailer: A portable dwelling that is towed by a motorized vehicle.

Turbidity: A measure of suspended solids in well water.

Turnover rate: The time required for all the water in a pool to cycle through the filter.

Two-pipe system: For jet pumps, an alternative to the packer-type ejector discharge method. One pipe draws well water through an ejector, and another pipe feeds the pressurized water to the jet.

Two-pipe system: A hydronic system that uses separate supply and return lines.

Uniform Plumbing Code® (UPC): The model code of the International Association of Plumbing and Mechanical Officials (IAPMO). It has been published since 1945.

Unit cost: In estimating materials used in a job, the total material cost divided by the number of units used.

Upfeed system: A type of recirculation system that supplies hot water as it travels up the system.

Upskilling: A type of training that provides someone with more advanced skills through additional education.

Variable-capacity system: A type of water pressure booster system that uses two or more pumps to provide water in response to peaks and lows in demand.

Venturi: A device that accelerates water and lowers its pressure by passing it through a narrow pipe opening. It is also called a diffuser.

Verbal communication: Messages that are sent using voice or other audible sounds. Also called *spoken communication* or *oral communication*.

Verifier: A person who inspects medical gas and vacuum systems and verifies that they meet all applicable standards and codes as well as manufacturers' specifications.

Vertical turbine pump: A pump used in water pressure booster systems where high-pressure and low-flow conditions exist.

Vibration isolators: Devices such as rubber pads and flexible line connectors that reduce the effects of pump vibration.

Visual communication: The process of using visual elements, such as photos, graphics, colors, and fonts to convey messages.

Volute: In centrifugal pumps, a geometrically curved outlet path.

Water hammer arrester: A device that absorbs the pressure surge that occurs when water flow is suddenly stopped in a system, known as water hammer.

Water pressure booster system: A plumbing installation that increases water pressure in the fresh water supply system.

Water supply fitting: A fixture that allows water to flow from a recirculation system back into a swimming pool.

Water table aquifer: An aquifer in which water flows toward bodies of water. Also called an unconfined aquifer.

Water table: The top of an aquifer, below which rock, sand, and soil are completely saturated with groundwater.

Well casing: A pipe section inserted into a well that maintains the shape of the well and prevents contamination.

Well head: A pointed rod used to create a driven well by being forced into the ground with a hammer.

Well: A hole in the ground that connects with an aquifer. Wells are equipped with pumps to lift water from the aquifer into water storage tanks.

Well-casing adapter: A well cover that protects the well casing from air leaks.

Work breakdown structure (WBS): A method of organizing the entire scope of work in a logical manner based on how the work will be executed.

Workers' compensation: Insurance that companies have to cover medical expenses, lost wages, and other costs for employees who are injured or become ill while completing work for their job.

Written communication: The process of sending messages or information using words, whether digitally or physically.

Zone control valve: In a hydronic system, a valve that allows the temperature in various parts of the building to be preset.

REFERENCES

02401 Business Principles for Plumbers

Be a Successful Building Contractor, Third Edition. 2004. R. Dodge Woodson. New York: McGraw Hill.

Construction Accounting & Financial Management, Fourth Edition. 2020. Steven J. Peterson. Hoboken, NJ: Pearson Education.

Financial Management and Accounting for the Construction Industry. 2023. New York: Construction Financial Management Association.

Financial Management in Construction Contracting. 2013. Andrew Ross and Peter Williams. Ames, IA: Wiley-Blackwell.

NASCLA Contractors Guide to Business, Law and Project Management, Basic 13th Edition. 2020. Phoenix, AZ: National Association of State Contractors Licensing Agencies (NASCLA).

Plumbing Estimating Methods, Third Edition. 2023. Joseph J. Galeno and Sheldon T. Greene. Kingston, MA: R.S. Means Company.

Small Business Accounting Simplified, Fifth Edition. 2010. Daniel Sitarz. Carbondale, IL: Nova Publishing Company.

46100 Introduction to Leadership

Architecture, Engineering, and Construction Industry (AEC), *www.aecinfo.com*.

The Center for Leadership Studies, *www.situational.com*.

Conflict Resolution Playbook: Practical Communication Skills for Preventing, Managing, and Resolving Conflict, Jeremy Pollack. 2020. Rockridge Press.

Construction Industry Alliance for Suicide Prevention, *www.ciasp.org*.

The Constructor Construction Encyclopedia, *theconstructor.org*.

Crucial Conversations: Tools for Talking When Stakes Are High, Patterson, Kerry, Joseph Grenny, Ron McMillan, and Al Switzler. 2002. Maidenhead, England: McGraw-Hill Contemporary.

Dare to Lead: Brave Work. Tough Conversations. Whole Hearts. Brene Brown. 2018. New York, NY. Random House Publishing. www.penguinrandomhouse.com.

Engineering News-Record, *www.enr.com*.

Equal Employment Opportunity Commission (EEOC), *www.eeoc.gov*.

The Five Dysfunctions of a Team, Lencioni, Patrick M. 2002. London, England: Jossey-Bass.

Mentoring for Craft Professionals, NCCER. 2016. New York, NY: Pearson Education, Inc.

National Association of Women in Construction (NAWIC), *www.nawic.org*.

National Census of Fatal Occupational Injuries (NCFOI), *www.bls.gov*.

National Institute of Occupational Safety and Health (NIOSH), *www.cdc.gov/niosh*.

National Safety Council, *www.nsc.org*.

Occupational Safety and Health Administration (OSHA), *www.osha.gov*.

Smart Choices: A Practical Guide to Making Better Decisions. John S. Hammond, Ralph L. Keeney, Howard Raiffa. 2015. Boston, MA. Harvard Business Publishing Education. *hbsp.harvard.edu*.

Society for Human Resources Management (SHRM), *www.shrm.org*.

Total Worker Health, *www.cdc.gov/niosh*.

United States Census Bureau, *www.census.gov*.

United States Department of Labor, *www.dol.gov*.

02403 Water Pressure Booster and Recirculation Systems

Efficient Building Design Series, Volume 3: *Water and Plumbing*. 1999. Ifte Choudhury and J. Trost. New York: Prentice Hall.

International Plumbing Code®, Latest Edition. Falls Church, VA: International Code Council.

Plumbing: Cold Water Supplies, Drainage, and Sanitation. 1994. Fred Hall. New York: Longman.

Plumbing: Hot Water Supply and Heating Systems. 1994. Fred Hall. New York: Longman.

Water Quality & Systems: A Guide for Facility Managers. Second Edition. 2020. Robert N. Reid. Gistrup, Denmark: River Publishers.

02404 Indirect and Special Waste

ANSI Standard 112.14.3, *Grease Interceptors*, Latest Edition. Washington, DC: ANSI.

Introduction to Wastewater Treatment Processes. 2014. Bruce Jefferson, Simon Parsons, and Elise Cartmell. New York: Wiley-Blackwell.

Water, Sanitary, and Waste Services for Buildings, Fifth Edition. 2001. Alan F. E. Wise and J. A. Swaffield. Boston, MA: Routledge.

02405 Hydronic and Solar Heating Systems

Hydronic Radiant Heating: A Practical Guide for the Nonengineer Installer. 1998. Dan Holohan. Bethpage, NY: Dan Holohan Associates.

Quick & Basic Hydronic Controls: A Contractor's Easy Guide to Hydronic Controls, Wiring, and Wiring Diagrams. 2001. Carol Fey. Littleton, CO: P.I.G. Press.

Repairing Hot Water Heating Systems. 2012. Robert VanNorden. Seattle, WA: Amazon Digital Services, Inc.

Solar Hot Water Fundamentals: Siting, Design, and Installation. 2011. Peter Skinner, et al. Albany, NY: E2G Solar.

02406 Codes

International Plumbing Code®, Latest Edition. Falls Church, VA: International Code Council.

Uniform Plumbing Code®, Latest Edition. Ontario, CA: International Association of Plumbing and Mechanical Officials.

Code Check: A Field Guide to Building a Safe House. 2000. Redwood Kardon, Michael Casey, and Douglas Hansen. Newtown, CT: Taunton Press.

Code Check Plumbing: A Field Guide to the Plumbing Codes. 2006. Michael Casey, Douglas Hansen, and Redwood Kardon. Newtown, CT: Taunton Press.

The Engineering Resources Code Finder for Building and Construction. 2001. Dennis Phinney. Anaheim, CA: BNI Building News.

02408 Private Water Supply Well Systems

Audel Technical Trades Series, *Water Well Pumps and Systems Mini-Ref. 2012.* Roger D. Woodson. New York: John Wiley & Sons, Inc.

Efficient Building Design, Volume 3, Water and Plumbing. 1999. Ifte Choudhury and J. Trost. Upper Saddle River, NJ: Prentice-Hall, Inc.

Finding Water: A Guide to the Construction and Maintenance of Private Water Supplies. Rick Brassington. Chichester, New York: John Wiley & Sons.

International Plumbing Code®, Latest Edition. Falls Church, VA: International Code Council.

02409 Private Waste-Disposal Systems

Advanced Onsite Wastewater Systems Technologies. 2006. Anish R. Jantrania and Mark A. Gross. Boca Raton, FL: CRC Press.

Anatomy of a Disaster. 2008. Washington, DC: Chemical Safety Board (video). *www.csb.gov/videos/anatomy-of-a-disaster*

International Private Sewage Disposal Code®, Latest Edition. Falls Church, VA: International Code Council.

A Builder's Guide to Wells and Septic Systems, Second Edition. 2010. New York, NY: McGraw Hill.

02410 Swimming Pools and Hot Tubs

International Plumbing Code®, Latest Edition. Falls Church, VA: International Code Council.

International Swimming Pool and Spa Code™, Latest Edition. Falls Church, VA: International Code Council.

02411 Plumbing for Mobile Homes and Travel Trailers

Code Check Plumbing: A Field Guide to Plumbing, Fourth Edition. 2011. Redwood Kardon, Michael Casey, and Douglas Hansen. Newtown, CT: Taunton Press.

International Plumbing Code®, Latest Edition. Falls Church, VA: International Code Council.

International Residential Code®, Latest Edition. Falls Church, VA: International Code Council.

The Manual for Manufactured/Mobile Home Repair and Upgrade. 2002. Mark N. Bower. Aberdeen, SD: Aberdeen Home Repair.

Recreation Vehicle Plumbing Systems. 2000. Reston, VA: Recreational Vehicle Industry Association.

NFPA 501, Standard on Manufactured Housing, Latest Edition. Quincy, MA: National Fire Protection Association.

02412 Introduction to Medical Gas and Vacuum Systems

ASSE/IAPMO/ANSI Standard #6010, Medical Gas Systems Installers, Latest Edition. Washington, DC: American National Standards Institute.

ASSE/IAPMO/ANSI Standard #6020, Medical Gas Systems Inspector, Latest Edition. Washington, DC: American National Standards Institute.

ASSE/IAPMO/ANSI Standard #6030, Medical Gas Systems Verifiers, Latest Edition. Washington, DC: American National Standards Institute.

ASSE/IAPMO/ANSI Standard #6040, Medical Gas Systems Maintenance Personnel, Latest Edition. Washington, DC: American National Standards Institute.

ASTM B88, Standard Specification for Seamless Copper Water Tube, Latest Edition. West Conshohocken, PA: ASTM International.

ASTM B280, Standard Specification for Seamless Copper Tube for Air Conditioning and Refrigeration Field Service, Latest Edition. West Conshohocken, PA: ASTM International.

ASTM B819, Standard Specification for Seamless Copper Tube for Medical Gas Systems, Latest Edition. West Conshohocken, PA: ASTM International.

Brazing, Second Edition. Mel M. Schwartz. 2003. Novelty, OH: ASM International.

CGA M-1, Guide for Medical Supply Systems at Consumer Sites, Latest Edition. Chantilly, VA: Compressed Gas Association.

The Copper Tube Handbook. 2010. New York: Copper Development Association.

NFPA 55, Compressed Gases and Cryogenic Fluids Code, Latest Edition. Quincy, MA: National Fire Protection Association.

NFPA 99, Health Care Facilities Code, Latest Edition. Quincy, MA: National Fire Protection Association.

NFPA 101, Life Safety Code, Latest Edition. Quincy, MA: National Fire Protection Association.

United States Pharmacopeia, Latest Edition. Rockville, MD: United States Pharmacopeial Convention, Inc.